RAILWAY
CONTROL SYSTEMS

Project Group

K W Burrage CEng FIEE FIRSE FCIT Chairman
F How MA CEng MIEE MIRSE
*J Waller BSc (Eng) CEng MIEE FIRSE
C I Weightman IEng MIRSE FPWI
R L Weedon General Secretary IRSE
*M E Leach BSc FIRSE General Editor

*Past President IRSE

The Institution, as a body, is not responsible for the views and opinions expressed by individual authors.

Editorial Note

The production of this book, which is a sequel to *Railway Signalling* published for the Institution of Railway Signal Engineers in 1980, has been managed on behalf of the Institution by a Project Group working under the chairmanship of K W Burrage, Director of Signal and Telecommunications Engineering, British Railways Board.

The Project Group has been particularly assisted in this task by the valuable co-operation it has received from:

EB SIGNAL (UK) Ltd
(formerly ML Engineering (Plymouth) Ltd)
GEC ALSTHOM Signalling
Westinghouse Brake and Signal Ltd

The Group also wishes to express its appreciation of the contributions made by the following authors, from which the text of the book has been prepared:

T Akehurst	A C Howker
R E B Barnard	J Lethbridge
R M Bell	J A Mason
B Botfield	G May
D N Bradley	A Porter
C R Bray	C H Porter
D C Bulgin	A J R Rowbotham
A H Cribbens	R C Short
D Fenner	D H Stratton
R J Fenton	E F Sutton
A J Fisher	D C Taylor
A R Hardman	K L Walter
B D Heard	

The greater part of the line diagrams in the book were drawn by Mrs Jennifer Hyde of Westinghouse Brake and Signal Ltd

RAILWAY
CONTROL SYSTEMS

compiled by a Project Group
of the Institution of Railway Signal Engineers
under the general editorship of
Maurice Leach

A & C Black · London

First published 1991
Reprinted 1993, 1996, 1999
A & C Black (Publishers) Limited
35 Bedford Row, London WC1R 4JH

ISBN 0–7136–3420–0

© 1991 Institution of Railway Signal Engineers

A CIP catalogue record for this book is available
from the British Library.

Apart from any fair dealing for the purposes of research or private
study, or criticism or review, as permitted under the Copyright, Designs
and Patents Act, 1988, this publication may be reproduced, stored or
transmitted, in any form or by any means, only with the prior
permission in writing of the publishers, or in the case of reprographic
reproduction in accordance with the terms of licences issued by the
Copyright Licensing Agency. Inquiries concerning reproduction outside
those terms should be sent to the publishers at the above named address.

Typeset by August Filmsetting, Haydock, St Helens
Printed in Great Britain by Antony Rowe Limited, Chippenham

Foreword

These words are being written some ten years after the publication of *Railway Signalling*, the first IRSE Textbook. In that time we have seen unprecedented growth in microprocessor-based systems, in particular with the introduction of solid state interlocking, a major advance in signalling technology on British Railways. We have also received requests to produce information on more conventional, long standing, systems which could not be accommodated within the original volume.

The Council felt it appropriate therefore to produce this companion volume, including reference to changes from the original volume, so as to complete the record of current practice on British Railways.

Whilst our Institution was formed, and is based, in the United Kingdom, more than 50% of the corporate membership is now from overseas administrations. It is our hope, indeed our intention, that further volumes will follow, covering the signalling principles and practice of many other countries.

If, by doing this, we can help further with the training of those new to our demanding profession, then our prime aim will have been achieved.

J Waller
President, Institution of Railway Signal Engineers

Contents and Illustrations

Foreword v

Preface ix

1 Recent Changes in Signalling Philosophy 1

1.1	SIMBIDS line fitted with TOWS 6
1.2	Special signs for SIMBIDS equipped lines 7
1.3	Layout for control table examples shown in **Figs. 1.4, 1.5, 1.6** and **1.7** 9
1.4	Typical control table for controlled signal — main class route 10
1.5	Typical control table for controlled signal — call-on class route 12
1.6	Typical control table for points 14
1.7	Typical control table for position light shunt signal 15
1.8	Typical control table for CCTV monitored level crossing 16

2 Solid State Interlocking 18

2.1	Duplication for safety — basic concept 19
2.2	Duplication for safety — a practical system 21
2.3	Triple redundancy technique used for interlocking processors 24
2.4	SSI basic structure 26
2.5(a)	The SSI interlocking 27
2.5(b)	Block diagram of SSI system 28
2.6	Interlocking software principles 32
2.7	Interlocking module software organisation 33
2.8	The interlocking program 34
2.9	Typical configuration of one trackside data link 36
2.10	Message format on the trackside data links 37
2.11	SSI long distance data transmission 39
2.12	Application of signal module 41
2.13	Application of point module 42
2.14	Internal structure of trackside interface module 44
2.15	Trackside module power interface isolation A – signal module B – point module 45
2.16	Trackside module voltage sensing and contact detection 47
2.17	Block diagram of SSI design workstation 49
2.18	Layout for SSI data examples (with subroute/suboverlap diagram) 52
2.19	Control panel: entrance/exit buttons 53
2.20	Control panel: other buttons/switches and indications 53/54
2.21	Route setting and locking and associated functions 54
2.22	Point associated data 55
2.23	Incoming functions and MAP data 56
2.24	Signal output data 57
2.25	Subroute/suboverlap release 58

3 Single Line Signalling 59

3.1	Key token instrument — general arrangement 65
3.2	Key token main section (signalman) instrument 68
3.3	Key token instrument commutator arrangement 69
3.4	Key token automatic operator instrument 70
3.5	Key token intermediate instrument 73
3.6	Key token system — Reed-type transmission 74
3.7	Key token system — transmission over public telecommunications lines 76
3.8	Simplified infrastructure: radio electronic token block system 78
3.9	Simplified infrastructure: fixed signals 79
3.10	Radio electronic token block: diagram of system 81
3.11	Radio electronic token block system: driver's and signalman's equipment 82
3.12	Simplified infrastructure: no-signalman remote key token system (trainman operated) 85
3.13	No-signalman remote key token system (trainman operated) 88
3.14	No-signalman remote token system: Reed-type transmission 89
3.15	Key token system: no-signalman remote instrument for use with physical circuits 90
3.16	No-signalman remote key token system (trainman operated): transmission over physical circuits 92

3.17	British Railways tokenless block system 93	
3.18	British Railways tokenless block system: transmission over physical line circuits 95	
3.19	British Railways tokenless block system: Reed-type transmission 96	

4 Immunisation and Earthing of Signalling Systems 99

4.1	Mechanism for generating conductive interference 100
4.2	Mechanism of effects of induced voltages 101
4.3	Generation of traction interference currents 103
4.4	Track circuit current paths through the traction return system 106
4.5	Arrangement of tail cables: local and line circuits 107
4.6	Mechanism of earth fault finders 112
4.7	Lightning strike voltage gradient 114
4.8	Lightning protection for a line circuit 115

5 Train Detection 117

5.1	TI21 basic track circuit block diagram 118
5.2	TI21 transmitter block diagram 118
5.3	TI21 receiver block diagram 119
5.4	Tuned area with one transmitter and one receiver — wiring schematic for normal power mode 121
5.5	General arrangement for a typical axle counter track system 122
5.6	Coupling between transducer, transmitter and receiver 123
5.7	Trackside electronics for an axle counter detection point 124
5.8	Schematic diagram of axle counter evaluation unit 125
5.9	Principal signals in the axle counter system 126
5.10	Axle counters in a multi-section configuration 127
5.11	Typical transponder system for train detection 129
5.12	Block diagram showing the track transponder and train-borne interrogator 131
5.13	Practical application of transponders on a simple double line section of track 132
5.14	Practical application of transponders on simple single line with passing loops 133

6 Level Crossings 134

6.1	Automatic half barrier crossing — layout 135
6.2	Open crossing with no controls — layout 135
6.3	Automatic open crossing locally monitored — layout 136
6.4	CCTV monitored remote barrier crossing — layout 138
6.5	Automatic half barrier crossing — bidirectional strike-in circuits 141
6.6	Automatic half barrier crossing — double line control circuits 142
6.7	Barrier and open crossings — road signal lighting and proving circuits, audible warning circuits 144
6.8	Automatic half barrier crossing — boom operating and indication circuits 145
6.9	Automatic half barrier crossing — indication circuits to signalbox 147
6.10	Automatic half barrier crossing — single line basic control circuits 148
6.11	Automatic half barrier crossing with signal regulation 150
6.12	Automatic half barrier crossing with signal regulation, continued 151
6.13	Automatic open crossing locally monitored — bidirectional strike-in circuits 154
6.14	Automatic open crossing locally monitored — control circuits 155
6.15	Automatic open crossing locally monitored — signal circuits 157
6.16	Automatic half barrier crossing locally monitored — bidirectional strike-in circuits 159
6.17	Automatic half barrier crossing locally monitored — double line control circuits 160
6.18	Automatic half barrier crossing locally monitored — rail signal circuits 163
6.19	Automatic half barrier crossing locally monitored 164

6.20	CCTV monitored remote barrier crossing — typical layout — signal and train approaching circuits 169
6.21	CCTV monitored remote barrier crossing — signal control and interlocking circuits 171
6.22	Manned barrier crossing — typical control console — indication and alarm circuits 172
6.23	CCTV monitored remote barrier crossing — control and indication circuits 173
6.24	CCTV monitored remote barrier crossing — miscellaneous circuits 175
6.25	CCTV monitored remote barrier crossing 176
6.26	CCTV monitored remote barrier crossing — barrier control circuits 178
6.27	Remote barrier crossing — boom operating and indication circuits 179
6.28	Crossing with miniature warning lights — control circuits 180

7 Equipment 183

7.1(a)	DC neutral line relay BR 930 183
7.1(b)	AC immune line relay BR 931 184
7.1(c)	Biased AC immune line relay BR 932 184
7.1(d)	Slow pick-up neutral relay BR 933 185
7.2(a)	Magnetically latched relay BR 935 186
7.2(b)	Polarised magnetic stick relay BR 936 186
7.3	Typical arrangement of banner repeating signals with right hand route 189
7.4	'Toton' position light speed signal 191
7.5	Clamp lock standard detection circuit 194
7.6	Train-operated points 196
7.7	Typical arrangement of lifting barrier unit 199

8 Operator Interface 201

8.1	The networks and systems of IECC 202
8.2	IECC general subsystem configuration 203
8.3	IECC signalman's display — typical overview 206
8.4	IECC signalman's display — typical detail view 207
8.5	IECC general purpose display 209
8.6	IECC keyboard layout and legends 210
8.7	ARS — illustration of principle 214
8.8	ARS — conflict assessment 216
8.9	ARS — timetable and predicted train graphs 217
8.10	Timetable processor 221
8.11	CTC office control machine face layout 231
8.12	CTC typical field station indications in control office 233

9 Signalling the Passenger 235

9.1	Average ratings of attributes of importance scale 236
9.2	Split flap indicator 237
9.3	CCTV monitor 238
9.4	Character generator-based system schematic 239
9.5	Digital-to-video converter-based system schematic 240
9.6	Dot matrix LED display 241
9.7	Liquid crystal display 243
9.8	Manual control unit 245
9.9	Schematic of long line public address system 246
9.10	Schematic of complex information system 248
9.11	Schematic of London Underground dot matrix LED information system 253
9.12	Proposed schematic of information generator for IECC systems 255

10 Automatic Train Protection 258

10.1	Speed supervision to prevent run-by 259
10.2	ATP system and architecture 261
10.3	Train subsystem architecture 262
10.4	Track subsystem architecture 263
10.5	Track-to-train transmission 264
10.6	Possible effects on headway 266
10.7	Effects of in-fill 267
10.8	Release speed 268
10.9	Typical telegram sequences 270
10.10	ATP indication, warning and intervention curves 273
10.11	ATP driver's display 276
10.12	Signal interface with SSI system 280

11 The Future 282

Index 287

Preface

As we enter the 1990s, there is a welcome resurgence of interest in the value and viability of rail transport, both in terms of the economics of urban, intercity and freight railway systems and also in the environmental aspects of rail compared with other modes.

The future for rail transport in the UK appears brighter now than it has done for decades and there are exciting prospects opening up for new railway projects like Channel Tunnel and for improvements and modernisation to existing railway infrastructure.

The signal engineer has an essential and vital role to play in this new era for rail transport, since with that professional engineer lies the responsibility for ensuring that control systems are provided which permit rail traffic to flow safely and efficiently.

The aim of this book is to furnish a reference work of the signalling technology in present use on British Railways. It is presented primarily for serious study by those intimately concerned with the design and provision of signalling systems. However, it is also hoped that it will be of general interest to the large and enthusiastic body of people who just believe in railways and rail transport.

Signalling technology is constantly evolving and therefore this work can only be up to date to the extent it was possible to include the latest developments at the time of final editing before going to press.

I wish to place on record my thanks and appreciation for the hard work and commitment of my colleagues on the IRSE Textbook Project Group who have worked so well with all those who have contributed to bringing this work to publication.

Finally it is my hope that students of signal engineering, practising professional signal engineers and also those with just a general interest in signalling matters will find this volume a treasured and valuable addition to their bookshelves.

K W Burrage
Chairman, IRSE Textbook Project Group

CHAPTER ONE

Recent Changes in Signalling Philosophy

Throughout the history of engineering, the engineer has had the problem of keeping up with changes in technology, changes in business requirements and lessons learned through experience. The railway signal engineer has not been immune to these, particularly in the last 20 years with the rapid development of electronics, including computers, and the changes in the way British Railways is managed as a business.

The existing signalling still relies on the driver to interpret the lineside signals and to control his train accordingly. This inevitably leads to the possibility of misunderstanding due to psychological effects which the engineer must be aware of, and he must design his systems to minimise the danger.

This chapter gives an insight into some of the changes in the philosophy of signalling and the reasons for these changes.

Junction Signalling

The basic concepts of junction signalling remain the same now as they were when *Railway Signalling*, the first volume of the IRSE Textbook (hereinafter referred to as the 'Textbook'), was written. The refinements that have taken place include the following:

Approach Release from Red
If a signal is approach released from red, this inevitably means that trains regularly approach the signal at danger and drivers get accustomed to the signal clearing for the diverging route, as they get close to the signal. This situation can 'condition' drivers to such an extent that on occasions when the signal does not clear as the driver expects, the train over-runs the signal with the obvious danger of an accident.

This problem can be minimised by allowing the approach released signal to clear as early as possible so that on occasions when the signal is not going to clear the driver has more time to realise this.

It might be thought that allowing the junction signal to clear as soon as the train passes the signal in rear, which will be showing single yellow, would be a solution to this problem. However this can cause other problems in that if the signal is allowed to clear when the train is a long way from it, the driver may see the change of aspect in the distance, but may not see the junction indicator. He may then assume that the signal has been cleared for the straight, fast, route and accelerate to an unacceptably high speed for the junction before sighting the junction indicator.

These two conflicting requirements can both be met by allowing the signal to clear as early as possible after the train has approached close enough for the driver to see the junction indicator. Since a position light junction indicator is considered to be visible at a distance of 738 m the signal will generally be allowed to clear when the train reaches this point. If, due to sighting problems, the junction indicator is not visible at the same time as, or before the main signal aspects, the clearance will be further delayed until the train reaches the point when the junction indicator is visible.

Conversely if the signal is visible for less than 738 m and the junction indicator comes into view before, or at the same time as the main aspects, there is no reason to prevent the junction signal from clearing as soon as the train has passed the signal in rear at yellow. This arrangement has the added advantage that the driver does not see the junction signal at red unless he should be stopping at the signal.

It was felt desirable at one time that in addition to the junction indicator, train drivers should also have some other form of warning that the junction was set for a diverging route. Therefore the signalling was arranged to give an AWS caution warning as the train passed over the AWS inductor for the junction signal. In order to achieve this, the junction signal was held at yellow (or double yellow in 4-aspect signalling) until the train had passed over the inductor.

This arrangement was found to cause trains to be delayed, particularly if the next signal along the diverging route was not far beyond the junction. At other places, it also resulted in the junction signal giving a yellow aspect when no such aspect should be displayed because, for example, the next signal was a distant signal several miles away and the correct proceed aspect at the junction could only be green.

Maintaining the junction signal at yellow until the train had passed over the AWS inductor has therefore been discontinued and the junction signal can now display the least restrictive aspect the conditions ahead allow as soon as it is cleared.

Approach Release from Yellow with Flashing Yellow in Rear

Following several years' experience of the use of flashing yellow aspects for junction signalling, the requirements have been amended to allow for greater flexibility in application.

The table of speeds given in the first volume of the Textbook for determining when to provide approach release from yellow in preference to approach release from red, has been totally abandoned. The provision is now based solely on operating expediency and this has led to some cases where low speed (eg 12.5 km/h) junctions have been provided with signals approach released from yellow where a large proportion of trains take the diverging route. The holding of the junction signal at yellow (or double yellow) until the train has passed over the AWS inductor has been discontinued for the same reasons given above under approach release from red.

The description of approach release from yellow given in Volume 1, is still correct, except that the junction signal is now prevented from displaying an aspect less restrictive than single yellow after the train passes the signal in rear, showing flashing single yellow only if: (i) the junction signal can be seen from a distance greater than 738 m; or (ii) the junction indicator is visible only after the main signal aspects; or (iii) the signal beyond the junction is at danger because of conditions ahead of the junction. If the signal is held at yellow for the first two of the foregoing reasons, it is allowed to show the least restrictive aspect which the conditions ahead permit once the train is within 738 m of the signal and the junction indication is visible.

In Volume 1, it is stated that if a train is within 277 m of the signal to the rear of the junction signal at the time the latter signal is about to clear, or the flasher unit for the former signal is faulty, then the junction signal will not be allowed to clear by approach release from yellow, but instead must be held at red and be approach released from red.

This is done to avoid the possibility of the driver, having received a steady single yellow on the signal to the rear of the junction signal, seeing at a distance the main proceed aspect on the junction signal and mistaking this for the latter signal cleared for the straight, fast route.

Consideration will show that provided the junction signal main aspects do not come into view before the route indication and cannot be seen at greater than 738 m, this danger should not occur and therefore these controls are no longer provided in such cases.

Junctions that are not Approach Released

If the speed of the junction is equal to the 'line speed' allowed for the fastest (straight) route, then a train travelling at full line speed may safely approach and travel through the junction without reducing speed. In such cases, therefore, the signal in rear of the junction can show green whilst the junction signal shows green with a route indication, more restrictive aspects being displayed only if required by conditions ahead of the junction.

At present any junction which can be travelled through at line speed or within 6.25 km/h of line speed is not approach released.

A change is being considered to allow a greater speed difference than 6.25 km/h where the junction signal is a considerable distance from the junction points. This may provide a means of signalling medium speed (eg 31 km/h) junctions on provincial lines with modest line speeds (eg 47 km/h) without the cost of providing flashing yellows. It will be necessary to ensure that a driver is given adequate time to reduce speed together with an AWS warning in this situation.

Possible Future Developments in Junction Signalling
The use of position light junction indications can cause confusion in certain installations where it has been found that drivers think an indication at position 4 is being displayed when in fact position 3 is illuminated. This problem is particularly prevalent at night. Similar confusion has been found between positions 1 and 6.

It is therefore proposed that if on a particular signal, position 4 is provided, position 3 must not be provided, likewise if position 1 is provided, position 6 must not be provided.

Concern has been expressed about the possibility of drivers forgetting which junction they are approaching with a flashing yellow sequence. It is proposed that if junctions are close together and of different speeds only one of them may be approach released from yellow and the other junction must be approach released from red.

Combining of Berth and Overlap Track Circuits

The berth and overlap track circuits of automatic signals may be combined as one track circuit, but not at a controlled or semi-automatic signal. Some regions of BR have allowed this practice in the past, particularly if the signal has a reduced overlap.

The combining of berth and overlap track circuits prevents both the interlocking and the signalman from being able to detect whether a train stopped on the track circuit has come to a stand in rear or in advance of the signal. This means that a train which stops a short distance in advance of the signal will not be detected as having passed the signal, and the route locking will not then be held after the approach locking has been released. If, as may happen, the signalman has cancelled the route because he thinks the train has stopped on the approach to the signal, the driver, having passed the signal at green, may then restart his train and proceed through the route which is no longer locked. This potentially dangerous situation is mitigated only by the rules and regulations required to be followed by signalmen and train crews.

Several diagrams and statements in the first volume of the Textbook reflect the former practice of combining berth and overlap track circuits. These situations in modern practice require separate overlap track circuits.

Release of Approach Locking

Volume 1 of the Textbook covers two methods by which the approach locking of signals may be released by the passage of a train. The standard method of operation adopted for use on BR is that the first track in advance of the signal should be clear following the simultaneous occupation of the first and second tracks in advance of the signal.

The reference states that another precaution taken where there are very long distances between stop signals is to increase the time interval to 4 min. A 'long distance' in this context is accepted to be in excess of 830 m between the signal displaying single yellow (in 2-, 3- or 4-aspect signalling) and the signal at danger.

The introduction of solid state interlocking (SSI) has enabled the time interval allowed for the release of approach locking to be individually set at no extra cost for each route from a signal and also for each class of aspect displayed for a route. This means, for example, that a signal having displayed a call-on aspect may release after only 30 s, whereas if it were part of a relay interlocking it would normally have used the same timing relay as that associated with the main aspect, with a release time possibly as long as 240 s.

Emergency Replacement of Automatic Signals

Early installations with automatic signals had no provision for replacement of these signals from the signalbox because this was regarded as expensive and unnecessary.

Following an incident during the 1960s, replacement facilities were provided on certain automatic and semi-automatic signals. After another incident during the 1980s, it has been decided on all future schemes to provide all signals with emergency replacement facilities. Fortunately the additional cost will be minimal with SSI, because all signal controls are located within the central interlocking at the signalbox.

Time of Operation Locking

This is the locking which is applied to facing points in an overlap if they are within the track locking distance of the protecting signal. BR has now introduced a standard track locking distance of 20 m.

Subsidiary and Shunt Signals

The proceed aspect displayed by both subsidiary and shunt signals is identical (ie two white lights), and the meaning to the driver is also identical for both. In other words he is to proceed cautiously towards the next signal or buffer stop being prepared to come to a stand short of any obstruction. Thus the use of route indicators to differentiate between a subsidiary call-on signal and a 'shunt ahead' no longer applies.

Subsidiary signals, except semaphore signals, are now referred to as position light signals associated with a main aspect, and colour light shunt signals are referred to as position light signals.

Route indicators are now provided in association with position light signals only when justified because:

- The physical layout of the different routes makes it essential to provide the indications, for example some routes are available to electric traction, others are not. The route indications may be 'grouped', for example all routes available to electric traction may have route indication 'E', those not available, indication 'D'.
- The signal has more than one route for which a position light signal can be displayed and at least one route leads to a 'limit of shunt' signal.
- The position light signal is associated with a main aspect for which a route indication is provided. A position light junction indicator if provided for use with the main aspect, is also used when the position light signal is cleared for the same route.

The former limit of shunt sign was made of glass and proved to be a favourite target for vandals. The new type of limit of shunt signal is similar to a standard shunt signal except that it displays two horizontal red lights and no white lights (see Chapter 7).

Point to Point Locking

The advantage of providing this form of locking is that if the route setting equipment is out of action and the signalman is operating the points using the individual point switches, the locking provides some protection against signalman error.

The disadvantage is that if a set of points fails to operate, other sets of interlocked points are unable to be moved from the signalbox until the failed set is repaired. This necessitates the moving of points by hand (ie a member of staff has to 'crank' the points on site and then clip them).

The delays to rail traffic are therefore likely to be increased because of point to point locking. Safety may also be reduced because of the relatively high probability of an incident occurring as a result of a misunderstanding between the signalman and the man working the points by hand. This risk is reduced if the signalman is able to continue to operate some of the points from the signalbox.

Therefore point to point locking is not provided in the most modern power signalling schemes. The omission of this form of locking does mean, however, that additional locking may be required elsewhere, because provision of point to point locking has in the past enabled the simplification of other types of interlocking.

'Warning' Class Routes (Delayed Yellow)

This class of route is referred to several times in the first volume of the Textbook. It should be emphasised that in modern practice, a 'warning' route only sets if a separate exit button is operated by the signalman to show that he wishes to select the warning route and there is the required overlap of 46 m minimum which is proved by a separate track circuit not combined with the berth track circuit.

Simplified Bidirectional Signalling

There are occasions, during engineering works and emergency working, when it is necessary to allow for single line working over one track of a two-track railway.

The traditional method of achieving this has been by introducing pilot working where the pilot has acted as a 'human token' for the single line working. This has the disadvantage that it takes time to organise and a sizeable number of staff to implement. It is therefore totally unsuitable for emergency use, such as allowing trains to pass a failed train, and when used for planned engineering works, causes considerable delays.

In order to provide a more satisfactory solution, bidirectional signalling has been provided in recent years between cross-overs. Initially this was full signalling complete with AWS equipment. The cost of this full bidirectional signalling was found to be so high that with the exception of a few very busy routes it proved to be uneconomical. A review of the requirements for bidirectional signalling produced what is known as simplified bidirectional signalling (SIMBIDS).

6   RECENT CHANGES IN SIGNALLING PHILOSOPHY

Fig. 1.1 SIMBIDs line fitted with TOWS

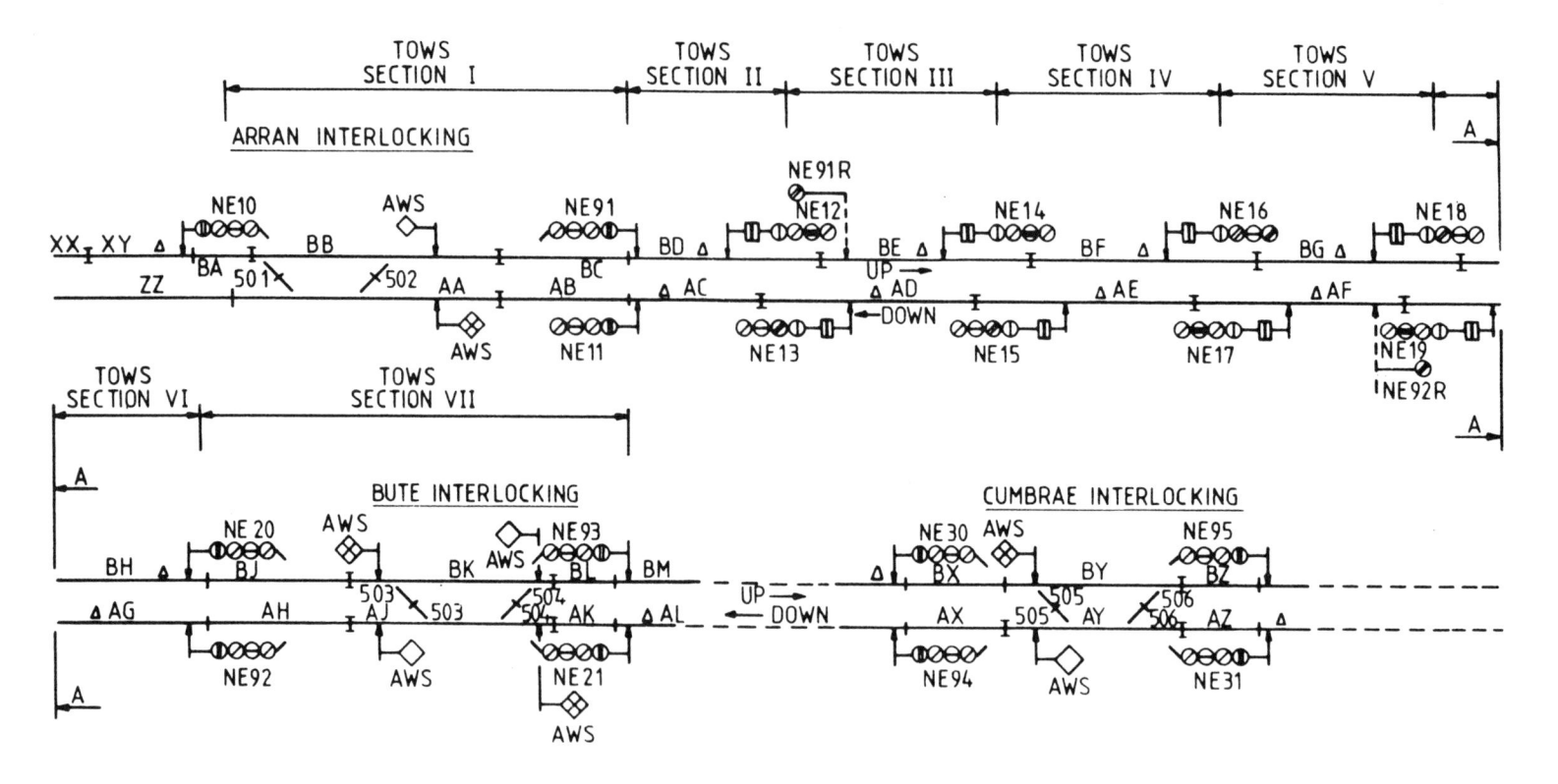

Fig. 1.1 gives a simple example of SIMBIDS. The following points should be noted:

- The 'wrong' direction signals do not have any AWS equipment.
- The 'right' direction signal AWS equipment is not suppressed for wrong direction trains.
- The wrong direction signals are normally mounted to the right-hand side of the line to which they apply.
- The fact that a train is entering or leaving a section on which SIMBIDS is provided is indicated to the driver by the provision of special AWS signs (see **Figs. 1.1** and **1.2**).
- The signals at the end of each SIMBIDS section can be approach released from red for all routes. If they are not, the fixed yellow signals shown in **Fig. 1.1** become yellow/green signals.
- The maximum speed allowed in the wrong direction is 44 km/h.
- Patrolman's lockout devices are provided to prevent wrong direction running whilst the patrolman is walking

along the track, unless a train operated warning system has been provided throughout the section for other reasons.
- Intermediate wrong direction signals may be provided where justified to reduce the headway and must be provided if the section is more than 6.25 km in length.

Staff Warning Systems: TOWS and ILWS

The increase in train speed and the introduction of more bidirectional lines has made the job of working on the track more difficult and hazardous, especially on curved lines.

The traditional method of warning staff has been to provide lookout men who sound whistles, sirens, etc, when a train is seen to be approaching. On curved lines with high speed working, this can require the provision of advanced lookouts who signal, with blue and white flags, to the lookout stationed near the men working.

The provision of lookouts and advanced lookouts can require as many as five staff from a typical gang of nine men. This can obviously cause problems and is expensive.

The provision of a fixed trackside train operated warning system (TOWS) has therefore been justified in some places. This consists of sirens placed on poles at 200 yard intervals through the section. The sections are chosen so that the warning given to staff is a minimum 25 s and maximum 90 s for trains approaching at line speed.

All clear is indicated by an intermittent so-called safe tone on the sirens; a warning is given by a continuous tone. The system is provided with switches at access points so that it can be switched on only when required, thus reducing the annoyance to local residents caused by unnecessary operation of the sirens.

Fig. 1.2 Special signs for SIMBIDS equipped lines

A typical example of a TOWS system on a SIMBIDS line is shown in **Fig. 1.1**. The circuits are arranged such that:

- A minimum of 25 s warning is given to staff for trains approaching at line speed when the staff are working at the beginning of a section, and a maximum of 90 s warning time when they are working at the far end of the section.
- The warning is given for trains approaching in both right and wrong directions.
- The warning ceases as the train leaves the section. This requirement means that on bidirectional lines the strike-in track circuit for one direction must be inhibited for trains in the opposite direction. This introduces a potentially dangerous situation since the right side failure of a track circuit can appear as a train receding from the section and hence can suppress the warning. This situation is prevented from becoming dangerous by only allowing the inhibition of a strike-in track circuit to be maintained for a short time. The inhibition becomes ineffective when a vital timing relay is de-energised.
- If a train is routed away from the TOWS section, the warning is inhibited.
- If a train is waiting at a signal which is regularly used to stop trains, the warning will be inhibited until the signal is ready to clear. The signal will then be held at danger until the warning has sounded for the required time, after which the signal will clear. The delayed clearance of the signal is not enforced when the TOWS system is switched off.

The TOWS system has proved useful in certain places but it does however have the following shortcomings:

- The system, when in use, emits a loud noise that makes it unsuitable in residential areas.
- There is no indication of which line the train is on so that on busy multi-track sections, staff have to stop work whenever a train approaches on any line when in many cases they could continue to work, provided that no train is approaching on the line they are working on or standing foul of.
- There is no special indication of a second train approaching: the warning tone continues until the second train passes. Unfortunately the warning tone also continues in any case until the first train leaves the section, and since this is the far more common reason for the warning tone sounding after the train has passed, it has been found that some staff have become 'conditioned' to assume this is always the case. Thus although the rules require staff to remain clear of the track until the warning returns to safe tone there is a risk that they may in fact resume work as soon as the first train has passed the work site, and some staff have, as a result, been killed by a second train approaching.

The inductive loop warning system (ILWS) is being developed to provide a staff warning system which avoids the shortcomings of the TOWS.

The principal advantages of ILWS are:

- The warning tones are only produced in the headphones worn by the warden (ILWS lookout) protecting the gang working. It can therefore be used in residential areas without causing annoyance.
- An advanced version of ILWS will be available for multi-track lines which will give audible and visual warnings to identify the lines on which trains are approaching.
- A distinctive 'second train coming' warning will be given if a second train passes the strike-in point before the first train has left the section.

- During engineering work when one line is occupied by engineers' trains that are slow moving or stationary, it will be possible to isolate the occupied line so that staff working near the engineering train will still get a warning of trains approaching on other lines which are still open to fast trains. To ensure that this is done safely, a special 'qualified safe tone' is emitted via the warden's headphones instead of the ordinary safe tone. He thus is constantly reminded that no protection is given by the ILWS for the line that has been isolated and that he is responsible for providing other means of lookout protection for movements of the engineering trains.

It seems likely that before any trains are allowed to run at speeds above 200 km/h the lines concerned will either require to be fully equipped with either TOWS or ILWS, or work on the lines will be banned during periods of high speed operation.

Control Tables

Volume 1 of the Textbook makes use of control table extracts to illustrate certain types of interlocking. Although useful for illustrative purposes, these do not reflect the BR standard formats for control tables. These are provided for routes, points, banner signals, auto-signals, distant signals, AWS, ground frame releases, ground frame controls, level crossings, TOWS, absolute block controls, electric token controls, tokenless block and mechanical interlocking.

It will be appreciated from the list above that the constraints of space allow only a limited number of examples to be included in this book.

The control tables shown in **Figs. 1.4–1.7** are all related to the layout shown in **Fig. 1.3**.

The format for controlled signals requires the use of a separate sheet for each signal route; likewise the point format is for one point number per sheet. The standard notes are shown as $ followed by a number.

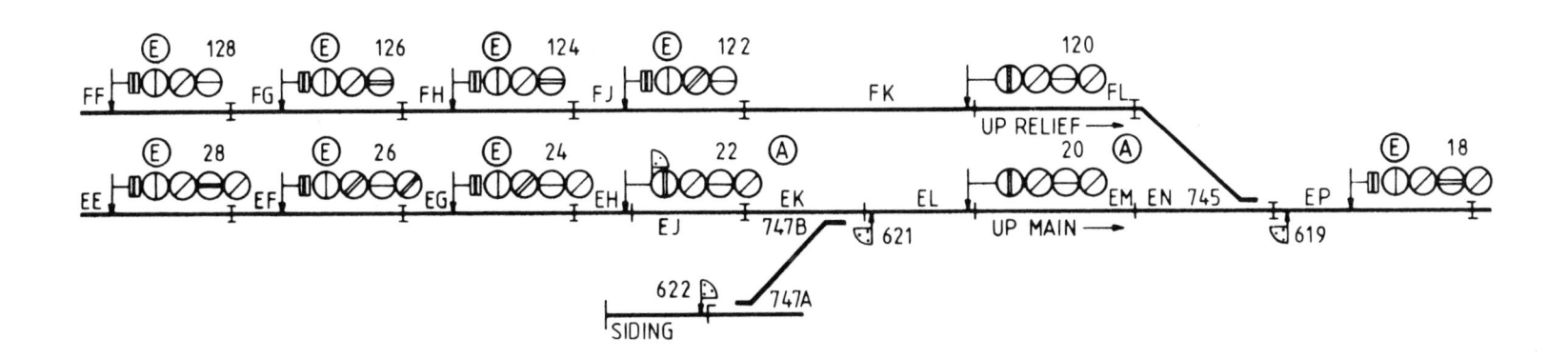

Fig. 1.3 Layout for control table examples shown in **Figs. 1.4, 1.5, 1.6** and **1.7**

	TRACKS CLEAR	TRACKS OCC.	POINTS SET AND DETECTED			SIGNALS ALIGHT	REPLACED BY TRACKS	SPECIAL CONTROLS
			and locked NORMAL	N or R	and locked REVERSE			
SIGNAL CONTROLS REQUIRE	EJ, EK, EL, EM, EN	EH$7	747, 745			20	EH occ, EJ occ	$1

	ROUTES NORMAL	ROUTE LOCKING NORMAL	TRACKS	POINTS SET OR FREE		TORR BY A/L FREE AND TRACKS CLEAR	SPECIAL CONTROLS
				NORMAL	REVERSE		
INTERLOCKING REQUIRES	22(C)		EL	747, 745		NOT PROVIDED	
	619						

	APPROACH LOCKED WHEN SIGNALS CLEARED AND APPROACH TRACKS OCCUPIED	SECTIONS IN REAR	A/L RELEASED BY SIGNAL ON	
			AND TRACKS or	SECS
APPROACH LOCKING CONTROLS	EH, EG, EF		EJ AFTER (EJ occ, EK occ)	180

	ROUTE IND.	ASPECT	EXIT SIG. AT	WITH JN.IND. AT	EXIT SIGNAL BANNER	TRACKS OCC.	ROUTE LOCKING	LOCKING MAINTAINED UNTIL ROUTE NORMAL AND TRACKS CLEAR or T.OCC. $8 SECS		APPLIED TO ROUTES
ASPECT CONTROLS		Y	R					EJ, EK, EL	EL 40	619
		YY	Y							
		G	YY>							
AWS	STANDARD CONTROLS PROVIDED									

Revisions	Drg. No.			
	Produced	Date	TYPICAL CONTROL TABLES	
	Checked	Approved		
	Microfilm No.		ROUTE:- 22(M) EXIT:- 20	Sht. No.
	Issued			

Fig. 1.4 Typical control table for controlled signal — main class route

Route Controls
(**Figs. 1.4** and **1.5**)

The '*signal controls*' section relates to the controls in the GR/HR circuits and the '*tracks occupied*' column refers to the requirements allowing the signal to display a proceed aspect. The '*replaced by tracks*' column relates to the disengager (GSR) track controls and any conditions for last wheel or other non-standard replacement.

The '*N or R*' column relates to points which are proved continuously, except when swung by forward route, conflicting route or point key, in which case the control is over-ridden for 5 s.

The '*interlocking*' section relates to the controls in the reverse lock relay (RLR) circuit. The '*TORR*' column relates to controls of the relay NR which is used to control RLR. The '*tracks*' column is used only when track controls are included in route selection, for example to select between a main or call-on route (**Figs. 1.4** and **1.5**). The '*routes normal*' column contains the routes from the same entrance signal, including facing shunt signals, which are not held by point locking. The '*route locking normal*' column contains the routes which, although initially locked out by point conditions, are freed as a train clears the points and are then directly opposing.

The bottom box which covers both the '*routes normal*' and '*route locking normal*' columns will contain directly opposing routes. When the table is used for an SSI interlocking, the route locking is applied immediately the route is set, and if the route locking is proved, the route does not need to be proved separately; therefore the bottom box is not used and directly opposing routes appear in the '*route locking normal*' column.

In the '*approach locking controls*' section, the '*approach tracks occupied*' column only includes track circuits back to the first controlled signal or signals in the rear which relate to the train approaching relay (TAR).

Each route for which the route in question is an exit will have the track circuits in that route separately listed in this table. This method of presentation simplifies the preparation, checking and testing of comprehensive approach controls through complicated point and crossing areas. The point conditions which may be required to bypass approach track circuits for movements not approaching the signal concerned are not specified. It should also be noted that it is not considered necessary to inhibit approach locking track circuits for movements proceeding away from the signal concerned.

The remainder of the approach track controls are identified in the '*sections in rear*' columns which relate to the ATSR relay, and the full controls will be shown on the route sheets of the signals identified. If the signals in rear are themselves not provided with comprehensive approach locking, then all of the controls are shown in the '*approach tracks occupied*' column; this has been shown for signal 20 (**Fig. 1.4**).

A space is allocated beneath the '*A/L released by signal on*' section heading for the addition of banner repeaters or distant signals when required.

The '*aspect controls*' section relates to controls in the FHR (flashing single yellow), HHR, FHHR and DR circuits. It is assumed that all route indications are required to be proved alight before the signal can clear unless a note is added in the '*route in*' column. The '*with jn.ind.at*' column is used for flashing aspect sequences and the '*tracks occ.*' column relates to controls required for stepping up to a less restrictive aspect on a signal already displaying a proceed aspect.

The '*route locking maintained until route normal*' section relates to any directly opposing locking initiated by this route. To obtain details of the route locking specified in the '*route locking normal*' columns of the '*interlocking*' section, it is necessary to refer to the route sheets of the routes as specified.

'*No, Standard or Special*' is added to the '*AWS controls provided*' box. Special AWS controls are specified on a special AWS control table. The route concerned and its exit button are identified as indicated on the sheet title block.

Fig. 1.5 Typical control table for controlled signal — call-on class route

SIGNAL CONTROLS REQUIRE

TRACKS CLEAR	TRACKS OCC.	POINTS SET AND DETECTED and locked NORMAL	POINTS SET AND DETECTED N or R	POINTS SET AND DETECTED and locked REVERSE	SIGNALS ALIGHT	REPLACED BY TRACKS	SPECIAL CONTROLS
EJ, EK	EH FOR T SECS	747			—	EH occ, EJ occ	

INTERLOCKING REQUIRES

ROUTES NORMAL	ROUTE LOCKING NORMAL	TRACKS	POINTS SET OR FREE NORMAL	POINTS SET OR FREE REVERSE	TORR BY A/L FREE AND TRACKS CLEAR	SPECIAL CONTROLS
22(M)		EL occ	747		NOT PROVIDED	
619						

APPROACH LOCKING CONTROLS

APPROACH LOCKED WHEN SIGNALS CLEARED AND APPROACH TRACKS OCCUPIED	SECTIONS IN REAR		A/L RELEASED BY SIGNAL ON AND TRACKS or	SECS
AS FOR 22(M)			EJ AFTER (EJ occ, EK occ)	30

ASPECT CONTROLS

ROUTE IND.	ASPECT	EXIT SIG. AT	WITH JN.IND. AT	EXIT SIGNAL BANNER	TRACKS OCC.
	PL				

AWS	STANDARD CONTROLS PROVIDED

ROUTE LOCKING

LOCKING MAINTAINED UNTIL ROUTE NORMAL AND TRACKS CLEAR or	T.OCC.	$8 SECS	APPLIED TO ROUTES
EJ, EK, EL	EL	40	619

Revisions		Drg. No.			
		Produced	Date	TYPICAL CONTROL TABLES	
		Checked	Approved		
		Microfilm No.		ROUTE:- 22(C) EXIT:- 20	Sht. No.
		Issued			

Point Controls
(Fig. 1.6)
One sheet is used for each set of points, divided between the controls required to move the points from normal to reverse (N→R) and those for moving the points from reverse to normal (R→N). In complicated areas, a separate sheet may be needed for each. Where the route required is normal with no consequent extended route holding, there is no entry in the '*route locking released by*' section (see entry for 619 route, **Fig. 1.6**).

The '*route locking released by*' columns are used for routes that directly lock the points one way and those that cause counter-conditional locking to be applied to swinging overlap points. It may be noted that these columns specify the conditions for releasing the controls applied under columns '*requires routes and route locking normal*' and '*routes and route locking normal*' when the controls in these columns are normal. If controls in the '*requires routes and route locking normal*' column are not accompanied by an entry in the '*route locking released by*' columns, this implies that the route normal only is required.

The '*time of operation locking*' section is used for the control of facing points by the berth track circuit when the points are close to the signal. Note that the '*when routes used*' column may be left blank if there are no signalled routes away from the points and all routes up to the signal are main class routes.

Notes Relating to Control Table Examples
(Figs. 1.4 and 1.5)
The entry marked $7 (**Fig. 1.4**) relates to a control effective only when the temporary approach release link is withdrawn. The entries for 745 points in **Fig. 1.4** refer to trailing points in the exit signal overlap and therefore do not appear in **Fig. 1.5** which relates to route 22(C).
The special control $1 in **Fig. 1.4** relates to the provision of automatic working controls. The entry for EL track clear within the interlocking controls (**Fig. 1.4**) is the control which provides the automatic selection between a main and a call-on class of route. EL occ is the equivalent control in **Fig. 1.5**.
The '*sections in rear*' columns are unused because signals 24 and 26 have no approach locking of their own and hence all the approach locking tracks are listed.
The '*route locking*' section refers to the locking that is applied to other routes when the route in question has been set, and lists the conditions required to release the locking. The time 40 s shown is notional and is only for illustrative purposes.
The entry 'EJ after (EJ occ, EK occ)' relates to the TASR relay and the time of 180 s (30 s in **Fig. 1.5**) relates to the AJR relay. For non-SSI interlockings, the time in **Fig. 1.5** would be the same as in **Fig. 1.4** (180 s) in order to minimise the number of timer relays required. The time of 180 s will vary with signal spacing and is therefore given only as an example.

YY > means double yellow or less restrictive aspect.

(Fig. 1.6)
The entry 'EL occ' 40 s relates to the fact that 745 points are called by 22(M) route as trailing points in the overlap of signal 20. Therefore if a train is proved to have occupied the berth track circuit of signal 20 for sufficiently long that it can reasonably be assured that the train has stopped, the overlap may be released.

The route locking tables list the track circuits up to, but not including, the deadlocking track circuit EN. The route locking relays [(UP) and (DN) USRs] proved in the point circuits will normally be EN(UP)USR, EN(DN)USR, and EN(UP)(O/L)USR.

(Fig. 1.7)
The note $5 relates to the provision of a track circuit over-ride button. The entry '621 or 621 off' is required because 619 must not be allowed to clear if 621 is out when it should display a red aspect, since 621 acts as the limiting signal for wrong direction movement.

The entry 'EP or EN after (EP occ EN occ)' refers to last wheel replacement.

14 RECENT CHANGES IN SIGNALLING PHILOSOPHY

Fig. 1.6 Typical control table for points

	REQUIRES TRACKS CLEAR	SET BY ROUTES OR GF	REQUIRES ROUTES AND ROUTE LOCKING NORMAL OR POINTS / GF	ROUTE LOCKING RELEASED BY			SWINGING OVERLAP REQUIRES		
				TRACKS CLEAR or	T.OCC. $8 SECS		TRACKS CLEAR and	POINTS SET OR FREE or	ROUTES AND ROUTE LOCKING NORMAL
POINTS CALLED N -> R	EN	120	20	EM					
			22(M)	EM, [EJ, EK, EL — — — — —EL	40]				
			619						
POINTS CALLED R -> N		20, 22(M), 619	120	FL					

TIME OF OPERATION LOCKING	EFFECTIVE	WHEN ROUTES USED	REL. BY	TRACKS CLEAR or	T.OCC. $8 SECS

Revisions		Drg. No.		TYPICAL CONTROL TABLES	Sht. No.
		Produced	Date		
		Checked	Approved		
		Microfilm No.		POINTS:- 745	
		Issued			

RECENT CHANGES IN SIGNALLING PHILOSOPHY 15

Fig. 1.7 Typical control table for position light shunt signal

	TRACKS CLEAR		TRACKS OCC.	POINTS SET AND DETECTED			SIGNALS ALIGHT	REPLACED BY TRACKS	SPECIAL CONTROLS
SIGNAL CONTROLS REQUIRE				and locked NORMAL	N or R	and locked REVERSE			
	EN, EM, (EL $5)		745				621 OR 621 OFF	EP OR EN AFTER (EP occ. EN occ)	

	ROUTES NORMAL	ROUTE LOCKING NORMAL	TRACKS	POINTS SET OR FREE		TORR BY A/L FREE AND TRACKS CLEAR	SPECIAL CONTROLS
INTERLOCKING REQUIRES				NORMAL	REVERSE		
			745			NOT PROVIDED	
	20, 22(M), 22(C), 622						

	APPROACH LOCKED WHEN SIGNALS CLEARED AND APPROACH TRACKS OCCUPIED	SECTIONS IN REAR	A/L RELEASED BY SIGNAL ON	
APPROACH LOCKING CONTROLS			AND TRACKS or	SECS
	EP		EN AFTER (EN occ. EM occ)	30

	ROUTE IND.	ASPECT	EXIT SIG. AT	WITH JN.IND. AT	EXIT SIGNAL BANNER	TRACKS OCC.	ROUTE LOCKING	LOCKING MAINTAINED UNTIL ROUTE NORMAL AND TRACKS CLEAR or T.OCC. $8 SECS		APPLIED TO ROUTES
ASPECT CONTROLS		PL						EN, EM		20
								EN, EM, EL	EL 40	22(M), 22(C), 622
AWS	NO CONTROLS PROVIDED									

Revisions		Drg. No.				
		Produced	Date	TYPICAL CONTROL TABLES		
		Checked	Approved			
		Microfilm No.		ROUTE:- 619 EXIT:- 621	Sht. No.	
		Issued				

16 RECENT CHANGES IN SIGNALLING PHILOSOPHY

Fig. 1.8 Typical control table for CCTV monitored level crossing

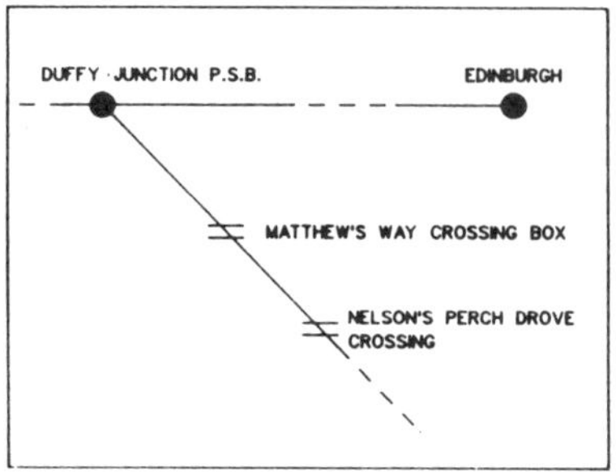

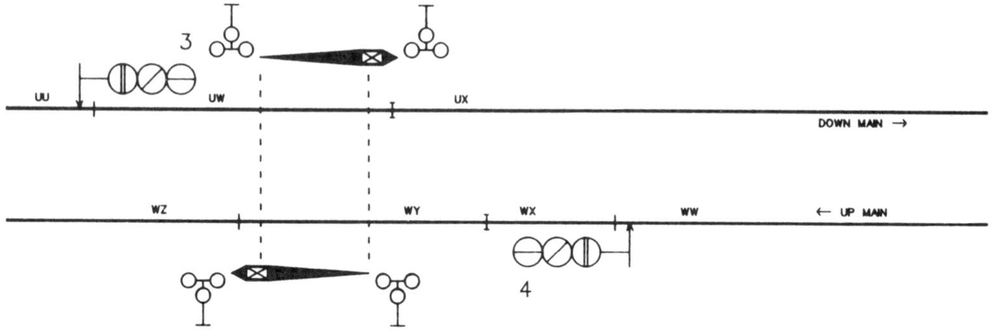

Control Table for CCTV Crossing
(Fig. 1.8)
This fig. shows the layout and controls of a CCTV monitored level crossing. The same type of control table format is used for manned, remote and trainman-operated full barrier crossings.

The example assumes that the signals are only controlled because of the crossing and would otherwise have been automatic signals. The approach locking controls are therefore provided, and shown, as part of the crossing controls and not the signal route controls.

The first three columns from the left relate to the commencement of the warning sequence and the subsequent lowering of the barriers. These columns are left blank because 'auto lower' facilities have not been provided.

The '*road light warning initiated*' columns relate to the provision of over-run track circuits which cause the road light sequence to commence automatically if a protecting signal is passed at danger.

The columns headed '*except when...*' relate to the disengaging controls applied to the track circuit controls when the train occupying the track circuits is heading away from the crossing.

The column '*inhib. by...*' is not used in the example, but is needed when shunt class routes read over the crossing and it is required to inhibit the operation of automatic raise.

The column '*local switches*' refers to the switches provided to allow the crossing to be operated from a control cabinet adjacent to the crossing during times of equipment failure.

CHAPTER TWO
Solid State Interlocking

Introduction

The development of digital electronic technology in the 1960s held out the offer of a cheap and flexible substitute for relays in performing the interlocking function. For a long time, however, electronics made no impact on signal control and interlocking because of the problem of complying with the very demanding safety standards.

During the 1970s the principles of the techniques described below for assuring safety with non-fail-safe components were developed, but it was not until the advent of the microprocessor that the solid state revolution in signalling became possible. There were now available mass produced devices with sufficient processing power to perform complex tasks such as signal interlocking or train control, whilst at the same time being able to carry out the redundancy management and self-checking logic necessary to ensure safe operation in equipment capable of being contained in a robust package suitable for operation at the trackside.

The objectives of British Railways in developing a solid state interlocking were to provide all the facilities available in existing relay interlockings at significantly reduced cost while maintaining high levels of safety and reliability. It was also considered important that the technology should be adapted to match the existing human resources of BR, so that with a minimum of specialist training, signal design and maintenance staff would be able to apply the new system.

These objectives were a major influence in shaping the SSI system. The configuration of concentrated central interlocking, vital lineside data link, and direct drive of signals and points from electronic modules was aimed at maximising savings in relays, multicore cables and equipment rooms. The safety and reliability targets determined the hardware redundancy strategies.

The need for signal designers, rather than software specialists, to be able to configure the system for each installation, dictated the requirement for the interlocking to be programmed by geographical data which is interpreted by the fixed program resident in the interlocking. The modularity of the system and its built-in diagnostic facilities were determined by the need for it to be maintained by existing technicians.

The design originated in work done at the British Railways Research and Development Division at Derby, and development was completed under a tripartite agreement between British Railways and the British signalling contractors GEC-General Signal Ltd and Westinghouse Signals Ltd. This provided for manufacture of SSI equipment by the contractors, subject to type approval by BR. Design and validation of programs was carried out by BR.

Safety and Availability Techniques

SAFETY BY REDUNDANCY

The remarkable processing power and low cost of modern micro-electronic systems make them potentially very attractive in process control applications, of which railway signalling is a specialised case.

Unfortunately, there is no conceivable way in which the components of a micro-electronic system can be regarded as inherently safe or can be shown to be safe by failure mode analysis. Such components are so complex that their modes of failure are effectively indeterminate. It is therefore impossible to use micro-electronic technology to construct equipment which satisfies the safety requirements of signalling without using some form of redundancy.

Redundancy may be defined as the provision of more physical resources than would be necessary to perform a function if perfect reliability could be assumed. The concept is not new, and much of the pioneering work on the behaviour of redundant systems was carried out in the 1950s. The application of redundancy to complex electronic systems is of course more recent, and its application to computer systems, particularly in the domain of real time process control, is very much a developing field of engineering. The remainder of this section will deal in some detail with the application of redundancy principles to the design of safe systems. Further applications of redundancy will be described in later sections.

Redundant System Design
Conceptually the simplest way of applying the principles of redundancy to the design of safe systems is to duplicate the equipment and demand identical results from the two subsystems. In the context of microprocessor-based equipment, such a system would consist of two processors connected to the same set of inputs and programmed to perform the same function, and a comparator whose task is to allow the system to operate only if the two sets of processor outputs are in agreement. At all other times, a predetermined safe set of output states must be enforced, and the comparator is of course required to perform this task in a safe manner.

The conceptually simple idea outlined above and illustrated in **Fig. 2.1** has a number of philosophical and practical shortcomings which it is instructive to consider.

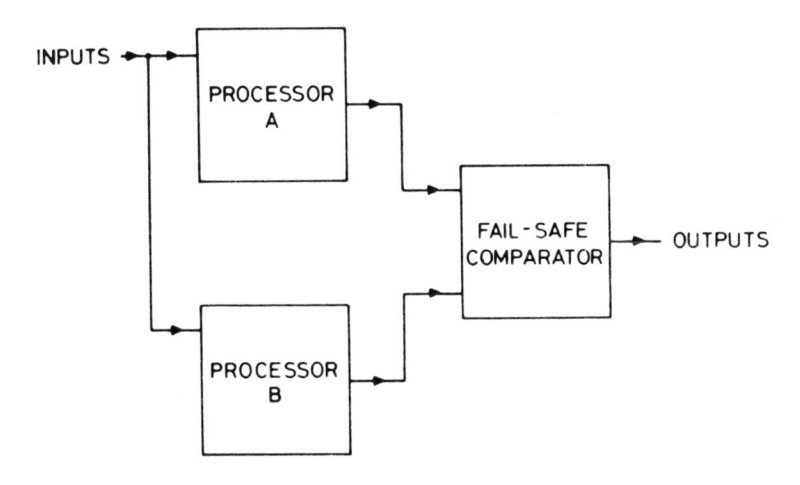

Fig. 2.1 Duplication for safety — basic concept

Redundancy Management

In ensuring that both processors agree and enforcing the safe condition if they do not, the comparator is performing a function which may be termed redundancy management. The need for the comparator to be inherently safe is an unattractive feature, resulting in an unhappy mixture of technologies as well as considerable expense. In a microprocessor-based system a much more elegant solution is possible, in which the comparison process is performed redundantly within the two subsystems and the safe output condition is enforced by means of special redundancy management hardware, which need not be inherently safe provided the shutdown mechanism is itself redundant and fully testable by both processors.

Common Mode Failures

If the two subsystems are identical in design, it is clear that design errors causing the system to carry out the wrong function would not be detected by the comparator. To avoid this risk, the designer can use different designs for the two systems (in which case the design is said to be diverse), or can demonstrate that the design is free of errors which might lead to unsafe behaviour. This choice applies to both hardware and software, and systems have been designed in which both are diverse. However, the penalties to be paid for this approach are considerable, as two development teams are required for both hardware and software, design freedom is restricted, and two types of equipment have to be maintained and stocked as spares.

Whether or not the design is diverse, no single fault or probable combination of faults must be capable of causing the same malfunction in each of the two subsystems. Care is required in the design of connections between the redundant system and the outside world, in the design of power supplies, and in providing effective electrical isolation between subsystems.

Fault Detection Coverage

In the system outlined above, faults are detected only when they manifest themselves as output disagreements. It is therefore possible for a combination of undetected faults to develop which may eventually lead to an unsafe failure. The system is said to lack sufficient fault detection coverage.

A major extension of fault detection coverage may be achieved by arranging for the two subsystems to exchange information about internal states; for example, the exchange of working memory contents is a powerful way of detecting failures at an early stage. It is also useful to exchange and compare the contents of program memory, as this will reveal failures affecting little used parts of the program, including those on which particular reliance may be placed in a failure situation. An additional benefit of such exchange procedures is that a microprocessor must be in a good state of health for them to succeed at all.

Self-testing also has an important role to play in fault detection, and may be applied to important areas of memory, to the central processor unit itself and to peripheral hardware. Although it is always possible to imagine ways in which a microcomputer might fail to detect an internal fault while continuing to appear perfectly healthy, this does not detract from the value of self-testing as a way of minimising the risk of undetected faults.

Some designs achieve safe operation using only a single processor with very elaborate self-testing arrangements. Such systems contain redundancy in the form of duplicated and testable input and output circuitry and diversity of information handling. Typically, the self-testing procedure is arranged to operate a vital relay which corresponds to the output comparator in the duplicated system.

Other designs seek to achieve perfect fault detection coverage by comparing the operation of the two subsystems at processor bus level. While providing a high level of hardware fault detection, this technique raises problems of lack of independence between the two processors, requires special hardware, and constrains the designer to make the two subsystems identical in every respect.

Divergence

In any parallel redundant system operating on a common set of input information, there is a finite risk that the two or more subsystems will not always agree about input states. This may be caused, for example, by lack of synchronism between the two subsystems or by uncertainty in reading an input in the course of changing state. Whatever the cause, such an event will generally lead to different output states, causing enforcement of the safe shutdown condition. This behaviour, which may conveniently be called divergence, is a serious threat to reliability and must be countered. The countermeasures, which may be called anti-divergence, need consist only of the exchange between subsystems of their current opinion of input states, together with a strategy for dealing with disagreements. The strategy may, for example, involve using previously agreed data or the use of the more restrictive data set.

Example: Safety by Duplication

Fig. 2.2 illustrates an approach to duplicated system design which embodies all the improvements on the basic concept outlined above. Comparison between subsystems takes place on two levels: (i) by exchange of internal states over a communication channel provided for the purpose; and (ii) by providing each processor with an independent set of inputs reflecting the final output states of the system. Redundancy management is handled through special hardware, by means of which either processor can enforce the safe output condition at any time. The redundancy management mechanism is itself redundant, providing each processor with more than one method of enforcing shutdown. It is also testable by both processors. Each subsystem incorporates a degree of self-testing, both of associated peripheral hardware and of important memory areas.

Fig. 2.2 Duplication for safety — a practical system

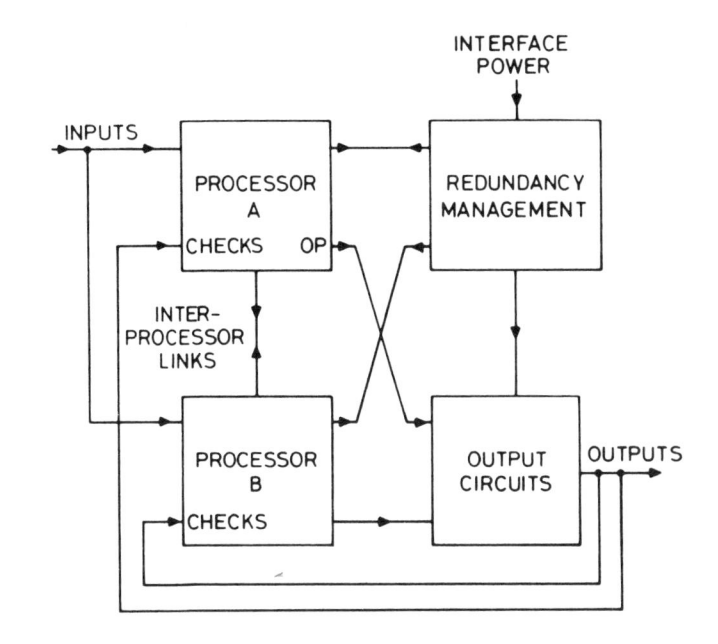

Exactly this type of architecture has been used in the trackside equipment for the SSI, in which solid state output switching devices are used to control signal lamps and point machines directly.

Calculation of Wrong Side Failure Probability

It is possible to find in the literature, various expressions for the reliability of redundant systems, including that of a duplicated system in terms of its probability of unsafe failure. Some are very simple, being based on the simple dual processor–comparator concept, and some are quite complicated, attempting to take into account the finer points of design and using rigorous probability theory. Unfortunately, mathematically exact solutions are of no consequence if very imprecise information is available to evaluate them, as is invariably the case. Therefore, a reasoned argument is presented stating first some critical assumptions, which leads to a figure for the probability of unsafe failure of a system such as that shown in **Fig. 2.2**.

The following assumptions are made:

- There are no design errors which might cause an unsafe failure which the system redundancy cannot detect or cope with. This is an optimistic assumption.
- That failures occur according to a Poisson distribution. This is optimistic, but not excessively so.
- That failures of the redundant parts of the system are independent. This is also optimistic.
- That all first failures, if undetected, would cause an unsafe malfunction of the system. This is pessimistic.
- That any subsequent failure would render the first, and itself, undetectable or leave the system unable to impose a safe state. This is very pessimistic.
- That the mean time to the detection of a failure is 1 s. This is a realistic figure for the type of system described.
- That the MTBF of a subsystem is one year. This is rather pessimistic.

It must be stressed that in any real situation, the making of such assumptions is unavoidable, even if they are not made consciously. It is one of the responsibilities of an approval authority to confirm that such assumptions are reasonable.

The argument proceeds as follows:

There are 8,760 hours in a year, and there are two subsystems. The probability of any failure in any 1 hour is

$$\frac{2}{8760} = 2.3 \times 10^{-4}$$

The probability that a second failure will occur before the first is detected (ie within 1 s) is

$$\frac{1}{8760} \times \frac{1}{3600} = 3.2 \times 10^{-8}$$

The combined probability of an undetected failure is therefore

$$7.4 \times 10^{-12} \, \text{hr}^{-1},$$

which is equivalent to 6.5×10^{-8} per annum

Clearly, whether or not all the assumptions are justified, figures like this give considerable confidence in the power of the technique.

AVAILABILITY

Many control systems are required to exhibit high availability, meaning that throughout the life of an installation, the proportion of the total time for which the system is not available for use must be very small. This requirement does not necessarily imply high reliability, since systems may be designed to tolerate faults without loss of performance. In complex electronic systems, fault tolerance is likely to be the only way of achieving very high availability.

Systems Requiring High Availability Only

In many applications there is no requirement for high integrity or fail-safe operation, and provided the availability requirement is met, it does not particularly matter what happens if the system should fail altogether. Under these circumstances, it is usual to employ a duplication technique in which one computer is active and on-line, while the other runs as a hot standby, meaning that it is also active and ready to take over from the first when necessary. The changeover will usually be automatic, with the object of minimising the disruption caused by a failure, and may be achieved using advice commonly known as a watchdog. The latter is a fault detection mechanism which expects to receive a periodic signal indicating that the subsystem is healthy and which effects the changeover if the signal ceases. Fault detection must otherwise rely on self-testing, since comparison of the two subsystems, while providing very effective fault detection, cannot identify which subsystem is at fault.

If it is required that the two processors agree in all respects (except for the immediate effects of the fault) at the instant of changeover, it is necessary to employ anti-divergence techniques to keep them in step during normal operation. If it is further required that the system should continue to operate while a failed subsystem is replaced, it will be necessary to update the new processor with the current state of the system, so that subsequently it can run in step with the on-line subsystem. Updating requires the current system state to be copied from the on-line subsystem to the new standby.

In the SSI system, the panel processors, which do not execute safety functions, are duplicated for high availability. Both processors run continuously in parallel. Their software includes anti-divergence and updating strategies to ensure that failure and replacement of a single processor will be invisible to the signalman.

High Availability Fail-Safe Systems

In applications where fail-safe behaviour and high availability are both required (electronic interlocking is an obvious example), the designer essentially has two alternatives. He can elect to use two fail-safe systems in a duplex high availability configuration as already described, or he can design a fault tolerant fail-safe system as a single entity. The former needs little elaboration, except to point out that considerable complexity will be introduced by the requirement for repair without interruption and that any process by which a new subsystem is updated must be proof against faulty data transfer. It happens that all the desirable features of such a system, which is in effect quadruplex, can be provided by a repairable triple redundant system.

24 SOLID STATE INTERLOCKING

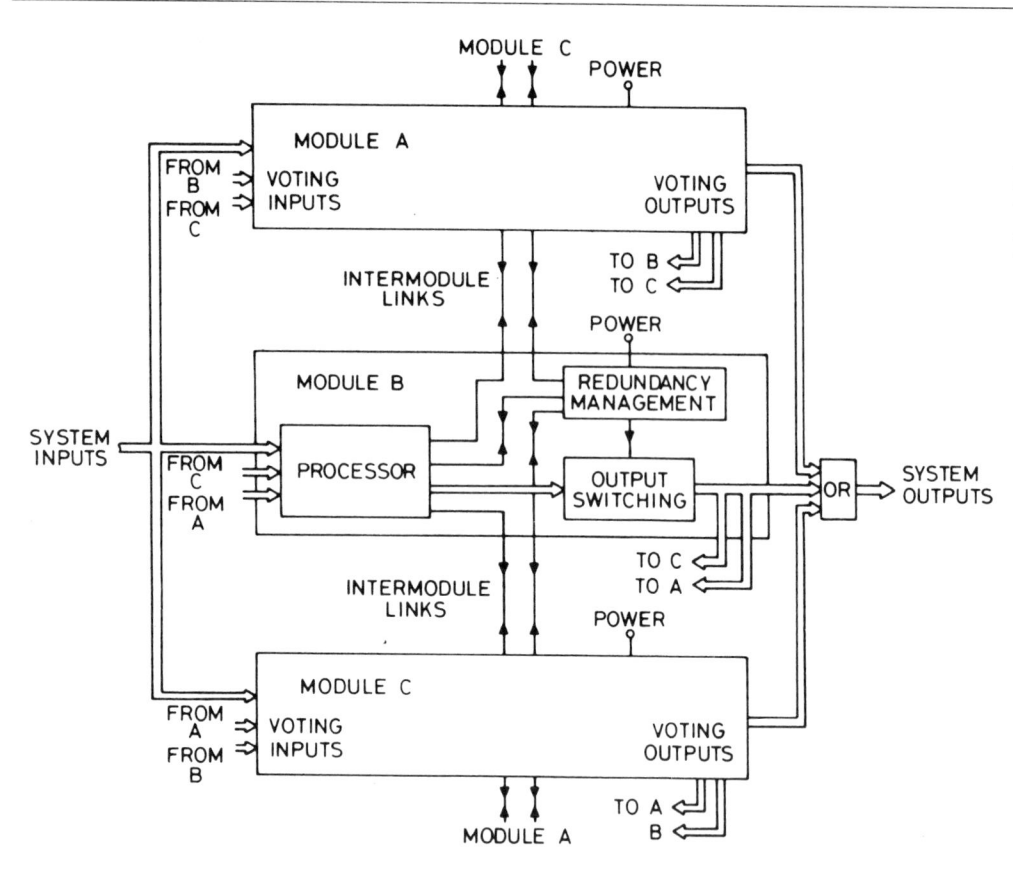

Fig. 2.3 Triple redundancy technique used for interlocking processors

Repairable TMR
The interlocking uses a technique called triple modular redundancy (TMR), which is, in many respects, an extension of the duplication technique described above. The majority voting process is performed redundantly within the three modules, both by monitoring output states and by comparison of program memory contents and selected system states. Each module (see **Fig. 2.3**) includes a redundancy management device which provides a redundant and testable mechanism for disconnecting a module in the event of a majority vote against it, and enforcing safe output states if no majority opinion exists. This mechanism can be activated by the parent processor acting alone, or by the other two modules acting together, and the design is such that an isolated or absent module appears to be voting against the other two.

As the safety level output of an interlocking consists only of the two data highway message streams, the maintenance of safe output states involves preventing the transmission of correctly coded data highway messages carrying incorrect information. Output from a disconnected module is prevented by the hardware design, but the first line of protection against incorrect data output is achieved by checking each transmitted bit and, if the data is incorrect, forcing the remainder of the message to be Manchester invalid.

Each subsystem has control over the redundancy management devices of the others in a way which ensures that the majority view prevails. The redundancy management devices provide the following facilities:

- Redundant and testable means of enforcing the isolation of the module in the case of a majority vote against it.
- Means of ensuring that the process of module isolation is irreversible, so that the possibility of further faults leading to an incorrect majority decision is eliminated.
- Automatic reconfiguration of the system at the time of module isolation, to become a duplicated system providing continued and undegraded safe operation without further fault tolerance.
- In the event of failure of the surviving duplicated system, means of enforcing an irreversible system shutdown with predetermined safe output states.

As the system described consists only of three independent modules, repair is simply a matter of removing the failed module and substituting a replacement.

When the replacement becomes active, it must be updated by the two operational modules by a transfer of system state information. In order to protect against the possibility of faulty information being transferred, identical information must be received from both operational modules.

SAFE DATA TRANSMISSION
Most industrial systems involve the transmission of information from one place to another, and railway signalling is clearly a prime example. It is obvious that system reliability depends upon the information received being the same as that sent, and in the context of a fail-safe system, it is also obvious that the probability of the information received being a corruption of what was sent must be very small. In present day signalling safety systems, vital reed circuits protect the information by using very narrow bandwidth circuits in which the signal strength is much greater than any interfering signal (the concept of energy redundancy) and by sending the information in the form of tone combinations (information redundancy). Even at the level of signalling line circuits, energy redundancy is used to ensure that information is incapable of corruption by interference.

With the advent of computer-based safety systems, the need to transmit safety information in serial data form arises. Techniques for the protection of data sent over serial telecommunications channels are of course very well developed. All of them involve the use of information redundancy, meaning that more information is transmitted than would be required if the communication channel were perfect. The additional information is mathematically related to the information to be protected in a way which enables certain types of data corruption to be detected or even corrected. A very wide range of coding methods is available, offering solutions to widely varying communication problems. It is not intended to go into any detail here, but merely to emphasise one or two essential facts about information coding.

Firstly, arbitrary levels of coding security may be obtained by applying arbitrary levels of coding, the penalty being the increase in channel space or transmission time required to accommodate the extra information.

Secondly, many standard codes are available which, in addition to detecting errors, will provide enough information to allow a limited number of errors to be corrected. Generally speaking, making use of this property increases the probability of reconstructing the information incorrectly, and its use in safety systems is not recommended.

Thirdly, it must be realised that whatever the level of coding security applied, the encoding and decoding is performed by hardware and software, and the safety of the processs depends upon the terminal equipment performing correctly. Terminal equipment must therefore also be safe.

The properties of standard codes are well documented, and a knowledge of the characteristics of the transmission channel will enable the probability of the code being broken to be calculated. In safety applications, account must be taken of the properties of the transmission channel under fault conditions, when its error statistics may be very different to the fault free condition.

System Description

OVERVIEW

The SSI system may be regarded as being divided into control centre and lineside subsystems (see **Fig. 2.4**).

At the heart of the control centre subsystem is the interlocking itself, which is a fail-safe multiple processor system which implements all the logic necessary to generate safety commands for lineside signalling equipment. Duplicated panel processors handle non-vital functions associated with the signalman's controls and indications, and a diagnostic processor interfaces with a technician's terminal shared by a number of interlockings to provide maintenance support facilities.

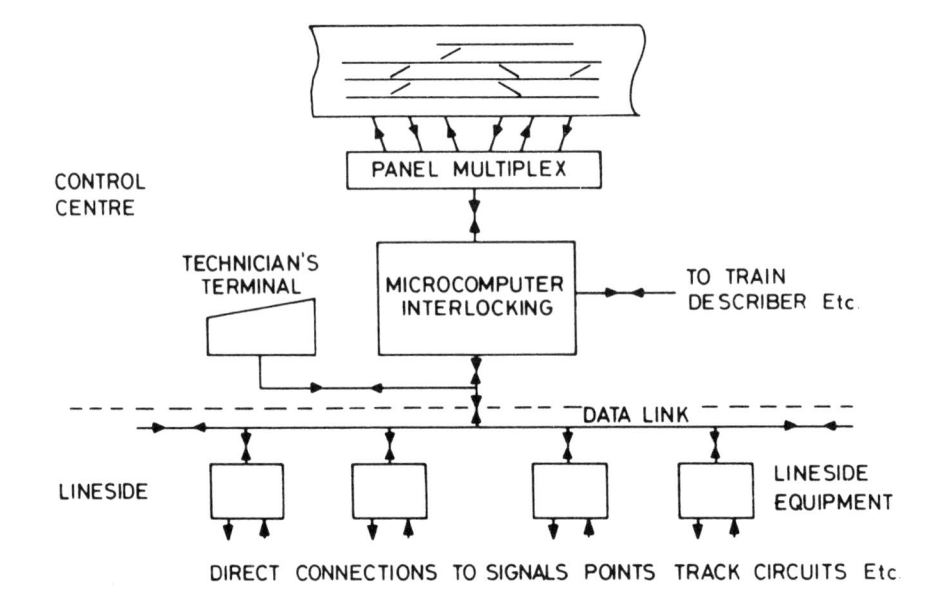

Fig. 2.4 SSI basic structure

A network internal to the control centre provides fail-safe communication between interlockings to enable information to be exchanged relating to adjoining signalling areas.

The lineside subsystem consists of the lineside data cable and trackside functional modules (TFMs) forming a fail-safe local area network which drives signals, points, and other signalling equipment according to command telegrams generated by the interlocking, and transfers back to the interlocking information from track circuits, points, and other equipment.

THE CONTROL CENTRE

The Interlocking Cubicle

Block diagrams of a typical SSI installation are shown in **Figs. 2.5(a)** and **2.5(b)**. At the control centre, the equipment for each interlocking is housed in a 19-inch cubicle. As will be seen, the extent of the area which can be controlled by one interlocking is primarily determined by the quantity of trackside equipment; in broad terms, one interlocking can control about 40 signals and between 20 and 40 sets of points.

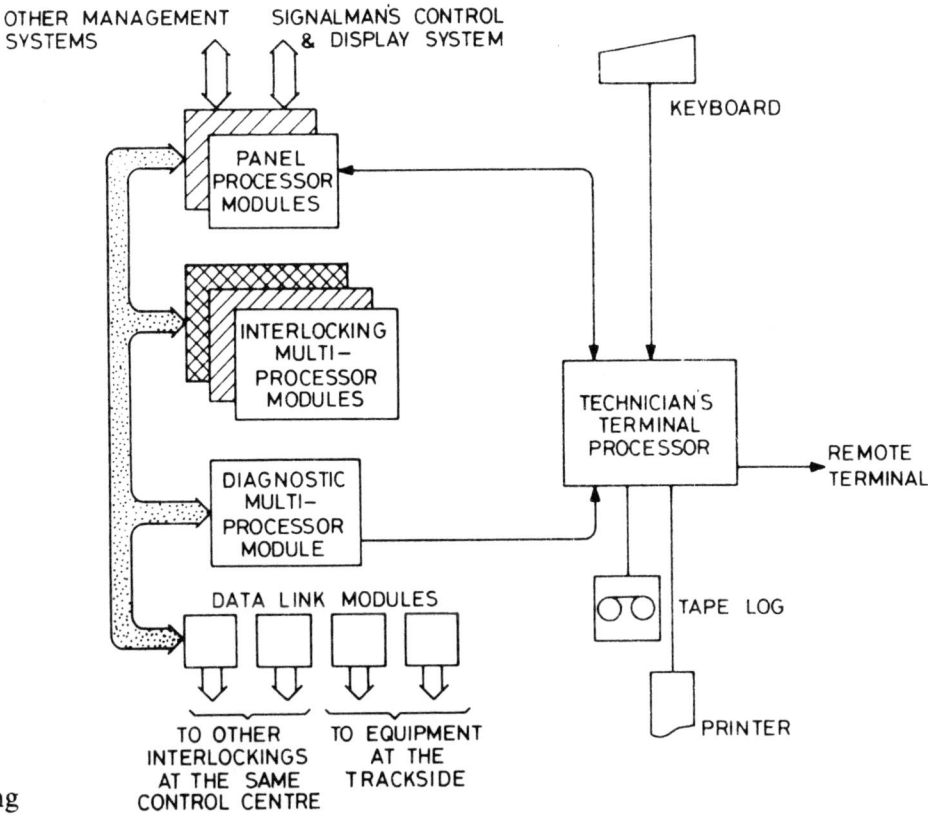

Fig. 2.5(a) The SSI interlocking

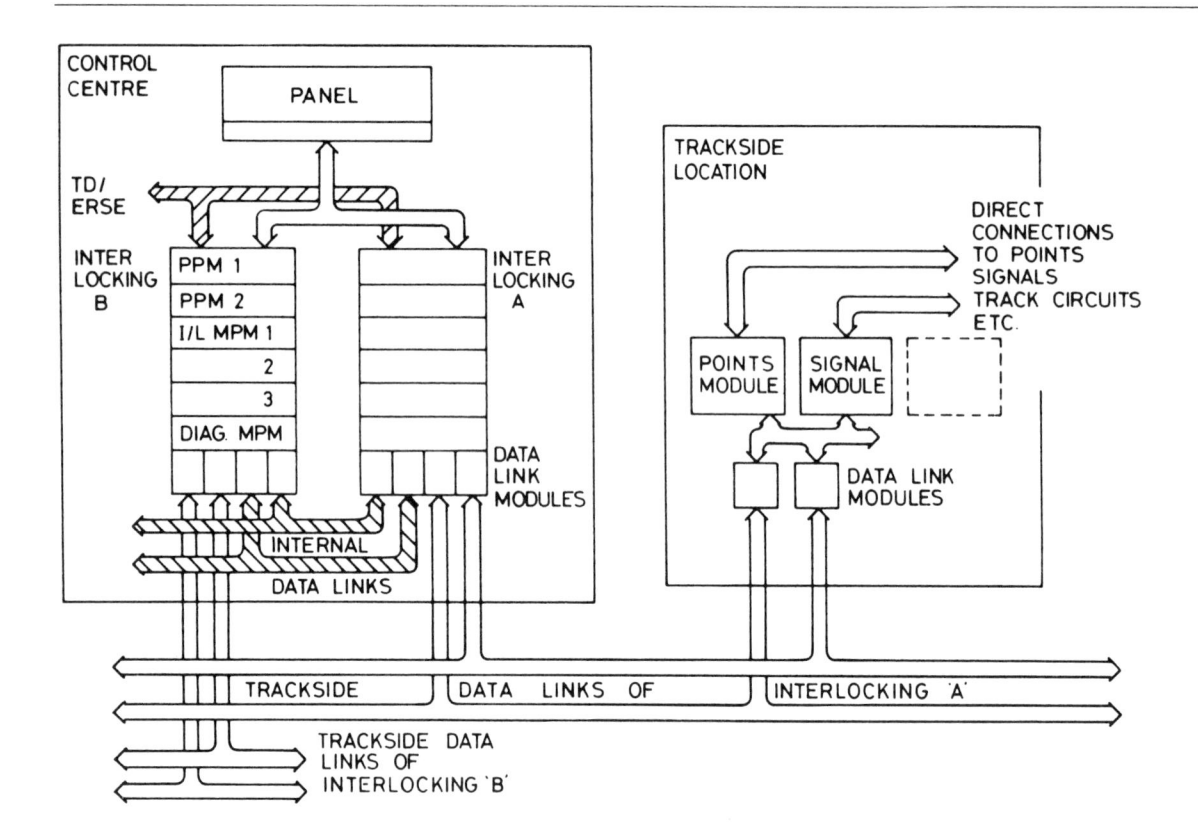

Fig. 2.5(b) Block diagram of SSI system

If the controlled area is larger, further interlockings are added as required. Together with the technician's terminal, which is housed in a further 19-inch cubicle of half height, this is the sum total of equipment required at the control centre; in bulk, it may be compared with the relay racks which would be needed for an electromechanical interlocking. Each interlocking cubicle houses the following equipment:

- Three interlocking multi-processor modules.
- Two panel processor modules.
- The diagnostic multi-processor module.
- Data link modules and power supplies.

The modules are fitted with standard multi-way connectors which carry all electrical connections, with coding pins. All links between modules are optically isolated to give immunity to electrical noise. The cubicle wiring to each multi-processor module includes an 8-bit binary address. For each bit, a wire link is made to either high or low voltage, to correspond to binary 1 or 0 as required. Five of these bits constitute the system identity, and identify the individual interlocking within the control centre. The remaining three bits are available to distinguish successive versions of geographical data (eg during modifications).

Each of the multi-processor modules and panel processor modules is fitted with an interchangeable memory module, containing the appropriate fixed program and geographical data in erasable programmable read-only memory (EPROM). A label on the memory module appears at the front of the mother module when it is plugged in, to identify its function and the date it was installed. It is usual to hold a ready-programmed spare memory module for every working module in a system, to minimise the time taken to replace a failure.

The interlocking equipment is designed for use in heated or unheated buildings without air-conditioning or other special provision. It takes power at 110 volts AC, the power consumption of a single interlocking cubicle being approximately 375 watts.

The Interlocking Multi-Processor Modules

The interlocking multi-processor modules (MPM) operate together as a two-out-of-three majority voting system. All three are identical and run the same program, in the course of which each checks itself against both its partners, comparing output states and memory contents in a continuous cycle. Each module is provided with an irreversible means of disconnection from the system following any component failure or divergence between itself and its partners. This takes the form of a security fuse circuit which supplies current to its outputs, so that if the fuse is ruptured, the module is unable to communicate with the outside world. Each module is able to trigger its own security circuit, and also to co-operate with either of its partners to trigger the security circuit in the third. All the security circuits are themselves tested regularly.

If, during the check, a module finds that it differs from both its partners, it triggers its own security circuit. If it agrees with one partner and they both differ from the third, it triggers the security circuit in the latter; if the attempt fails, it then triggers its own security circuit. This sequence of events ensures that a faulty module is always disconnected. The remaining two partners continue to operate as a two-out-of-two redundant system with no reduction in the level of safety. If a second fault were to occur before the failed module was replaced, both the remaining modules would isolate themselves, and a system failure would ensue. Hence it can be seen that in order to ensure high availability, the maintainer must replace a failed module as quickly as possible.

In each module, two subsidiary processors handle communications with the trackside data links, and a third handles communications with both internal data links. Each of these processors has its own fixed program which resides in a memory within the module.

The Panel Processor Modules

The panel processor modules are duplicated in order to provide fault tolerance. Both run the same program. Both receive all inputs, and they alternate at driving their outputs in normal operation. Hence, if one should fail, the other is able to continue normal operation without interruption. Again, rapid replacement is necessary to maintain availability and minimise the chance of the remaining module failing before the first failure is replaced. A new module is brought on-line without interrupting the working of the other.

The modules handle all information flowing between the interlocking and the signalman's display system, which may be an entrance/exit panel, a system based upon visual display units (VDUs) or the Integrated Electronic Control Centre (IECC).

Controls are received in the form of entrance and exit button operations, point key operations, etc, and are processed to generate route requests and controls for specific functions. These are passed to the interlocking multi-processor modules for further processing. Indications of all functions are received from the interlocking multi-processor modules, and drives are generated for the signalman's display in the form required.

The panel processor modules provide additional ports for general purpose communications between the SSI system and other management systems such as train describers, automatic train reporting and electronic route setting, as required.

Panel Multiplexers

An installation using an entrance/exit panel requires a panel multiplexer. Typically this takes the form of a microcomputer system which scans all the buttons and other controls on the panel, generating a serial data stream which it sends to the panel processor modules. It receives indications from the panel processor modules in serial form, and drives the panel lamps and other indications in accordance. Transmission to and from the panel processor modules is continuous, over separate pairs of wires. For high availability the panel multiplexer may be duplicated, with manual or automatic changeover. Both multiplexers then receive all inputs, one or the other driving the outputs at any given time.

The Internal Data Links

The number of functions which are required by more than one interlocking in the same control centre varies greatly from one application to another. At the boundaries between interlocking areas, there are normally routes having one end in each area as well as track circuits which are required by both interlockings. Other vital functions to be passed between interlockings include signal aspects for aspect sequencing, and point positions.

Communication between interlockings is over the internal data links, which are duplicated in order to give high availability. Each is a baseband serial link operating at a data rate of 10 kbaud and each interlocking is connected to it by data link modules (see below). Each interlocking in turn sends a Hamming coded telegram containing its 5-bit system identity, 128 bits of data, and parity bits, in the form of Manchester II code (see below). These two levels of coding combined give the required high level of integrity. The interlocking then listens to the telegrams sent by the other interlockings in turn, and extracts information which it requires from them.

The Diagnostic Multi-Processor Module

A comprehensive range of diagnostic aids is built into the SSI system to facilitate rapid location of any faults.

The TFMs and interlocking MPMs test themselves very thoroughly in the course of the safety checks implemented by their redundancy management software. In the case of the TFM, if a discrepancy indicating a failure internal to the module is detected, the module is shut down irrevocably by blowing its security fuse, and ceases to reply to the interlocking telegrams. If a discrepancy occurs in the input or output circuitry of the module, then the module isolates its inputs or outputs, but continues to reply to the interlocking telegram, indicating its condition by means of status bits in the reply telegram. The TFMs also report on the condition of the data links. If they are not receiving sufficient good messages over a particular link, they signify the condition of the link by means of further status bits in the reply telegram.

The diagnostic multi-processor module is a standard unit identical to the interlocking multi-processor modules, fitted with a memory module containing the diagnostic program, and is the link between the interlocking and the technician's terminal. It monitors all messages on the trackside and internal data links, extracting all changes of state of controls and indications and noting any failure of signals or points to respond to controls. It decodes the status bits in the messages received from the equipment at the trackside, in order to diagnose faults in individual trackside modules. It also analyses faults in the trackside data links themselves, and is able to locate them by noting simultaneous loss of transmission from groups of trackside functions over one or both links. Fault messages are generated and sent to the technician's terminal for display to the maintainer.

Diagnostic information relating to the status of the central interlocking equipment is put into the format of a command telegram and is transmitted over the lineside data link with the identity of address zero, which is not used for any TFMs. This is also interpreted by the diagnostic processor and used to generate messages to the technicians.

The diagnostic processor also recognises all changes of state of the signalling information and passes these on to the technician's terminal to be logged on tape. A full record of the recent activity of the system is thus available on a tape which can be interrogated off-line in the course of investigating failures or incidents.

The Technician's Terminal
The technician's terminal is housed in a separate cubicle of half height, and can serve up to six interlockings in the same control centre. The cubicle houses the technician's terminal processor, with a dual tape unit and a modem, and a printer with keyboard which is associated with it. The technician's terminal receives information from the diagnostic processor modules in the various interlockings, and also monitors all messages between the interlockings and the signalman's display system, to provide the following facilities:

- It prints the fault reports generated by the diagnostic processor modules in the various interlockings. Each report is accompanied by the time of occurrence, and is printed in plain language. Intermittent faults are detected and reported as such, in order to avoid unnecessary repetition of the same fault report. The modem enables a duplicate printer to be connected at a remote location.
- It keeps a log on tape of all changes of state in each interlocking, as advised by the diagnostic processor modules. The dual tape unit is buffered, so that the tape can be changed over without any data being lost. Each record on tape includes the system identity of the interlocking to which it applies and the time of occurrence. The tapes can be analysed on the design workstation, providing a complete analysis of the performance of the system which is available for subsequent investigation if required.
- It enables the maintainer, through the keyboard, to interrogate the interlockings in order to check the states of signals, points and 11 other functions, and to check the transmissions on the various data links.
- It enables the maintainer to apply certain restrictive technician's controls, locking individual routes, points and track circuits out of use on command through the keyboard. Each control is stored by the appropriate interlocking, and remains in force until it is removed through the keyboard. If the system performs a warm start after a power failure, the technician's controls are retained in memory along with all other vital data. If the power failure is prolonged, they must be re-entered, with all other controls.
- Lastly, it enables the maintainer to enter certain other commands via the keyboard, including those to start an interlocking up, and to reset the system clock (which is generated by the technician's terminal processor). The keyboard is normally locked, and the access for any purpose is under strict password control.

INTERLOCKING SOFTWARE
The SSI system depends on its software for the safety and availability of its hardware and data transmision, the correct execution of signal controls and interlocking, and the effectiveness of its diagnostic aids.

The software in the interlocking MPM is subdivided into programs providing initialisation, redundancy management (ie performing hardware safety checks and comparisons, implementing safety and availability strategies), interfacing with panel processors and communications links, and data transmission and reception.

At the heart of the interlocking software is the interlocking functional program. This generates command telegrams interpreting the signalling controls for the interlocking area encoded in the geographical data held in read-only memory (ROM), according to the image of the state of the railway held in random access memory (RAM) and continually updated by the indication telegram received from the trackside (see **Fig. 2.6**).

The geographical data must be written, compiled and checked for each installation, and in effect customises the SSI system to that particular location, whereas all the remaining software is common to all installations and does not need to be rechecked. The source data consists of a series of statements forming what is in effect a program written in a special purpose language, which is interpreted by the interlocking functional program.

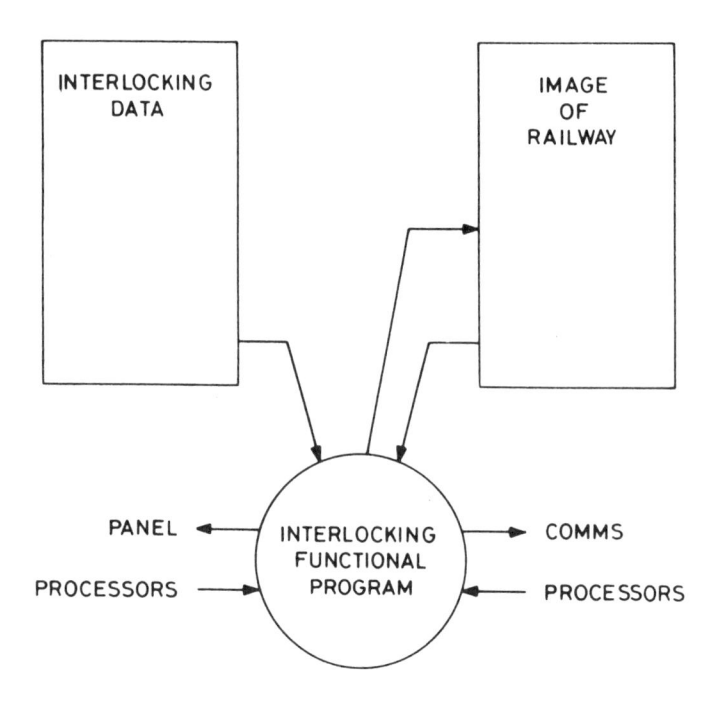

Fig. 2.6 Interlocking software principles

SOLID STATE INTERLOCKING 33

Fig. 2.7 Interlocking module software organisation

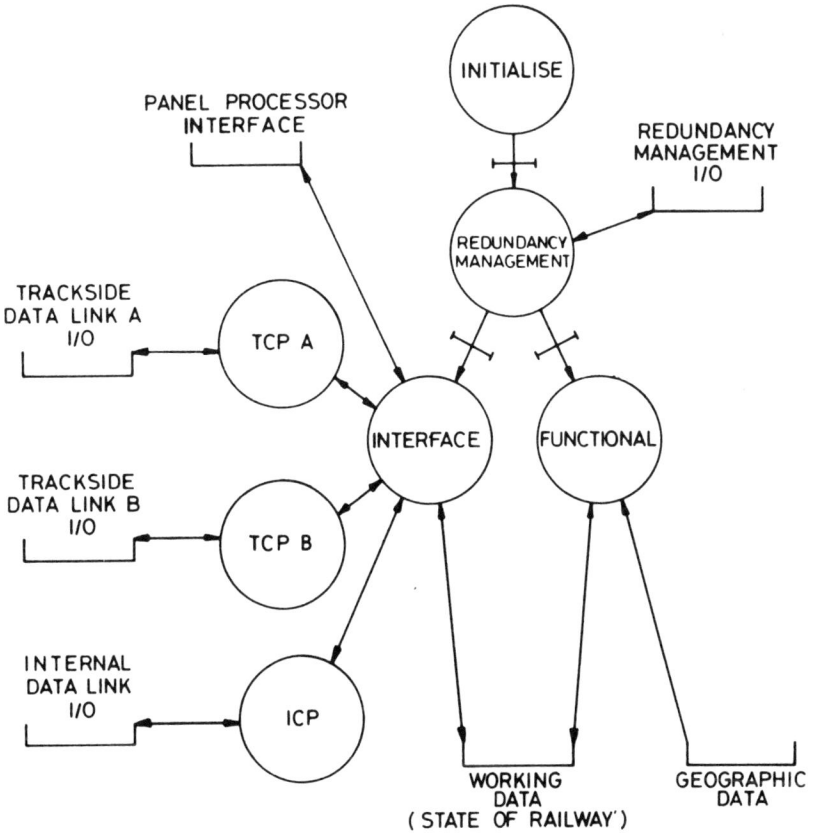

[Diagram: Interlocking module software organisation, showing INITIALISE, REDUNDANCY MANAGEMENT (with REDUNDANCY MANAGEMENT I/O), PANEL PROCESSOR INTERFACE, TCP A (with TRACKSIDE DATA LINK A I/O), TCP B (with TRACKSIDE DATA LINK B I/O), ICP (with INTERNAL DATA LINK I/O), INTERFACE, FUNCTIONAL, WORKING DATA (STATE OF RAILWAY), and GEOGRAPHIC DATA.]

The relation between the programs, data areas, and input and output channels of the MPM is shown in **Fig. 2.7**. Dedicated processor chips are provided to run the trackside communications programs (TCP) and internal communications program (ICP). The other programs are all run on the main (or functional) processor within the MPM.

Fig. 2.8 illustrates the operation of the interlocking program. During the basic minor cycle shown, the interlocking sends a message to one trackside functional module over the trackside data links, and receives a reply from it. The minimum time for this cycle is 9.5 ms. The time taken to exchange messages with all 64 possible addresses in turn defines the

34 SOLID STATE INTERLOCKING

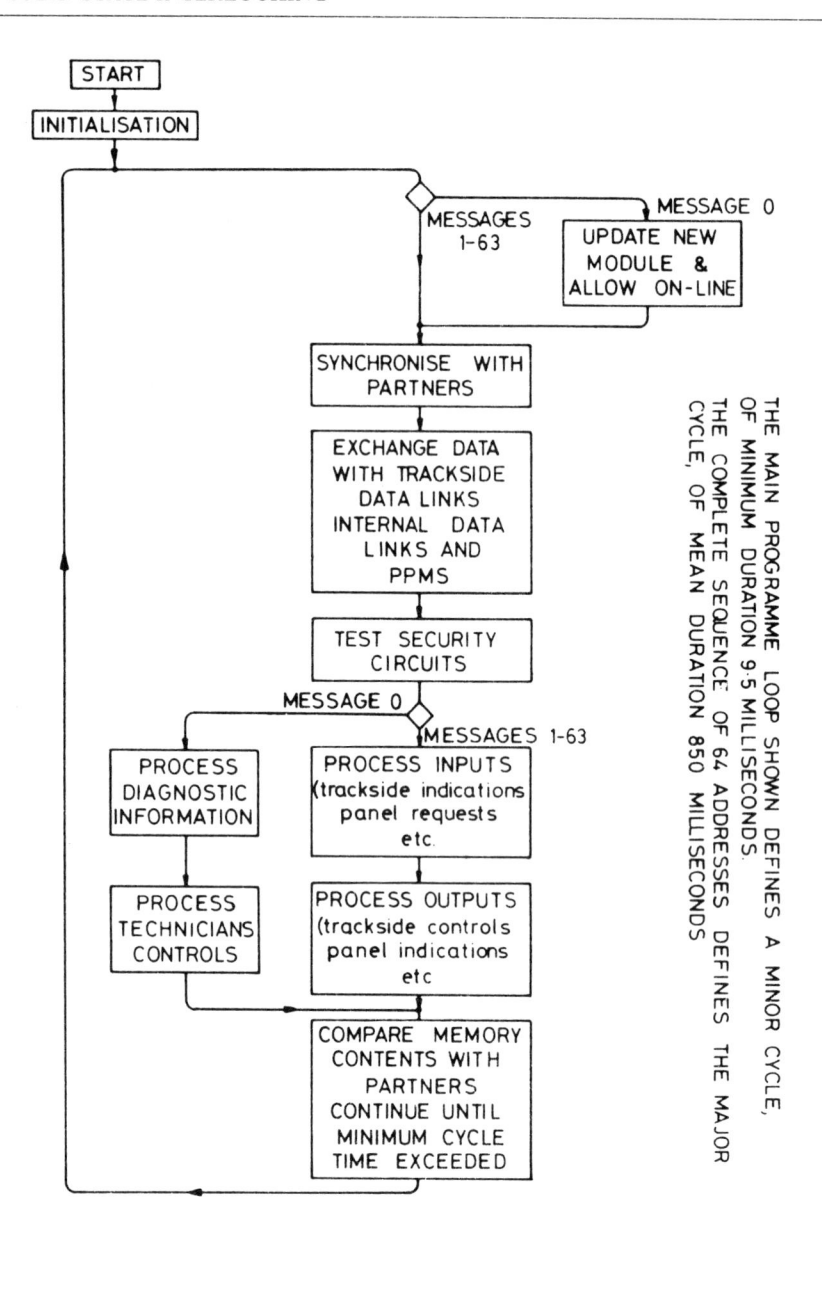

Fig. 2.8 The interlocking program

major cycle of the system, the average duration of which is 850 ms. The number of addresses available for trackside functional modules is 63, address zero being used for communication with the technician's terminal.

When power is first applied, the module executes the initialisation program, which enables it to start up in one of a number of possible modes, as follows:

- After a prolonged loss of power, the triplicated system starts up from cold. All controlled functions and all the signalman's indications are in their most restrictive states. When the three modules agree, the indications are set to their true states as reported by the trackside data links. Only then can controls be accepted.
- If the duration of the loss of power was less than about 6 ms, the states of functions are retained in memory in each module. If the modules are in agreement, the system performs a warm start, resuming operation without resetting controls or indications.
- If the module finds that only one other module is present, it co-operates with it to start up as a two-out-of-two system.
- If two other modules are present and already running, they provide the newly powered module with information to enable it to bring its memory up to date, and it joins them to run as a two-out-of-three system. This happens without any interruption of the operation of the system.

During each minor cycle, blocks of information are exchanged not only with the equipment at the trackside, as already noted, but also with other interlockings via the internal data links and with the panel processor modules.

If the indications received are agreed by the three modules, they are processed in accordance with the geographical data and are used to update the image of the state of the railway which each module holds in its memory. Panel requests and other controls are processed in accordance with data, and controls of trackside functions are set ready to be sent. If no information is received from a particular function (as might result, eg, from an equipment failure in the field), then after a predetermined interval that function is set in memory to its most restrictive state.

In both the interlocking and the trackside modules, the duplicated or triplicated processors contain identical programs, so there is no in-built protection against program design errors. A major effort of analysis and testing was therefore necessary to validate the software and ensure freedom from error.

The principles and techniques currently applied by British Railways to the validation of safety-related software seek to prove consistency between the various levels of design through the following procedures:

- Functional analysis, proving equivalence and completeness of the design with respect to the requirements specification.
- Structural analysis, proving modularity and path accessibility and providing a framework for later semantic analysis.
- Modular analysis, testing program modules against predicted behaviour.
- Information flow analysis, relating processes to variables and detecting illegal or absent variable references.
- High-level semantic analysis, demonstrating that the separately validated processes combine to satisfy the overall requirements of the system.
- Timing analysis, applied to processes with safety-critical timing constraints.

THE TRACKSIDE DATA LINKS

All signals, points, track circuits and other trackside signalling functions are connected to the interlockings which control them by trackside data links. Each link works over an individual screened twisted-pair copper cable having a normal impedance of 100 ohms, the equipment at each location being connected to it in parallel as shown in **Fig. 2.9**. The entire data link system is duplicated, from the control centre to every location at the trackside, in order to enable normal operation to continue in the event of a complete failure of one system. The data link cables may be laid in troughing, or may be ploughed into the ground. They are the only signalling cables required along the railway apart from power cables and local tail cables.

Message Format

Each data link is a baseband serial link working at an information rate of 10 kbaud. Transmission in both directions is in the form of telegrams, each of which consists of 30 bits of 100 microseconds duration each, together with a synchronisation pattern at the start and a terminator, as shown in **Fig. 2.10**. The direction bit in each telegram indicates whether it was sent by the control centre or by a trackside module. The address is in two parts, the first five bits representing the system identity and the remaining six bits the module address.

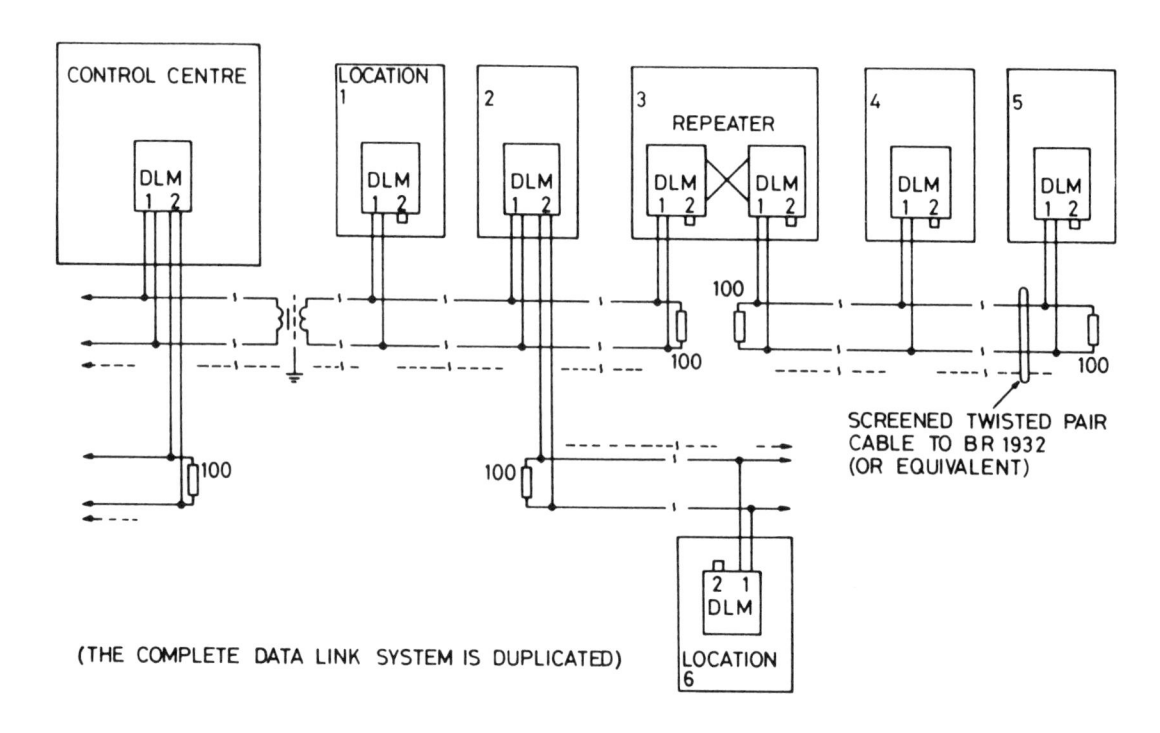

Fig. 2.9 Typical configuration of one trackside data link

Fig. 2.10 Message format on the trackside data links

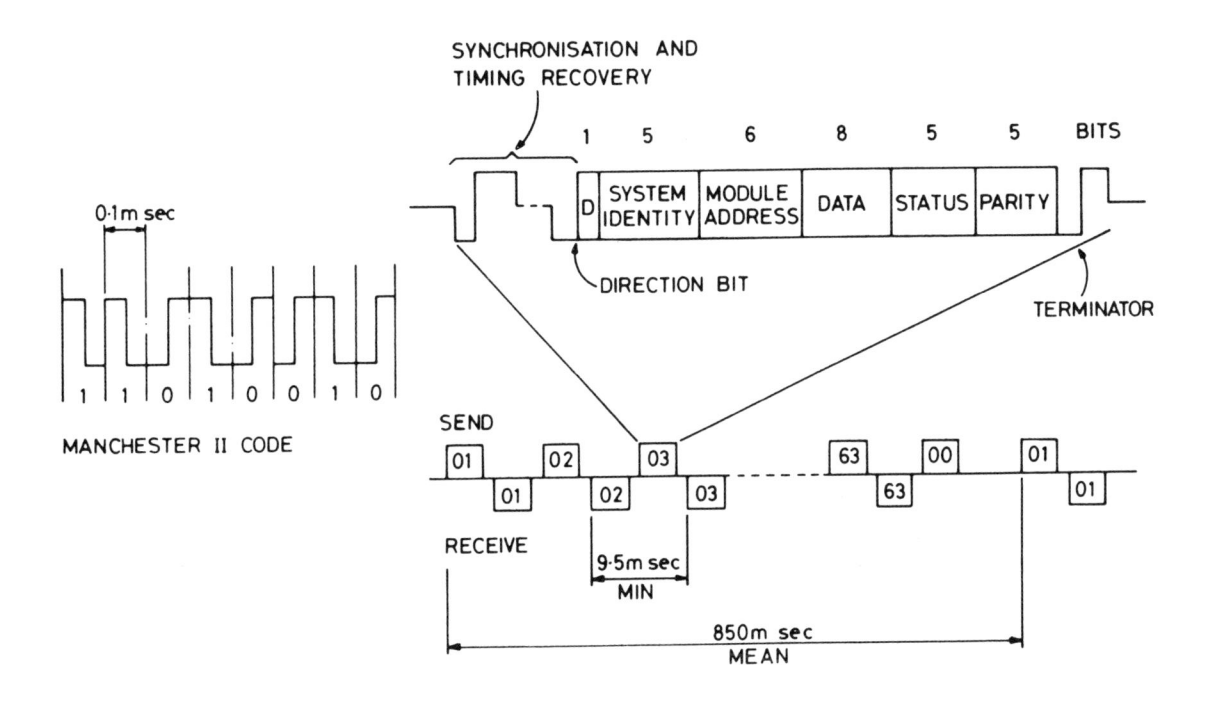

The eight data bits carry the signalling information content of the telegram. The five status bits carry diagnostic information from the trackside modules to the interlocking, for decoding by the diagnostic processor module. Five parity bits are generated by a truncated (31, 26) Hamming code. The telegram is sent in the form of Manchester II code, in which binary 0 and 1 are represented respectively by positive and negative-going voltage transitions on the line, as shown in **Fig. 2.10**, and at baseband (ie not modulated on to a carrier).

Protocol
Each interlocking at the control centre has its own physically separate trackside data link system, all telegrams on which carry its individual system identity. The interlocking sends out control telegrams to each of the 64 possible trackside module addresses in turn. When a trackside module receives a telegram carrying its own module address, it sends a reply telegram. In this way all the addresses are polled in turn, the average time taken to complete the cycle of 64 being 850 ms

and the minimum interval between successive telegrams sent by the interlocking 9.5 ms. Every address is always polled, whether or not equipment is actually fitted at that address. Only 63 are actually available for trackside modules, as the interlocking sends status information to the technician's terminal via the diagnostic processor module in the time slot corresponding to address zero.

Data Link Modules

The data link modules connect the interlocking and the trackside modules to the trackside data links. Each includes a transmitter and line driver with six input ports, and a receiver with six output ports. It is connected in parallel with one data link, two line ports being provided in order to allow for branching. The interfaces with SSI modules are optically coupled for noise immunity. Data link modules are of the same robust construction as TFMs for service in trackside apparatus cases. The same type is also used in the interlocking cubicle, in order to simplify spares holdings.

The interlocking is connected to the internal data links in the same way, an additional pair of data link modules being fitted in the interlocking cubicle when they are in use.

System Considerations

The distance over which the trackside data link can work without repeaters is limited by delay and other factors to 10 km. For operation over greater distances, repeaters are employed. Two data link modules are connected back to back to form a fully bidirectional repeater, receiving distorted telegrams in either direction at the end of section and retransmitting them with amplitude and waveform restored. A maximum of four repeaters may be fitted in any one link, at a maximum spacing of 8 km.

Each section of each link is terminated by means of a resistor, the value of which depends upon the characteristics of the cable used.

The Long Line Link

As already seen, the trackside data links are limited to a maximum range of 40 km from the control centre by considerations of attenuation and distortion. The area of control may be extended beyond this limit by the use of the long line link. This enables the vital controls and indications messages to be sent over channels in a standard telecommunications system, in order to extend the trackside data link in effect to distances of up to several hundred kilometres (see **Fig. 2.11**).

For this mode of operation each trackside data link requires one 64 kHz voice channel in a standard 30-channel pulse code modulated (PCM) telecommunications system using a 2 MHz carrier. A special SSI module, the long distance terminal, connects the SSI equipment to the PCM terminal via a contradirectional interface to CCITT recommendation G.703.

At the control centre, a pair of long distance terminals in the interlocking cubicle replaces the data link modules for the trackside data links. At the trackside, further long distance terminals connect the local PCM terminal to trackside data links which extend along the railway in the normal way. The maximum range of the long line link itself is limited by the time delays through the system, which depend upon the characteristics of the telecommunications system, but is generally up to several hundred kilometres.

The long distance terminal is a dual processor system which receives telegrams from the SSI equipment at the standard data rate of 10 kbits/s and reads them into an internal store. It adds an 11-bit control centre identity and retransmits the telegrams to the PCM equipment at the higher rate of 64 kbits/s. The higher data rate allows the telegrams to be sent to the PCM terminal in a form having a high degree of redundancy, giving the level of protection against corruption necessary to permit them to be sent over the open telecommunications network. The long distance terminal at the receiving end checks the received data and strips the identity before reading the telegram out of the SSI. These operations introduce an

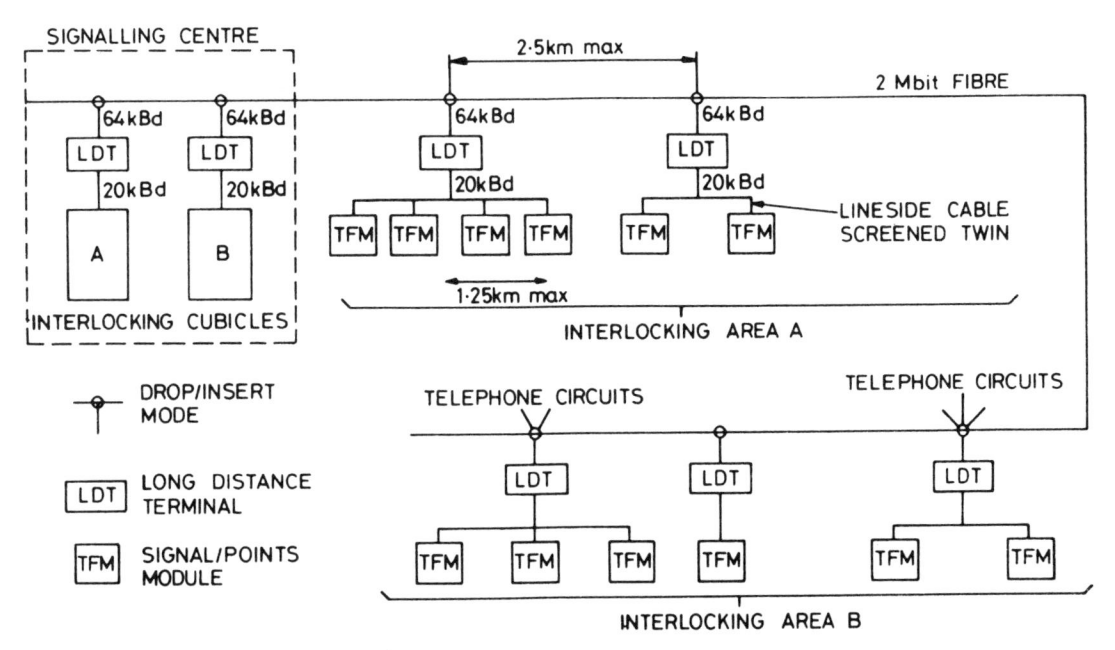

Fig 2.11 SSI long distance data transmission

overall delay into the system, and so reply telegrams are delayed by one address; after the interlocking has sent a telegram to address two, the reply it receives is from address one, and so on, if the long line is in use. The appropriate mode of operation of the interlocking is selected by a statement in the geographical data for that interlocking.

The long distance terminal is designed to the same standards as other trackside equipment, for housing in lineside apparatus cases. As in the case of the data link module, the same type is used in the interlocking cubicle.

TRACKSIDE FUNCTIONAL MODULES

The SSI system controls the equipment at the trackside by means of trackside functional modules, of which two types are in use at the time of writing — the point module and the signal module. They may be thought of as local multiplex terminals for the trackside data links, each providing eight vital outputs and eight vital system inputs. With the data link modules, they are housed in lineside apparatus cases of the conventional pattern, and are exposed to extremes of temperature and humidity, as well as high levels of vibration, industrial atmospheres and other hazards. They are therefore of robust construction conforming to standards laid down by the British Railways Board for equipment for service in a severe lineside environment.

They are connected to the trackside data links by pairs of data link modules, as shown in **Fig. 2.5**. Each requires one input and one output port of each data link module, so that up to six can be connected to one pair at a location. Each has a unique module address, the six bits of which allow 64 possible addresses. Address zero is not available, and the remainder define the maximum number of trackside functional modules which may be controlled by a single interlocking as 63.

Dual Processor System

When a module receives a telegram with its own address, it sets its outputs in accordance with the data bits, and sends a reply telegram in which the data bits repeat its inputs. These vital functions are supervised by two microprocessors operating as a two-out-of-two system to maintain the same level of integrity as the interlocking. (This arrangement lacks the fault tolerance of the interlocking, but since an individual trackside functional module controls only a small number of functions, the effect on overall system availability is small.) Both processors run the same program as in the interlocking, and a security fuse circuit provides an irreversible means of shutting the module down. The security fuse is not accessible from the outside of the module, and can only be replaced in the workshop.

The module accepts telegrams from either trackside data link. In normal operation, it switches between the two regularly, so that the availability of both links is known at all times, and reply telegrams are sent on both links by one or other processor. Should one link fail, the module continues to operate normally over the other alone, and uses the status information in the reply telegrams to advise the interlocking of the loss of reception.

Module Address and System Identity

The module acquires the two parts of its address externally, and so all modules of each type are interchangeable. The module address is read from an arrangement of links in the apparatus case wiring. The system identity is read from the first telegrams received after power is applied, and it is stored in a non-volatile memory, which retains it during power supply interruptions. It can be reset only by the technician at the depot. All telegrams on a trackside data link system of an interlocking carry the same system identity, and different interlockings have different identities. In this way, the system is protected against the effects of cross-talk in the event of cable faults, for example in cases where data link cables from different interlockings run parallel to one another.

The Signal Module

The signal module has eight AC outputs at 110 volts, and each is suitable for driving a signal lamp directly, as shown in **Fig. 2.12**. There are two current sensing return paths for lamp proving purposes, with a range of current sensitivities to suit alternative loads such as a single 24-watt signal lamp, a multi-lamp route indicator or an AWS magnet.

The six system inputs are used for track circuits and other functions. A coded message is generated by the module and is used to detect whether the contacts to be sensed are closed or open. Two more inputs are used internally, to repeat the lamp proving indications back to the interlocking.

The two processors of the module check each other's operation in a continuous cycle. They test every output circuit by switching it for a period of milliseconds and monitoring the voltages which appear in the circuit, and they test the security fuse circuit.

Fault tolerance is provided by limiting the effects of failures as far as possible. If the processors detect a failure in an output circuit, they remove power from all the outputs of the module, but continue to repeat indications. Red-retaining feeds then supply power to the most restrictive aspects of the signals controlled. Conversely, if an input failure is detected, the outputs may remain available; in this condition, the processors set every indication in the reply telegram to its more restrictive state. In each case, the status information in the reply telegram informs the interlocking of the state of the module. If, however, a fault causes the processors to diverge or to lose control of the outputs, the security fuse circuit is triggered. All outputs including the red-retaining feeds are then isolated, and the module ceases sending reply telegrams to the interlocking.

Loss of communication with the interlocking by a healthy module causes reversion to the red-retaining feeds, and the signals show their most restrictive aspects.

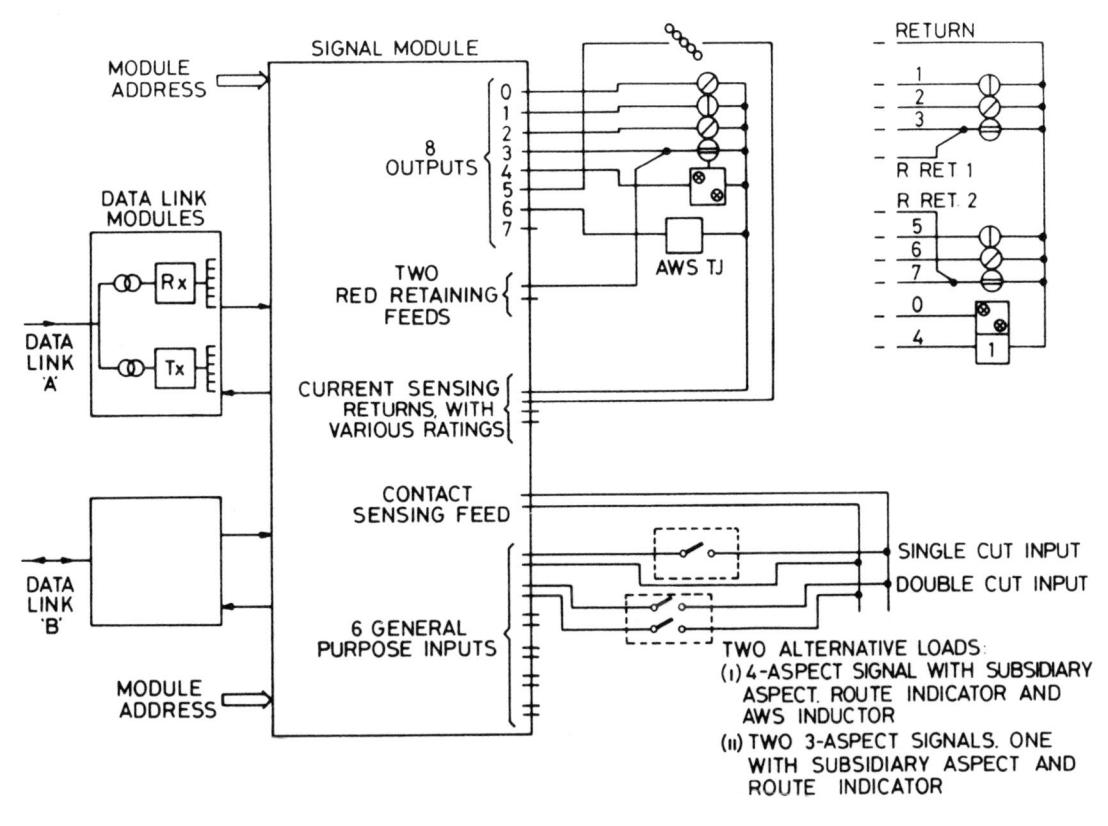

Fig. 2.12 Application of signal module

The Point Module

British Railways standard clamp lock point mechanism is electro-hydraulic. To throw the points, a hydraulic pump motor is first started, and then one of two solenoid valves for the normal or reverse position is opened. After the stroke is complete, the pump motor is stopped to avoid damage.

The point module controls clamp locks directly, as shown in **Fig. 2.13**. It receives an AC power supply at 140 volts and provides two independent groups of DC drives at 120 volts, each group consisting of two pump motor drives with a normal and a reverse valve drive. Each group can control two points working together, their valve drives being connected in parallel. It will be seen therefore that the module can control two independent groups, each containing two points.

The module uses one input for normal and reverse detection of each of the two groups of points, generating a coded signal which is passed over the detector contacts. In addition, four system inputs are used for track circuits and other indications, using the coded signal for contact sensing as in the case of the signal module.

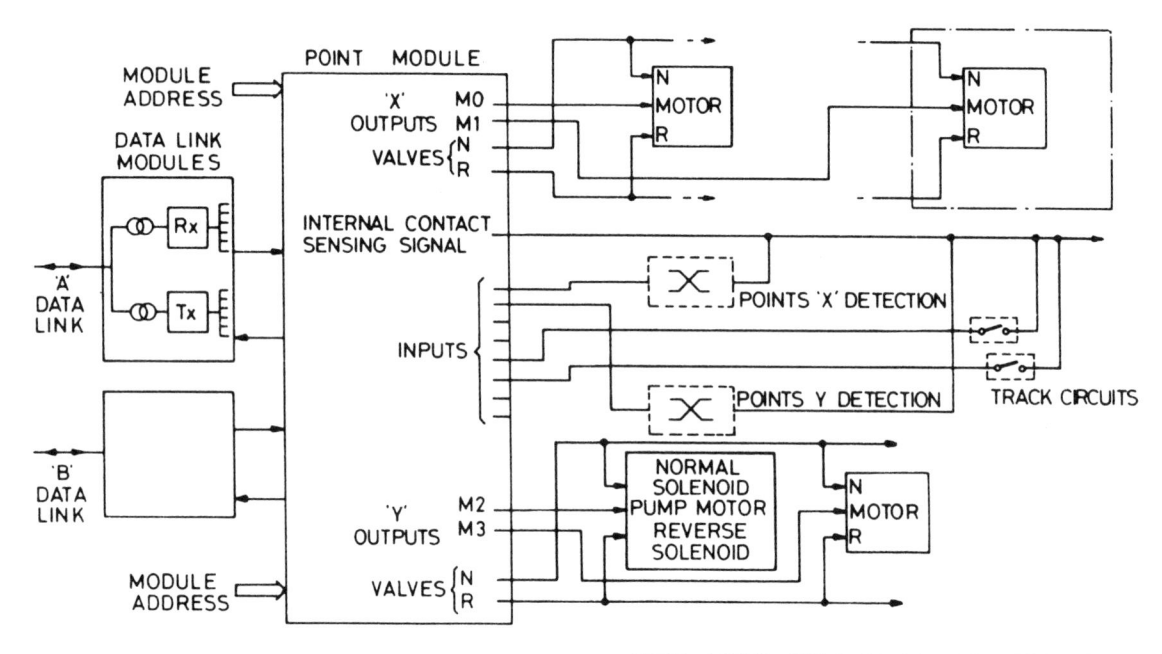

Fig. 2.13 Application of point module

After throwing the points, the interlocking removes its control as soon as the detection contacts are closed. If they should fail to close for any reason (eg if debris prevents the point blades closing), the module cuts off the motor drive after 8 s. If the detection contacts close and then open subsequently, the module drives the points again.

The behaviour of the point module under fault conditions follows the same principle as that of the signal module. Each of the groups of outputs may be isolated separately following a failure, leaving the other fully functional; indications, including point detection, continue to be sent back to the interlocking. If the security fuse circuit is triggered, all points are locked as last set, and the module ceases sending reply telegrams. If communication with the interlocking is lost, all points are locked as last set.

Trackside Interfaces

Two types of trackside functional modules are required to carry out the following functions:

- To manage the system redundancy to ensure safe behaviour under fault conditions.
- To receive and decode data highway messages addressed to the module, and to assemble, encode and transmit replies; to diagnose and report data highway unreliability and impose safe conditions when communication is lost.
- To detect the states of external switch or relay contacts.
- To control and monitor the state of power level outputs and the power interface fault protection mechanism.

In addition, the signal module is required to monitor signal lamp current and to ensure the illumination of the red signal lamp under fault conditions. The point module is required to detect the position of the points being controlled and to control the switching of a high current non-safety interface to the hydraulic pump motors.

Safe control of the modules is maintained using the dual processor technique described previously. Although the technique is capable in principle of satisfying the stringent safety requirements of trackside interface modules, translating these principles into a robust mechanism for switching AC at relatively high power levels presents considerable difficulties. The solutions adopted for the two types of module reflect subtle but fundamental differences in safety requirements, and the wish to maintain a degraded performance in the presence of faults where this can be achieved without compromising safety.

The internal structure of both types of trackside interface module is illustrated in **Fig. 2.14**. The design aims to divorce the management of processor redundancy from the safety management of the power output switching function. Only the failure of one or both processors, or an irreconcilable difference of opinion between the two, causes total (safe) failure of the module. Other failures cause only partial loss of function and are reported to the diagnostic processor in the control centre.

Power Interface Safety Management

The two essential distinctions between signal and point modules are that signal lamps are normally energised for long periods, whereas point machines are energised only for the time taken to move the points; and that it is permissible to allow a signal lamp to be incorrectly energised for short periods (up to 100 ms), whereas no transient energisation of a point machine is allowable under any circumstances. Power interface safety management for the signal module therefore aims to provide a means of enforcing safe output states in the event of an incorrectly energised output being detected, whereas for the point module, the aim is to maintain electrical isolation of the safety outputs at all times other than when the points are required to be moved.

The essentials of the technique are illustrated in **Fig. 2.15**. The signal power interface is protected by two normally open

Fig. 2.14 Internal structure of trackside interface module

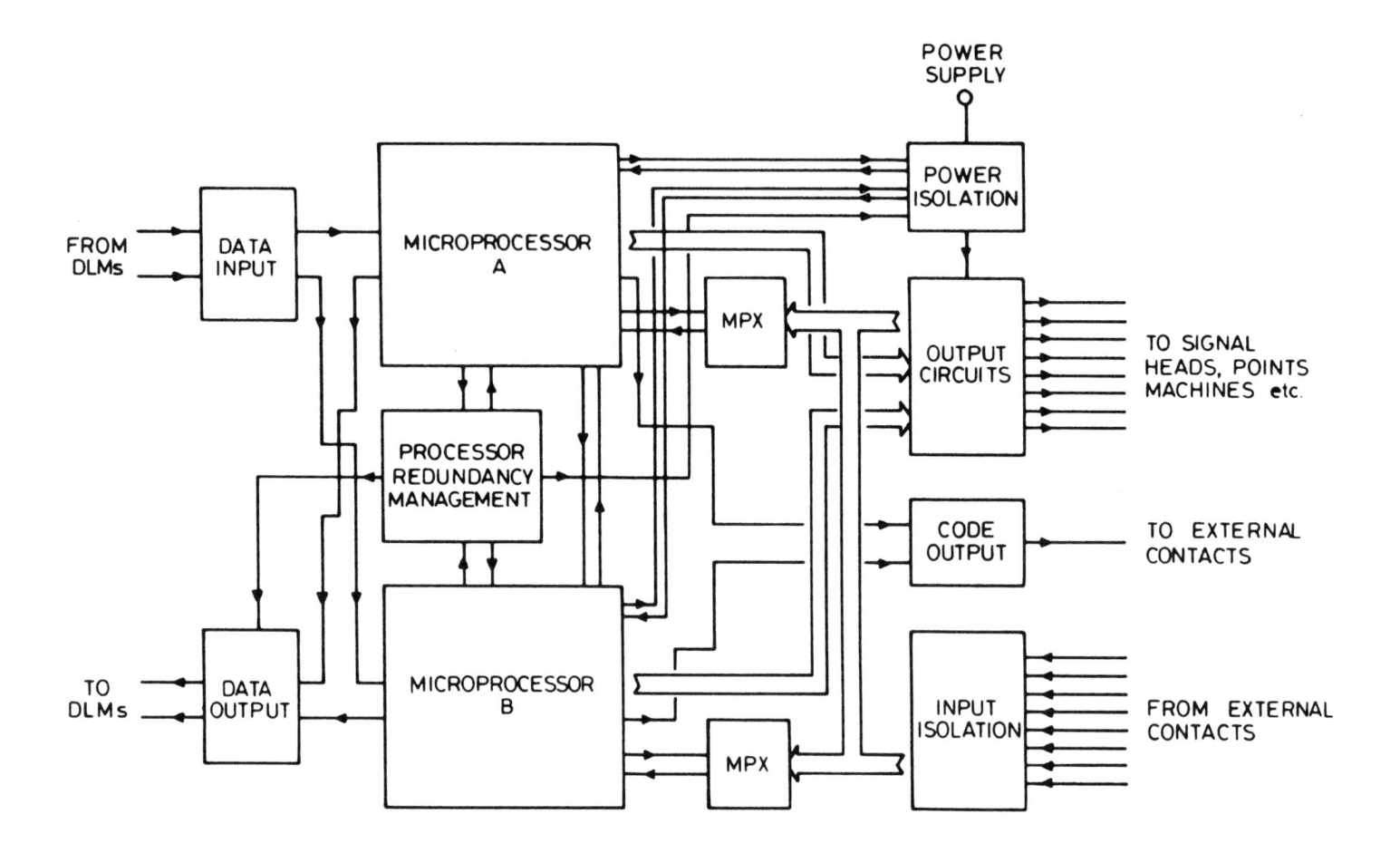

Fig. 2.15 Trackside module power interface isolation
A–signal module   B–point module

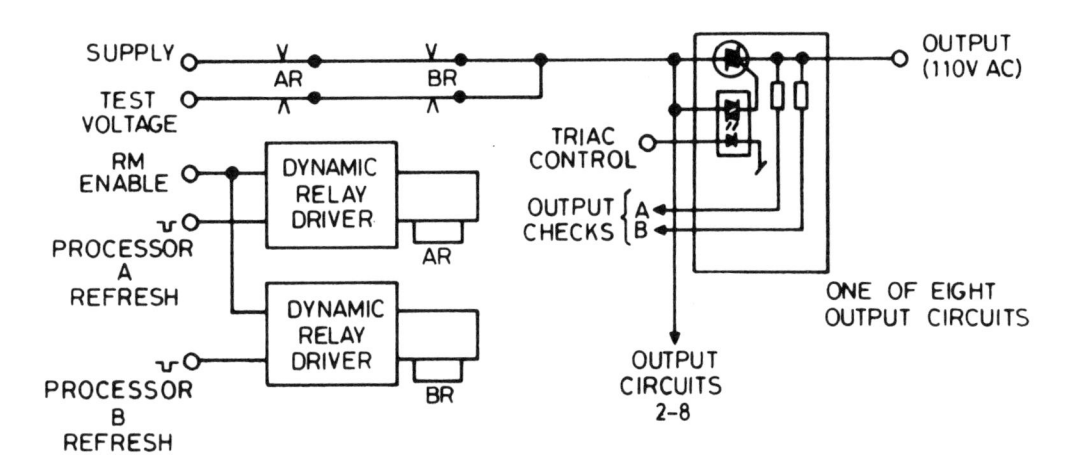

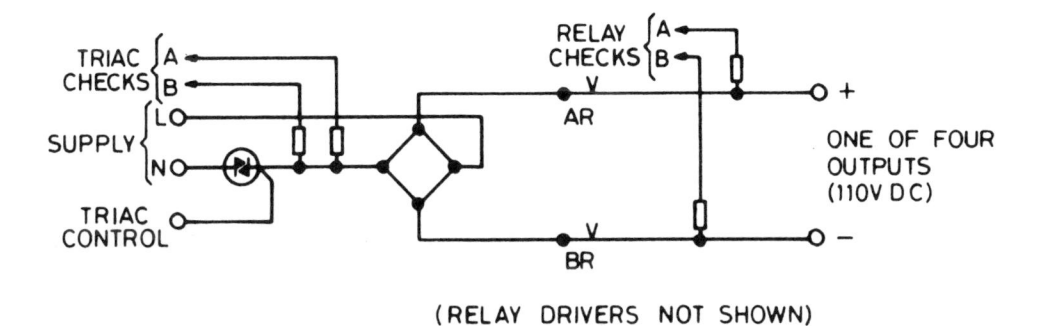

miniature relays placed in series and independently energised by the two processors. Energisation of the relays requires the presence of an enabling supply from the processor redundancy management circuit and the application of a dynamic refresh at intervals of not more than 100 ms. These relays are normally energised, supplying power to an array of triacs which switch 110 volts AC under processor control. The safe output condition is imposed by attempting to turn off all triacs and halting the dynamic drive to the relays. The ability to de-energise the protection relays is tested periodically, as is the ability to control each triac. Any fault results in the power interface being permanently disabled.

The safety outputs of a point module are isolated from the controlling triac by the normally open contacts of two relays, independently controlled by the two processors. In order to energise an output, it is necessary to close both relays and turn on the triac. At all other times (moving a set of points typically takes 2–3 s) the condition of the triac and relay contacts is constantly monitored and, if any of these three lines of defence is suspect, the associated points interface is declared unusable and the fault reported. The integrity of the monitoring arrangements is regularly tested by deliberately operating the triac or one of the relays.

The philosophy adopted is therefore one of testable redundancy, in that there is always more than one mechanism available for the processor to maintain the integrity of its power interface, and each mechanism and the means of testing it, is regularly tested for availability.

Voltage Sensing and Contact Detection
Power interface safety management depends on the ability to sense supply frequency voltages, and in particular to determine whether the voltage at a module output exceeds a certain amplitude. This function has been implemented in a way which integrates conveniently with mechanisms for sensing signal lamp currents and the state of external contacts. The principles are illustrated in **Fig. 2.16**.

The voltages to be monitored are passed through a potential divider and are added to a DC reference. Currents are passed through a suitable small resistance. The resulting small amplitude 50 Hz waveforms are sampled by an analogue multiplexer and compared with an offset reference which establishes a threshold level, the comparator output being read by the processor. Each voltage waveform is sampled in this way 32 times over the 20 ms supply cycle, and the number of samples exceeding the threshold is interpreted in terms of voltage amplitude. The two independent sets of hardware are arranged such that the two processors respond to opposite half cycles of the waveform. The integrity of the voltage sensing hardware is to a large extent checked automatically during normal operation; however, additional checks are carried out as part of the periodic testing of the isolation mechanisms.

External contacts are sensed by passing a coded signal through them. A Manchester-coded pseudo-random sequence is generated by the processors and its transformer coupled to the external circuit. The switched return is optically coupled into the two multiplexing circuits used for voltage sensing and subjected to the same sampling and comparison process. The code rate is such that sampling of the 32 Manchester half bits is synchronous with, and interlaced with voltage sampling.

Voltage sensing, current sensing and contact detection are carried out at the program cycle rate, typically every 50 ms.

Scheme Design and Data Preparation

DESIGN CONSIDERATIONS
All the interlocking equipment is located at the control centre. If trackside equipment is located more than 10 km away, repeaters are required in the trackside data links, at intervals not exceeding 8 km. With repeaters, equipment can be controlled at a range of up to 40 km.

Fig. 2.16 Trackside module voltage sensing and contact detection

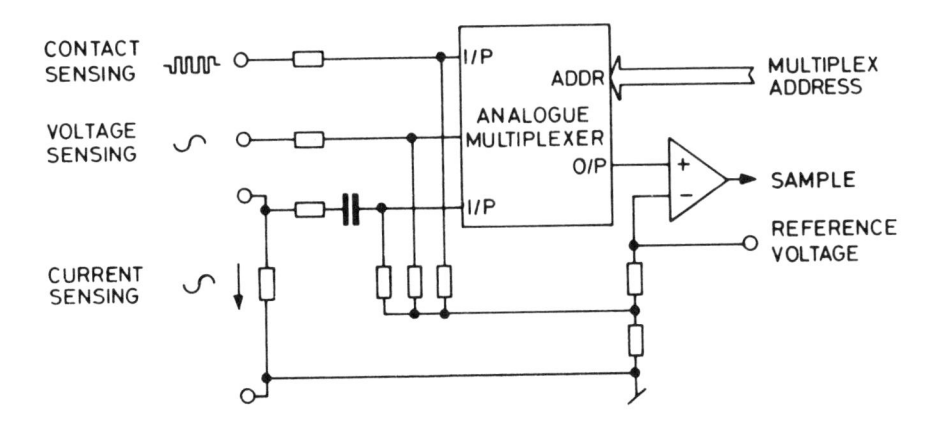

Allocation of the various trackside signalling functions to trackside function modules enables the total number of modules required to be assessed. This involves decisions on grouping of points, signals, track circuits and other functions. Each interlocking can control up to 63 trackside functional modules, and so the number of interlockings required can then be decided, it being prudent to provide spare capacity in each to allow for future modifications. When provision is made for the technician's terminal, the accommodation required at the control centre can be assessed (note that a large scheme with more than six interlockings would require more than one technician's terminal). Trackside functional modules are allocated to interlocking areas on a geographical basis, depending upon operating and other requirements. The cubicles for the interlocking and the technician's terminal, with their wiring, are of standardised design, as are lineside apparatus case layouts and wiring details for various numbers of trackside modules.

Fig. 2.9 shows a typical arrangement of the trackside data link, between 8 and 16 km in length. The link branches at location 2, and there is a repeater at location 3. An isolating transformer is shown between the control centre and location 1.

ALLOCATION OF TRACKSIDE FUNCTIONAL MODULES

Each of the outputs from the signal module can drive one signal lamp or an equivalent load, so that one module can drive (eg) two 3-aspect signals, or one 4-aspect signal with subsidiary aspect, two route indicators and an AWS electro-inductor. The module can generate flashing aspects; if any are required, a link is made in the apparatus case wiring and they are called up in the interlocking data (see below). Each of the six general purpose inputs can be used to sense one contact, or one pair of contacts in a double-cut circuit, of a track circuit or other relay. All remote functions, including non-vital indications such as train ready to start buttons, are brought in through the inputs of signal or point modules, and no other remote control system is required. The maximum length of leads from the module to the equipment it controls is 200 m.

The point module can drive up to four sets of clamp locks in two independent groups, as already explained. The maximum length of cables from the module to the clamp lock is 200 m. When point machines are in use, interface contactors are provided, circuit details depending upon the type of point machine and its power requirements.

Once the number of trackside functional modules at a location is known, power supply and cabling requirements can be worked out, as can the number of apparatus cases necessary.

THE DESIGN WORKSTATION

Scheme design and testing are carried out in the office, using the design workstation. As shown in **Fig. 2.17**, it consists of the following elements:

- Workstation computer.
- Simulator.
- Two colour display systems.
- Programmers for EPROMs.
- Communications network.

The workstation computer is a desk-top minicomputer with a high capacity store on magnetic disc, and allows several simultaneous users.

The simulator is a minimal SSI interlocking, containing a single panel processor module and also three multi-processor modules, of which one is configured as an interlocking module, one as a diagnostic module and the third as the special simulator module. They are all fitted with random access memory (RAM) in place of the usual EPROM, and geographical data created on the workstation computer is loaded into them automatically. The simulator module is able in effect to simulate the behaviour of the whole of the trackside equipment, replying to the control telegrams sent by the interlocking and sending meaningful indications in the reply telegrams in accordance with its own geographical data.

The state of all trackside functions as perceived by the simulator is shown on one of the colour displays. The other colour display communicates with the panel processor module, and simulates the signalman's control and display system. A device such as a mouse or a trackerball, which enables a cursor to be moved around the screen, is used for entering controls through either display system.

When design and testing are complete, the EPROM programmers are used to prepare sets of memory modules for installation on site.

The design workstation is used as follows:

- For each interlocking, a database is created which includes the track plan and screen displays, location areas and the allocation of signalling functions to point and signal modules. A detailed allocation of functions to the outputs and inputs of point and signal modules is carried out, a module allocation chart being produced for each module.
- Working from the control tables, geographical data is prepared for each interlocking. Independent checking of this source version of the data is carried out, the work-

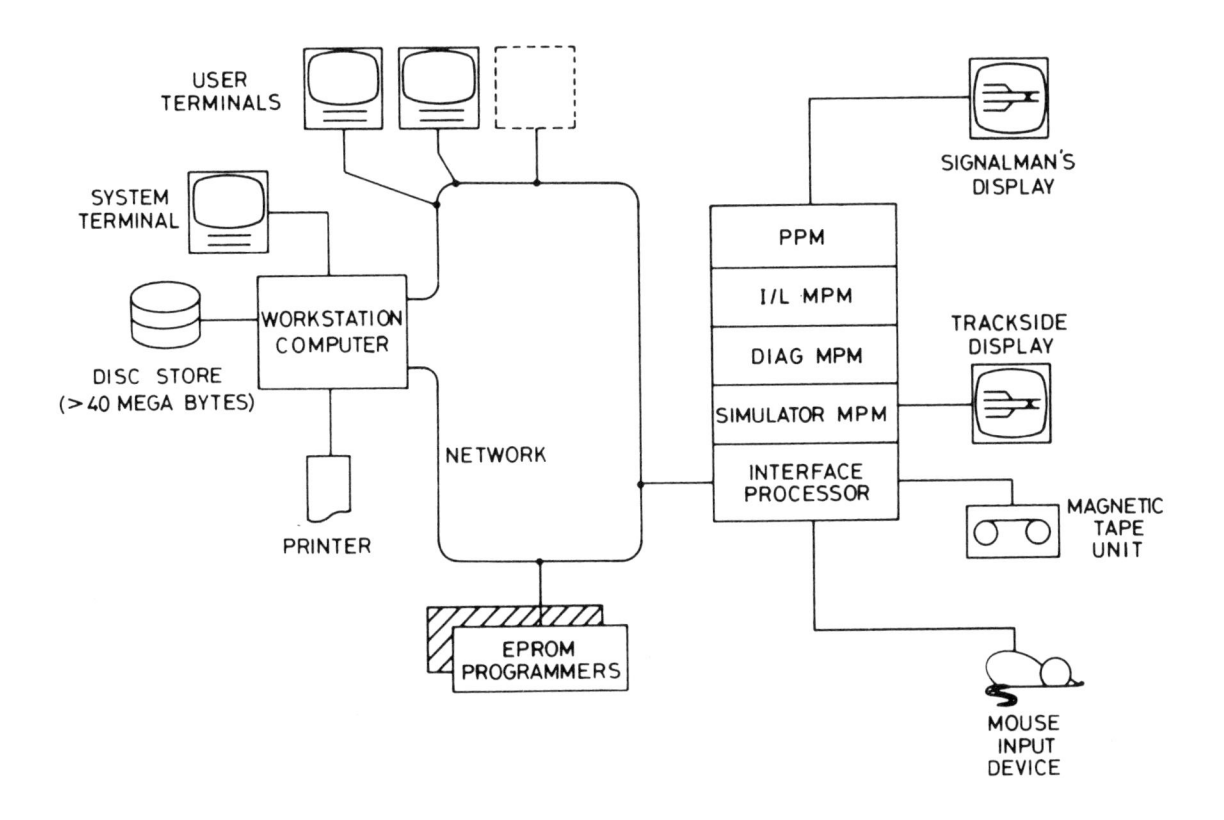

Fig. 2.17 Block diagram of SSI design workstation

station being equipped with a printer to provide copies for this and other purposes. This checking is a vital part of the applications work, its status being equivalent to that of circuit checking for relay interlockings.
- After checking, the data is compiled and loaded into the simulator. This and other operations are carried out on command from the system terminal.
- Functional testing of the compiled data against the control tables is then carried out, using the simulator and its display systems.
- When testing is complete, the data is loaded into memory modules. A decompiler program in the workstation computer allows SSI data to be regenerated from the EPROM contents in the memory modules. This decompiled version is compared with the data originally prepared, either manually or by the workstation computer, in order to ensure that the data installed on site corresponds exactly with the checked source version. Additional sets of copies of the EPROMs are prepared, to be kept in archive store.

When modifications to an existing installation are required, new data for the revised layout is prepared and tested in the office, using the workstation, and is installed in sufficient memory modules to equip all the working interlockings affected. If there are stage works, data for all stages may be prepared and held in readiness. When the time comes to carry out the modification work on site, the interlockings are switched off, the memory modules are exchanged, and the interlockings are switched back on again. They are then ready for service after a functional test to prove the modifications.

In addition to testing the data in the design workstation simulator, it is also possible to test the data making use of the real central interlocking and control panel or IECC hardware. An on-site simulator based on the technician's terminal processor can simulate the messages from trackside modules and other interlockings. This is particularly useful in the integration testing of the control centre and trackside equipment, when the simulator's replies to nominated trackside data link telegrams can be suppressed to enable the corresponding real trackside functional module to be connected and exercised.

In order to provide on-site simulation, one interlocking MPM in each central interlocking is converted to a simulator MPM by exchanging its memory module and inserting an adaptor cable between the MPM and its plug coupler in the cubicle wiring. The technician's terminal is adapted to provide simulator display and control facilities by the addition of a colour VDU and trackerball, and by loading it with programs and geographical data to control the display.

DATA PREPARATION

Data preparation is carried out by signal engineers, using the workstation. It consists of turning the information contained in the control tables into a form which the system can understand. There are two stages: first the information is expressed in a notation which the user can handle readily; then it is compiled to generate the machine code required by the processors.

Each interlocking requires data for its interlocking multi-processor modules, panel processor modules and diagnostic multi-processor module. The simulator also requires data for each interlocking to be tested. The workstation generates the data for each module from a single set of files.

The identity files include lists of all signalling functions by type. Only functions declared here are recognised by the system. The names of the files start with a code which identifies the interlocking; AB for example:

AB.TCS	—	track circuits
AB.SIG	—	signals
AB.PTS	—	points
AB.ROU	—	routes
AB.FLG	—	flags (see below)
AB.ELT	—	elapsed timers
AB.QST	—	route requests
AB.BUT	—	signalman's panel switches and buttons
AB.IND	—	panel indications

SSI Data Examples

Following allocation of the various signalling functions at the trackside to point and signal module outputs and inputs, preparation of data to configure each SSI to the required layout of signalling and to the control tables can commence.

Source data is prepared in a number of files, written in a near English format and entered into a SSI design workstation. It is then bench-checked and operated on by compilers which do lexical and syntactical checks and produce an object code version of the data, in a form acceptable to the programs in the various host modules. This code is then down-loaded into the simulator part of the workstation for testing, and when satisfactory, EPROMs are programmed with the object code for installation on site into the host modules.

These modules are the interlocking and diagnostic multi-processor modules (MPMs) and panel processor modules (PPMs). Interlocking MPM source data files are:

IPT — input functions from the trackside and other interlockings

OPT — output functions to the trackside and other interlockings, including signal aspects, approach locking and route release

FOP — flag operations, ie subroute and suboverlap freeing and other miscellaneous functions

PRR — panel requests, which includes route availability, setting and locking, point keying, etc

PFM — points free to move, which contains all point locking

MAP — geographical data map of layout used for looking for approaching trains for approach locking release and TORR (train operated route release) purposes

The first three files are processed by the module program once per major cycle: the other files are accessed as required.

PPM data files are PSD, PPD, PTD, POD (panel, signal, points, track and other data respectively) and PBK (panel button and key switch data), the first four being regularly processed and mainly used for controlling indications to the signalman.

The diagnostics MPM data is compiled from the point and signal module allocations together with the data link map DIA file and is assessed as required. This allows the MPM to deduce faults which are then passed on to the technician's terminal.

For each signal, set of points, track circuit, route, subroute/suboverlap, etc, identified by lists in identity files, memory is allocated and this is used by each interlocking MPM to record the state of the railway. This can be updated from a number of sources, namely information received from the trackside (lamp proving, track circuits, point detection) or other interlockings, actions of the signalman (button pulled, route set, locking of subroutes/suboverlaps, point controls, point key switch position, signal auto-mode); subsequent program action (train in section proving, free of approach locking, signal stick, aspect control, aspect displayed), or technician's bars (points disabled, route barred, aspects disconnected). Timers are also provided for each track circuit and signal and for miscellaneous purposes. Most of the above memory is copied across to the PPMs which also have memory to record the states of panel indications and button/switch contacts.

Thus the data dictates the actions of the various module programs on this memory and the output of the interlocking to the trackside, the signalman and to other interlockings. Arrangements of data range from general purpose constructs such as conditional statements where the program is advised of every action (eg as in the PRR file), through to specialised data types where, for example, just a track circuit or an incoming telegram bit from the trackside is identified; from this the program updates the memory accordingly including restarting track circuit timers, delaying track circuit clearance, etc (see IPT file).

Punctuation such as '"$.' is provided to guide the program as to the end of various data constructs, with logicals '(or) if then else' having their usual meanings and '@ { }' being data pointers, ie pointing to labels (*) within the data. These last items allow the programs to deviate from their normal serial progress through sections of the data.

52　SOLID STATE INTERLOCKING

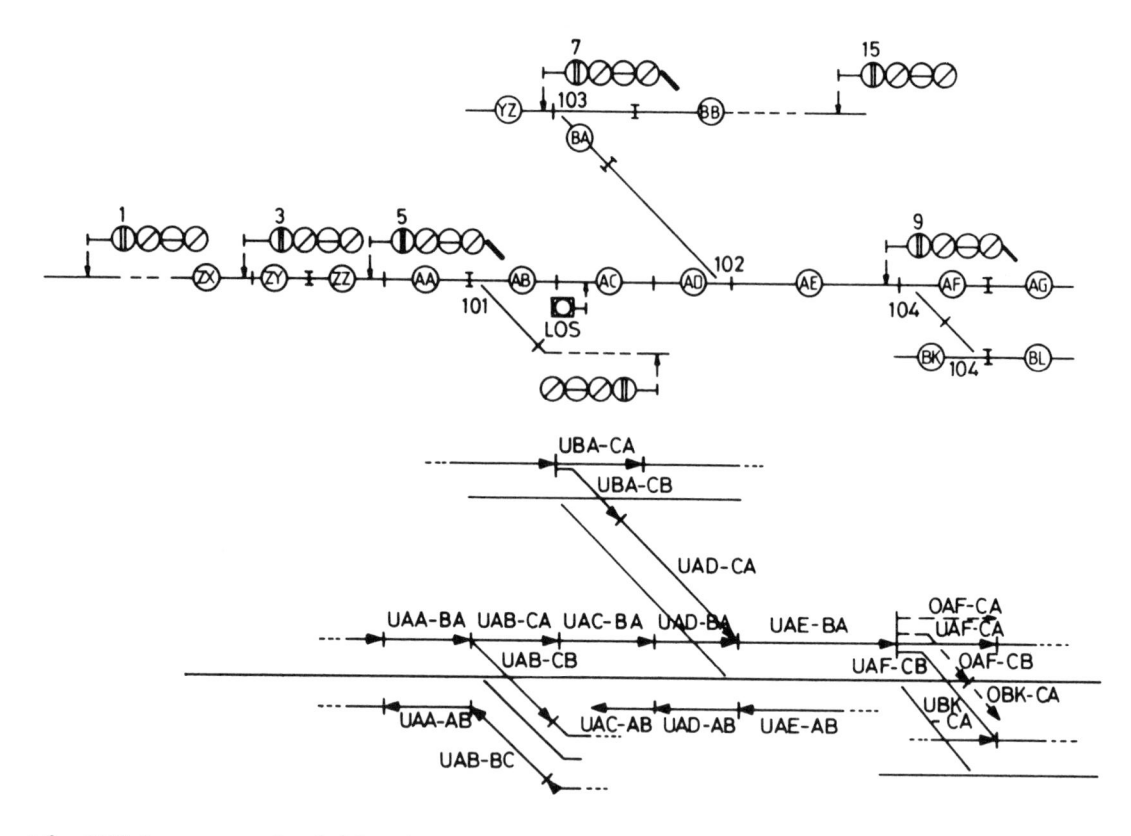

Fig. 2.18　Layout for SSI data examples (with subroute/suboverlap diagram)

Fig. 2.18 shows the layout used for the data examples. This is the same layout used for the relay interlocking circuit examples in the first volume of the Textbook, to allow comparisons to be made. Subroutes and suboverlaps have been added to this fig., these being similar to the route sticks used in a relay interlocking. Extra comments (preceded by /) have been added to the data examples to assist readers in understanding the data.

Setting Up a Route – Panel Data
The PPMs receive changes of state of button contacts from the NX panel, and data in the PBK file is used to link associated entrance and exit button operation and send panel requests for route setting to the interlocking MPMs. The operation of 5 entrance and 9 exit buttons can be followed through the data above with entrance button being flashed and panel request QR5A generated (see **Fig. 2.19**). If the route

Fig. 2.19 Control panel: entrance/exit buttons

Fig. 2.20 Control panel: other buttons/switches and indications. (c) overleaf

```
/ PRR File
/ Typical route setting data.
*QR5A      / tests for availability of route
              if R5A a , P101 cnf , P102 cnf , P103 cnf ,
              ( P104 cnf or P104 crf ) UAA-AB f , UAC-AB f ,
           / commands for setting & locking route
              then R5A s , P101 cn , P102 cn , P103 cn ,
              @OL104Q , UAA-BA 1 , UAB-CA 1 ,
              UAC-BA 1 , UAD-BA 1 , UAE-BA 1 ,
              S5 clear bpull .
           / subroutine for setting overlap, called from @OL104Q
*OL104Q    ( if P104 cr then OAF-CB 1 , OBK-CA 1
              or if P104 cnf then P104 cn , OAF-CA 1
              or P104 cr , OAF-CB 1 , OBK-CA 1

*QXS5      S5 set bpull .   / 5 entrance button pulled
*QAUTO5    if R5A s then S5 set auto .
                           / 5 auto button pushed
*QXAUTO5   S5 clear auto ./ 5 auto button pulled
```

(a)

```
/ PBK File – Buttons & key switches
/ Other typical items in first button list.

B5 (A) F s ,    = QAUTO5    / Auto button 5 pushed
B5 (A) FM s ,   = QXAUTO5   / Auto button 5 pulled
B102N s ,       = QP102QN   / Points key switch 102 moved to N
                B102C s ,   = QP102KC /  "   "   "   "   C
B102R s ,       = QP102QR   /  "   "   "   "   "       R

/ PSD File – Signal Indications
1S5 , I5DG s , I5RG sf $ . / Indications are lit, flashed
    / or extinguished according to state of aspect code and
    / lamp proving, button pulled and free of approach
    / locking (foal) bits in signal memory.
```

(b)

```
/ PPD file – Point key indications
1P102    I102N s , I102R s , I102(OC) sf $ . / Indications
    / are lit, flashed or extinguished according to
    / state of detection and control bits in
    / memory as appropriate.
/ PTD File – Track and route light indications
/ Track circuit example without connections
1TAC , UAC-BA f
    IACT s     IACU s
$.
```

is subsequently set, then as the state of the associated route memory is copied across from interlocking MPMs to the PPMs, the entrance button will be lit steadily through data in the POD file above.

Other Panel Data

Other examples of panel data to translate making or breaking of panel button and switch contacts into panel requests are shown in **Fig. 2.20**. Also included are examples of panel indication data where the state of signals, points, track circuits, routes, subroutes, etc, copied across from the interlocking MPMs memories to that of the PPMs, are then tested in combination as required to determine which panel indications should be lit.

(c)

```
/ Track circuit example with connections
1TAD , UAD-AB f , UAD-BA f , UAD-CA f
   I102CT s    I102CU s              / Common segment
UAD-AB f , UAD-BA f   I102NT s       / Normal segment
I102NU s , P102 cdn , I102NKU s
UAD-CA f   I102RT s                  / Reverse segment
I102RU s , P102 cdr , I102RKU s
$ .
/ Indications lit or extinguished dependent on state of
/ track circuits, subroutes and point controls and detection
/ as appropriate
```

Where IECC systems are employed controlling SSIs, the PPM data is not required, this being provided in a different form within the IECC system. Thus the PPMs merely act as communication ports sending state of the railway information to the IECC system and receiving panel requests for passing on to the interlocking MPMs.

When the interlocking MPMs receive panel request QR5A for setting route 5A, the appropriate section of data in the PRR file is processed (see **Fig. 2.21**). Tests are made for route not barred by technician (R5A a), points controlled to, or free to move to, the required position (eg P101 cnf) and any directly opposing route locking free (opposing subroutes free). If all are true, the route memory is set, normal or reverse point control bits are set, subroutes and suboverlaps are locked to hold the route and the button pulled memory is cleared. The example shown of a data subroutine (carry out commands at ...) is used by routes up to 9 signal to set an available overlap, although with this simple overlap example, the overlap data does little more than lock suboverlaps for controlling route lights in the overlap. Alternative commands are provided to avoid unnecessary swinging of the overlap and allowance has been made for initial start-up conditions when

Fig. 2.21 Route setting and locking and associated functions

```
/ PRR File
/ Typical route setting data.
*QR5A        / tests for availability of route
             if R5A a , P101 cnf , P102 cnf , P103 cnf ,
             ( P104 cnf or P104 crf ) UAA-AB f , UAC-AB f ,
             / commands for setting & locking route
             then R5A s , P101 cn , P102 cn , P103 cn ,
             @OL104Q , UAA-BA l , UAB-CA l ,
             UAC-BA l , UAD-BA l , UAE-BA l ,
             S5 clear bpull .
/ subroutine for setting overlap, called from @OL104Q
*OL104Q      ( if P104 cr then OAF-CB l , OBK-CA l
             or if P104 cnf then P104 cn , OAF-CA l
             or P104 cr , OAF-CB l , OBK-CA l
*QXS5        S5 set bpull .   / 5 entrance button pulled
*QAUTO5      if R5A s then S5 set auto .
                             / 5 auto button pushed
*QXAUTO5     S5 clear auto ./ 5 auto button pulled
```

neither point control bit is set (equivalent to neither point NLR/RLR relay being latched up).

Also shown are data examples of processing panel requests for 5 entrance button pulled and 5 auto button pushed and pulled.

Point Associated Data
Next a test such as P102 cnf in **Fig. 2.21** is carried out, if the control bit in points memory for that position is not set, then data in the PFM file is accessed (*P102N in **Fig. 2.22**) to test whether the points are free to move.

First comes automatic tests for points not keyed or disabled against the move and then the data is processed; this

Fig. 2.22 Point associated data

```
/ PFM file
/ Points-free-to-move data

*P102N    TAD c , ( TBA c or P103 cdn ) , UAD-CA f
*P102R    TAD c , TAC c , UAD-BA f , UAD-AB f

/ PRR file
/ Point key requests

*QP102QN    P102 kn , P102 xkr        / Key to N
            if P102 cfn then P102 cn .

*QP102QR    P102 kr , P102 xkn        / Key to R
            if P102 cfr then P102 cr .

*QP102KC    P102 xkn , P102 xkr.      / Key to C

/ OPT file
/ Point output data

/ Address 21 – Points Module
*021 'P102 G s 76 $ .

/ The state of the points memory control bits
/ is copied into output telegram bits 7 (normal) and
/ 6 (reverse), this being interpreted by point module at
/ address 21 to drive this set of points to the required
/ position.
```

```
/ PRR file
/ Overlap swinging subroutines

*P104QN    P104 cn
           if OAF-CB 1 then OAF-CB f, OAF-CA 1
                      if P104 dr then EP104 = 0
*P104QR    P104 cr
           if OAF-CA 1 then OAF-CA f , OAF-CB 1 , OBK-CA 1
                      if P104 dn then EP104 = 0

/ Where a set of points is a facing connection in an overlap
/ eg points 104, then route setting or point keying requiring
/ to swing the overlap have a data subroutine substituted for
/ the usual points command to move the points in order to
/ swing the overlap. Time EP104 is used in 5 signal aspect
/ controls (see Fig. 2.24) to suspend proving of 104 points
/ during overlap swinging.
```

typically includes tests for track circuits clear, free of route/overlap locking, etc. The program then resumes processing the data in the PRR file with a true or false result from this excursion into the PFM file.

The other usual way of moving points, ie point keying, is implemented through point key requests, examples of which are shown in **Fig. 2.22**. (Special requests used for static updating of point key position after interlocking or panel start-up are omitted for clarity.) The point key memory is unconditionally updated but point controls are only changed if free to do so, similar to route setting.

Data in the OPT file then translates the state of the points normal and reverse control bits into telegram bits in the outgoing message to the point module controlling the points. The 8 s drive cut-off and motoring facility are part of the functions of the module, so that no data has to be provided for these features.

Incoming Functions from the Trackside

The state of bits in incoming telegrams from signal and point modules are used to update the various memories. Examples are shown in **Fig. 2.23** of special data constructs for signal lamp proving and route indicator proving, track circuits and point detection and conditional statements for a train ready to start plunger.

Special features include the setting of a signal aspect code to its most restrictive condition if a module indicates that it has disabled its outputs (red-retaining mode), and also a delay feature on a track circuit going from occupied to clear, where three consecutive messages reporting track clear are required before the track circuit memory is changed to clear.

A latch memory (used for miscellaneous functions) is used to store the plunger having been operated and to drive (via PPM data) the flashing indication to the signalman. The latch is unset by a second conditional statement when the signal is cleared (less restrictive than red) or the berth track circuit is cleared.

```
/ IPT File
/ Address 19 – Signal Module
    *I19   "S5 7  / Signal 5 lamp proving on bit 7
           "S5 rip 6
           /       Signal 5 route indicator lamp proving on
                   / bit 6
           "TAA 3 / Track Circuit input on bit 3
                  if G s 4 then L5TRTS s
                  / Train ready to start plunger on bit 4
                  if (TZZ c or S5 set <r) then L5TRTS xs
/ Address 21 – Point Module
    *I21   "P102 76 / Points 102, N&R detection on bits 7&6
           "TAD 2 . / TC input for TAD on bit 2
```

```
/ Map File
           .... / Look back to 9 via 104N
           #TAG
           if P104 cdr then pass
*TCAFD    #TAF # S9
*TCAED    #TAE #TAD / Look back from 9
           if P102 cdn then  ^ TCACD
           if P102 cdr, P103 cdn then pass
           if P102 cdr then  ^ TCBAD / via 102R
           fail
*TCACD    #TAC
           if P101 cdr then pass
*TCABD    #TAB #TAA #S5
*TCZZD    #TZZ #TZY #S3 / Look back from 5
*TCZXD    #TZX ....     / Look back from 3
           ....  / Look back to 7 via 103N
           #TBB
           if P103 cdr then pass
*TCBAD    #TBA #S7
*TCYZD    #TYZ .... / Look back from 7.
```

Fig. 2.23 Incoming functions and MAP data

MAP Data

MAP data is arranged geographically, and is accessed from other files to look for approaching trains, usually for comprehensive approach locking release or train operated route release (TORR) purposes. Labels (*) are provided to give jump to points within the file and entry and exit points from other files. MAP searches start at a defined entry point and then work through the data until a pass or fail is encountered, or a defined exit point is reached, this last event being declared a pass. Tests include track circuits (fail if not clear) and signals (pass if on and free of approach locking) (usually referred to as foal). Sections of typically MAP data are shown in **Fig. 2.23**.

In this example (see **Fig. 2.24**) module 19 outputs 7 to 2 are allocated to 5 signal red, bottom yellow, top yellow and green aspects, route indicator and AWS inductor respectively.

Here the route select type of the special data construct for signal 5 is used. If no route is set, the output telegram bit for the red aspect is set, otherwise the appropriate data subroutine for the route set is processed. First in this are the train in section proving track circuits (first occupied and cleared, second occupied) used in approach lock release with the train in section proving code updated accordingly to record the stage reached. Second are the two track circuits which when occupied with signal off unset the signal stick. The aspect control data contains all tests (except any approach release or route indicator proving) for the signal to clear, including automatic tests for entrance button not pulled and aspect not disconnected by technician, and if all are true the aspect control bit is set.

The tests in temporary approach data have to be satisfied if the technician applies this control. Aspect sequence data first includes data for putting the signal to red; if the signal can clear, the aspect sequence data which advises of the signal ahead, the number of aspects and the telegram bits which refer, is then processed and the correct aspect is calculated, the appropriate telegram bits are set and approach locking is applied. The telegram bit indicated in AWS data is set if the signal is at green and alight.

Fig. 2.24 Signal output data

```
/ Address 19 – Signal Module
*019    'S5              / Start of special for S5
        R5A s , @R5A     / Route Select Data
        R5B s , @R5B
        G s 7.           / Set red aspect (teleGram set bit 7)
        $ .              / End of special and telegram
*R5A    #TAA #TAB .      / Train In Section Proving data
        #TZZ #TAA .      / Single Stick Unsetting data
                         / No Route Step Up data
        if S5 set stick , TAA c , TAB c , TAC c , TAD c , TAE c ,
        P101 cdn , P102 cdn , P103 cdn , S9 set 1p , TAF c
        ( P104 cdn or P104 cdr or EP104 < 8 , S5 set ascon )
        ( P104 cn or P104 cr , TBK c ).
                         / Aspect Control data
        if TZZ o .       / Temp. Appr. Control data
        .                / No Route Indicator data
        G s 7            / Aspect Sequence data
        S9 seq 4 , G s 7654 .
        G s 2 .          / AWS data
        S5 alt > 240 .   / Appr. Lock. Rel.data (not comp.)
        R5A xs       .   / Route Release data (no TORR).
                         / End of data for R5A
*R5B etc . . . . . .
/ To illustrate use of MAP data
/ Excerpt from data for *R9A
S9 alt > 240 if {TCAED , }TCZXD , }TCYZD .
                         / Approach Lock Release (comprehensive)
R9A xs   if {TCAED , }TCZXD , }TCYZD    .
                         / Route Release (includes TORR)
```

In approach lock release data, the approach lock release time is defined. Program action automatically examines the state of button pulled, aspect control and free of approach locking bits, train in section proving code and approach locking timer, to determine which actions of freeing approach locking, starting and stopping the timer and resetting the train in section proving code should be carried out. Signal 5 is shown approach locked when cleared, but in the data shown for signal 9, a test list including a map search is shown, and if the entrance button is pulled and the map search results in a pass, then the approach locking will be freed.

The route data includes the route(s) to be released and if TORR is not provided, as in this example for signal 5, then the route will be released if the entrance button is pulled and the approach locking is free. If a test list for a search for approaching trains is included, then TORR will be operative and subject to being free of approach locking, the signal stick unset, the signal not in auto-mode and no trains approaching, the route will then be released.

Final program actions include setting the signal stick if the entrance button has been pulled or the signal is in auto-mode, and cancelling auto-mode if the entrance button has been pulled.

Fig. 2.25 Subroute/suboverlap release

```
UBA-CB f   if   R7B xs , TBA c   .
UAA-BA f   if   R5A xs , R5B xs , TAA c   .
UAB-CA f   if   UAA-BA f , TAB c   .
UAC-BA f   if   UAB-CA f , TAC c   .
UAD-CA f   if   UBA-CB f , TAD c   .
UAD-BA f   if   UAC-BA f , TAD c   .
UAE-BA f   if   UAC-CA f , UAD-BA f , TAE c   .
OAF-CA f   if   R5A xs , R7B xs, TAF c ,
                ( TAE o > 90 or UAE-BA f )   .
OAF-CB f   if   R5A xs , R7B xs , TAF c,
                ( TAE o > 90 or UAE-BA f )   .
OBK-CA f   if OAF-CB f
```

The usual sequential sectional release of route locking is provided, as shown in **Fig. 2.25**. Safeguards against bobbing track circuits are provided by the delay in the interlocking in accepting that a track circuit is clear, but additional precautions are taken where automatic route setting is provided by additionally requiring (own track circuit clear for 15 s or next track circuit occupied).

CHAPTER THREE

Single Line Signalling

Introduction

The United Kingdom's early railways were for the most part built with two unidirectional lines of way. Double lines with left hand running were normally designated up and down lines, with the up line usually leading to the principal town on the system; single bidirectional lines were often built where traffic was expected to be light. In recent years, it has been the practice to convert existing double lines to single line where traffic permits in order to benefit from the much lower cost of maintaining the permanent way. This, in conjunction with the modernisation and automation of level crossings and the more recent application of low cost signalling techniques, has dramatically reduced the cost of operating rural railways.

The considerations to be taken into account in deciding the system of operation are: (i) the timetable, with stopping patterns, headway requirements and crossing and connectional information; (ii) the required line speed with the ideal speed of passing loops; and (iii) the type of motive power or coaching stock to be used, and whether the stock is captive to the service, or if through working is required.

From this information can be determined the position of passing loops, although in practice it is usually found that these are located by geographical considerations, such as the position of stations, level crossings and gradients. Also determined are the form and length of the passing loops, the layouts to be adopted at the extremities of the single line, the speed over the pointwork at crossing stations, and the requirement for left or right hand running, or for the loops to be bidirectional to allow for high speed crossing or overtaking.

General Philosophy

Unlike double lines, on which it is only necessary to protect against rear collisions between succeeding trains, on a single line there must also be protection against collisions between trains travelling in opposite directions. To achieve this, there must be direct interlocking between the starting signals reading into each end of the single line section. This interlocking must then be maintained until the movement has passed through the section and clear of the single line at the other end. Furthermore, once a train has entered the section at the starting signal, this signal may then not be cleared for a succeeding train until the initial train has either passed out of the single line section and has cleared the overlap of the home signal, or is clear of the overlap of the first of any intermediate signals which may be provided for headway purposes. The same applies to the opposite direction.

It may not be practical to apply a direct interlock between opposing signals, in which case the train must carry a unique token as an authority to occupy the section. This token may be a single staff, or may be part of a multiple token system in which only one token may be available at any one time. The operation of single lines carrying passenger trains is governed by the requirements of the Department of Transport, which are reproduced below.

Requirements of the Department of Transport for the Operation of Single Lines

On single lines used for passenger traffic, arrangements must be made to prevent opposing movements as well as to maintain a space interval between following trains; as a rule, this will be some form of token as described in Methods I to IV below, with regulations to the effect that no engine or train may travel upon the single line, beyond the protection of the signals at a crossing place unless its driver is in possession of the appropriate token. Alternatively, a non-token method of operation depending only upon the obedience of signals, as in Method V, may be adopted where specially approved. In each case a formal undertaking must be sent to the Department of Transport, stating the method of operation which will be adopted.

METHOD I: TRAIN STAFF AND TICKET
With this system a train staff and a set of train tickets, of paper or metal, are required as tokens for each section. No train may leave a crossing place unless the train staff for the section through which it is about to travel is at that crossing place, and no other driver may proceed into a section except for the purpose of shunting within the protection of signals, unless he has the train staff in his possession, or has seen it. So long as movement through a section takes place alternately in opposite directions, the driver of each train must carry the train staff as his authority. But if two or more trains require to travel in succession through a section in the same direction, before any train passes through in the reverse direction, the driver of each train except the last must be shown the train staff and be given a train ticket as his authority to proceed. The train staff will then be carried through the section by the driver of the last train of the series. The train tickets must be kept in the signalbox or booking office in a locked box which can only be opened by a key forming part of the appropriate train staff.

Removal of the train staff must relock the ticket box. The absolute block system must be used to ensure a proper space interval between trains. A single line section may be divided into two or more block sections by intermediate block posts with the necessary signals, in order to increase its capacity for following movements.

METHOD II: DIVISIBLE TRAIN STAFF
In this modification of Method I, the train staff has one or more detachable portions, each serving as a train ticket and is so labelled. The tickets must remain attached to the train staff and travel with it, except when removed for issue to the drivers of the earlier trains of a series travelling through a section in the same direction, to whom the train staff must be shown. The direction of movement through a section must not be reversed unless the train staff and all its tickets are at the crossing place from which the next train will start. In other respects the routine to be followed is the same as in Method I, and the absolute block system must be used to ensure a proper space interval between following trains.

METHOD III: ONE TRAIN WORKING
With this system there must not be more than one train upon the single line or any section of it at one and the same time. The train staff for the line or section must be carried on the train, and no train tickets may be used. Block instruments are not necessary.

METHOD IV: ELECTRIC TOKEN
The instruments at opposite ends of a section in which the tokens are kept, must be so arranged that after withdrawal of a token from one of them, a second token cannot be obtained from either instrument until the one already withdrawn has been transferred to the instrument at the far end of the section, or returned to the instrument from which it was taken. Separate block instruments are unnecessary, but if token instruments are provided at intermediate block posts between crossing places, they should be interlocked to prevent the simultaneous issue of tokens for movement in opposite directions.

The line wire connecting token instruments should preferably have continuous insulation as a precaution against accidental contact with other wires; a continuous uninsulated return wire may be necessary to prevent interference with the instruments by earth currents. Auxiliary token instruments, to be operated by station staff or enginemen and controlled by the main instruments for the section, may be used where necessary to avoid delay in obtaining a token from a signalbox, or surrendering it.

Intermediate token instruments similarly controlled may be provided at outlying sidings in order that the main instruments may be restored to normal, and through running resumed, when a train working at the siding has been shunted clear of the running line and while the token carried by it is locked in the intermediate instrument. A bank engine token may also be used where specially approved, to enable an engine to return from an intermediate point in the section, to the station from which it started, after assisting a train from the rear.

Tokens and train tickets must be clearly marked with the name of the section for which they are valid; those for adjoining sections must differ in colour or in shape, or both. Only the person authorised to do so may hand the token to a driver or receive it from him, or show the train staff to him under Methods I and II.

Points leading to a siding in a section must have some locking device which can be released by insertion of the token for the section, including the train tickets used in Method I if of metal. The points must be left normal for through running before the token can be removed.

Special signalling arrangements may be made to enable crossing places to be closed when traffic is light. These must include some interlocking device to make the short section tokens unobtainable when the long section is in use, and vice versa.

Where motor trolleys are used for track maintenance, arrangements are necessary to ensure that they are clear of the line at derailing points before the token for the section can be obtained. Similar arrangements for controlling the issue of tokens may be required to enable permanent way work to be carried out in the section without protection by flagmen.

It is desirable to control a signal leading to a section so that it cannot be cleared unless the appropriate token has been obtained from the token instrument under Method IV or the staff is available for issue to the driver under Methods I and II.

METHOD V: DIRECTION LEVER AND TRACK CIRCUIT
This non-token system requires continuous track circuiting through the section. The signals for entering the section must be so controlled that they cannot be cleared unless all track circuits in the section are unoccupied, and unless released by the signalman at the far end; his release must be repeated each time the signals are cleared and there must be a control to ensure that they are put to danger behind each departing train. Precautions against disregard of signals may be required in the form of detonator placers working with them, or of facing trap points.

OTHER METHODS
Other methods of operation, or combination or adaptation of the foregoing methods to meet special conditions, may be approved; full particulars of the arrangements to be adopted have to be submitted before installation.

Most modern systems fall into the last category, or are a combination of the systems allowed above. However the DoT requirements, although a little dated, have been included because it can be seen in the foregoing descriptions how these basic requirements are achieved by methods using modern technology (radio and data interchange), or by the application of modern techniques to established equipment (token instruments linked by multiplexing techniques).

The following are the systems commonly used to operate single lines in the United Kingdom:

- Token, staff, and tablet systems which generally conform to Method IV of the DoT requirements. Staff and tablet systems are now rare; however, electric key token systems are increasingly used in conjunction with simplified infrastructure. The systems available are:
    - Signalman-to-signalman token working for through single line sections. This can be further subdivided into systems where the communication between the instruments is by a multiplex system using the public telecommunications network or by means of a fail-safe data-carrying radio system, which can additionally be used to establish a speech channel with the train crew.
    - 'Signalman to automatic operating' token working for terminal branch lines upon which it is required to have more than one train at a time.
    - No-signalman remote key token working, (NSTR) where the train crew operate the instruments under telephone instructions from the controlling signalman. This is a true low cost system.
    - Radio electronic token block (RETB) in which the token is in the form of data which is passed by a fail-safe radio system from an interlocking to a train-borne instrument. This is a true low cost system.
    - Telephone electronic token block. Similar to RETB except that the data is passed over the public telephone system. (This system is under development.)
- Train staff and ticket, and divisible train staff systems, generally conforming to Methods I and II of the DoT requirements. This is not normally used, except in the form of an emergency back-up for no-signalman remote token working and radio electronic token block during a system failure.
- One train working. Single train staffs are used for one train working on light traffic branch lines. These generally conform to Method I of the DoT requirements.
- Track circuit block, which generally conforms to Method V of the DoT requirements. This requires a lineside cable route and power supply, and continuous track circuits through the section. However it is a most versatile system, as it depends entirely upon the obedience of fixed signals, and therefore does not impose any speed reduction for the exchange of tokens. It is also convenient if intermediate automatic signals need to be provided for headway purposes. It can be divided into the following subsystems:
    - Both ends of the single line are controlled from one signalbox by the same interlocking. This is only used where the section is short, and only the normal interlocking circuits are required to prevent opposing movements.
    - Both ends of the single line are controlled from one signalbox through separate interlockings. Special circuitry is required to maintain the interlocking of opposing routes in the two interlockings.
    - Both ends of the single line are worked from separate signalboxes by means of direction levers or switches. The opposing interlocking is carried out by a direction lever in the receiving signalbox, which is only free to be pulled if the section is clear. This operation transmits a release to the sending signalbox to allow a train to travel through the section. The direction levers are not interlocked, but safety is achieved by the release circuit using a common line for both directions. Once operated, the direction lever cannot be normalised unless the section is clear after the train has passed through, or the signalmen have co-operated to effect a cancellation.

- This system is similar to the previous one, but the track circuits are normally de-energised, and are only energised to prove the section clear initially when it is required to operate the direction lever. In this system, the track circuits are energised from the sending end to the receiving end to allow the receiving signalman to reverse the direction lever; once this is operated, the track circuit is maintained energised from the receiving end to the sending end to allow the sending signalman to clear the starting signal. Once the train has passed through the section, the track circuit is again energised from the sending end to the receiving end to prove the section clear and allow the direction lever to be normalised. This system is truly economical in that the track ciruits work on the prevent shunt, and thus can be very long. The release circuit only requires one pair of lines.
- This is as for the previous system but there is no release circuit, the system being entirely dependent on track circuit operation. One lineside circuit is required for cancellation purposes.
- Tokenless block is classed as 'other systems' under the DoT requirements. Like track circuit block, these systems depend entirely upon the obedience of fixed signals, thus avoiding the need to slow down to collect a token. Such systems do not require the section to be fully track circuited, and they do not provide a formal maintained interlock between opposing signals at each end of the section. Safety is achieved by the operation of block instruments giving a release to the starting signal. Once the release is given, a cancellation is possible if the train has not entered the section, but once the train is detected as having done so, it must also be detected as having vacated the section (normally, but not necessarily, at the opposite end to which it entered). It is usual to detect the passage of the train by treadles, which give a more positive indication of the actual presence of a train than a track circuit. Tokenless block can be divided into the following subsystems:
  - BRB tokenless block, which is available in two forms, one requiring four physical lineside circuits, and the other using a multiplex system to allow the transmission of the four required channels over a single lineside circuit. It is a true low cost system.
  - Direction lever tokenless block, which is a hybrid system, interlocking opposing signals by the use of a direction lever in the receiving signalbox. This direction lever can be normalised at any time until the train enters the section. Thence the train must be detected as having vacated the section before the direction lever can be normalised. The system requires three lineside circuits.
  - Scottish Region tokenless block, which is the only system which is a true simulation of token working, but without the physical token. The information is transmitted between the block instruments using intermittent polarised pulses. The move can be cancelled by co-operative action between the signalmen at any time until the train enters the section. Thereafter, if the train does not completely pass through the section, it is necessary to use a sealed release to cancel. It requires one lineside circuit, the block bell being included within the system.
  - Lock and block uses an instrument and circuits very similar to the double line block system, with the exception that the block commutators may only be operated in one direction, and are electrically locked. The release circuit is established via the commutators of both block instruments, the receiving end requiring to be at *line clear*, and the sending end requiring to be at *normal*. Once a release is established, the block line for the opposite direction is disconnected. The

system may not be normalised until the train has operated a treadle on vacating the section, thus allowing the commutator to be placed to normal. If the move is cancelled, it is necessary for the signalmen to co-operate to reset the instruments by operating mechanically interlocked cancelling switches.

Constraints of space prevent the full description of all systems mentioned.

Electric Key Token Working

GENERAL DESCRIPTION OF APPARATUS

Main Section (Signalman) Instrument
(Fig. 3.1)
The instrument is provided with a token magazine, capable of holding approximately 30 tokens. (The total number of tokens for a particular system must be capable of being accommodated in one instrument.) The magazine leads to a keyway through which the token is passed by way of a configuration notch to be placed in a further key receptacle attached to an interlocked commutator. By turning the token, the commutator can be rotated in an anticlockwise direction. The commutator is fitted with an electrically locked ratchet which prevents it from being rotated more than 90 degrees in an anticlockwise direction. Operation of the electric lock enables the commutator to be rotated a further 90 degrees in the anticlockwise direction, thus allowing the token to be withdrawn. The commutator can be rotated 180 degrees in the clockwise direction at any time by a free token being inserted into the instrument, providing that the token is of the correct configuration.

Configuration is achieved by cutting a slot at one of four positions along the outer edge of the token (A, B, C, D). The slot aligns with a corresponding projection in the keyway, which will thus only pass a key of the correct configuration. (Other special configurations are available.) Adjacent sections have tokens of different configurations and colours.

The instrument is fitted with a galvanometer to indicate the passage of an operating current and a plunger to operate a block bell and to give a release to other instruments. A three-position needle indicator may also be provided. (Some systems do not require the indicator, as its correct operation requires additional actions on the part of the signalmen, other than that required to obtain a token.)

The operation of the instruments is electrically simple in that the line circuits are polarity-conditioned by the commutator on each instrument. The lock can only be energised if the instruments are 'in phase', ie the lineside circuit is energised at the correct polarity via the commutators of both instruments being in the correct position. In some cases the lock is polarised, but more usually it is operated by a local circuit controlled by a polarised line relay.

Intermediate Instruments (No-Signalman)
These are placed at an intermediate siding ground frame in between token stations. The train crew use the section token to unlock the ground frame to give access to the siding; once the train is entirely clear of the running line, and the ground frame has been normalised and locked, the token can be placed into the intermediate instrument. This operates a commutator, which reverses the line circuit and places the main section instruments back into phase. Thus trains can now be passed using the main section instruments. Once the train is ready to leave the siding, and providing that the main section instruments are in phase, the signalmen at both ends of the section can simultaneously release the intermediate instrument, thus allowing the train crew to obtain a token, and regain access to the main line. This intermediate instrument may be designed so that it can only take one key; this is then known as a 'single capacity' instrument.

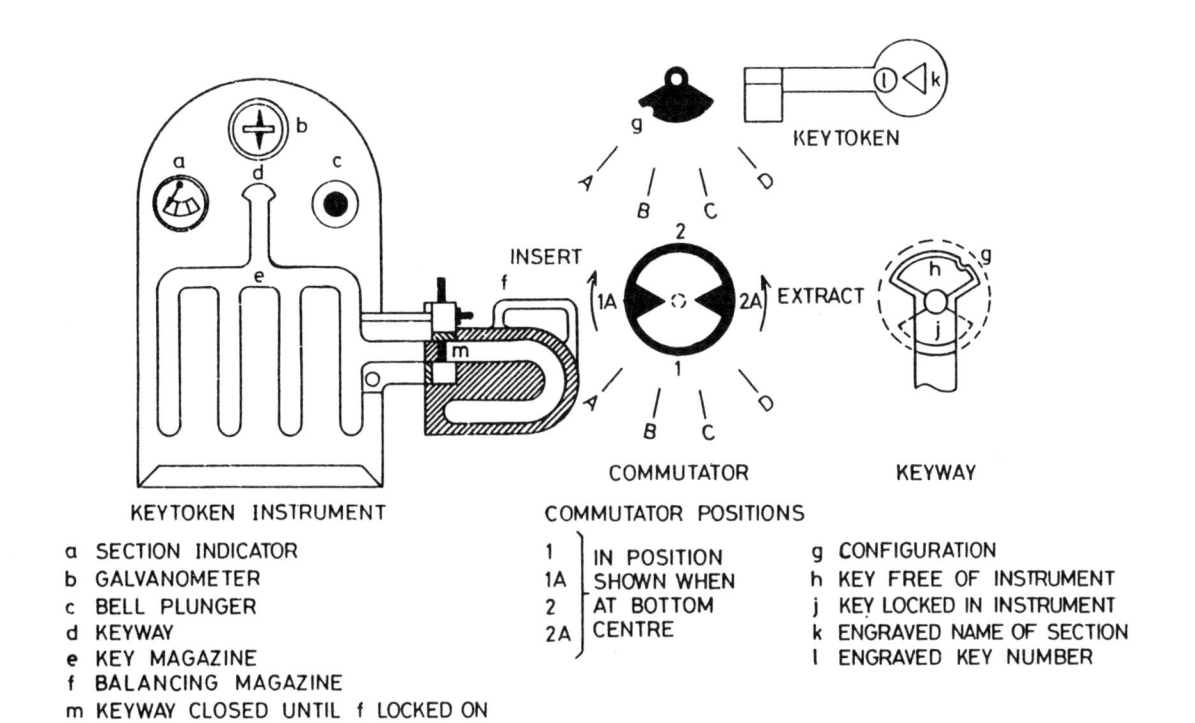

Fig. 3.1 Key token instrument — general arrangement

KEYTOKEN INSTRUMENT
a SECTION INDICATOR
b GALVANOMETER
c BELL PLUNGER
d KEYWAY
e KEY MAGAZINE
f BALANCING MAGAZINE
m KEYWAY CLOSED UNTIL f LOCKED ON

COMMUTATOR POSITIONS
1 ⎫
1A ⎬ IN POSITION SHOWN WHEN AT BOTTOM CENTRE
2 ⎬
2A ⎭

g CONFIGURATION
h KEY FREE OF INSTRUMENT
j KEY LOCKED IN INSTRUMENT
k ENGRAVED NAME OF SECTION
l ENGRAVED KEY NUMBER

It is possible to use an intermediate instrument for the remote issue or receipt of tokens at a place some distance from the signalbox, and where it would be inconvenient or cause delay for the signalman to do this in person. (See auxiliary instruments, below.)

Auxiliary Instruments (No-Signalman)
These are used for delivering or receiving a token at a distant point, where the signalman is unable to do this personally. Auxiliary instruments are usually a separate pair of simplified instruments which are of the same configuration as the main section instruments, but normally stand out of phase. When a signalman requires to despatch a train, he obtains a token from the main section instrument, which is then out of phase, and places the token into the auxiliary instrument which puts them in phase. This enables the train crew to withdraw a token from the remote auxiliary instrument, and proceed through the section.

Automatic Operator Instruments (No-Signalman)
These are used where one signalbox has complete control of a terminal branch line, and where the traffic pattern requires succeeding trains to use the branch. The automatic operator instrument is fitted with a special relay set, so that the sig-

nalman can instruct the remote instrument to transmit a release to the instrument at the signalbox. Once a train has passed beyond the limits of the single line section, the train crew insert the token into the remote instrument, which places the system back in phase. The signalman can then, either, with the co-operation of the train crew, give a release to the automatic instrument for a train to proceed towards the signalbox, or cause the automatic instrument to issue a release for another train to pass on to the branch.

Intermediate and auxiliary and automatic operator instruments are physically similar to the signalman type, but are without the galvanometer or plunger. The indicator shows locked or free, to advise the train crew when it is possible to obtain a token.

Any of the above instruments may be adapted to be operated by the train crew, but be electrically released from a distant signalbox, as would be the situation at a branch line joining a main line which was supervised from a large signalling centre.

No-Signalman Remote Instruments (Train Crew Operated)
These are used for sections which are not under the direct control of a signalman. The operation of the instruments is entirely carried out by the train crew, under the telephone supervision of the signalman. The instrument is fitted with a plunger which is used by the train crew to initiate the release of a token. There is also an indicator which may be in the form of a lamp, to show that a release has been successfully obtained.

Balancing Magazine
If there is a predominance of traffic in one direction, so that all the tokens finish up at one end of the section, the main instruments can be fitted with a balancing magazine (see **Fig. 3.1**). This is able to hold a number of tokens and is mechanically locked to either of the instruments for the section applicable. To balance the number of tokens between the ends of the section, the signalman to whose instrument the balancing magazine is attached, slides as many tokens as it is required to transfer into the magazine. The magazine is then detached from the instrument, which closes the keyway between the instrument and magazine so that tokens cannot be liberated from either the magazine or the main section instrument. The magazine is then sent by a convenient train to the far end of the section. The signalman there receives the magazine, attaches it to his instrument, and moves the tokens into the main instrument magazine. The tokens in the balancing magazine are not an authority to the driver to proceed through the section. With no-signalman remote token working, magazines are required to balance the receiving and issuing instruments.

EXPLANATION OF PHASE
As already explained, the operation of the instruments is dependent upon the polarity of the received current, which is reversed by the commutator every time that a token is either withdrawn or inserted. The commutators have positions 1 and 2, at which the keyways are open, and positions 1A and 2A, which conform to 1 and 2 in polarity, but at which the keyways are closed and the lock relay is connected to the line. The reversal of polarity takes place between 1A and 2, and 2A and 1. Initially with the commutators in position 1 and 1A, the instruments are said to be in 'even phase', and with the commutators at positions 2 and 2A, they are in 'odd phase'. It is, however, not possible to assign a commutator position to a particular phase, unless there are only two instruments on the system, because of the action of intermediate commutators. Instruments are known as being 'in phase', when they are both at 'even phase' or both at 'odd phase', at which withdrawal of a token is possible.

SINGLE LINE SIGNALLING

OPERATION OF SYSTEM

Main Section Instruments
(Figs. 3.2 and 3.3)
Consider an up train to proceed from signalbox Y (sending end) to signalbox Z (receiving end).

If able to accept, the signalman at Z holds in the plunger, transmitting an in-phase current. The indicator C coil operates the stop pin to 'out', preparing for *train coming from*.

At Y, the galvanometer deflects, upon which the token is turned anticlockwise through 90 degrees. The commutator completes the lock relay circuit, which because the received current is in phase, operates the relay in the correct direction to energise the commutator lock. The indicator B coil operates the stop pin to 'in', preparing for *train going to*. The token is then turned a further 90 degrees anticlockwise past the operated lock, which reverses the commutator and liberates the token. Thus the incoming current is pole-changed, reversing the galvanometer and energising the indicator A coil with an out-of-phase current, which operates the needle past the stop pin to *train going to*.

At Z, the signalman notes the galvanometer normalise, and then redeflect, at which the plunger is released. Similarly at Y, the signalman notes the galvanometer normalise, and operates the plunger momentarily, thus transmitting an out-of-phase current.

The indicator A coil at Z operates the needle to the stop pin 'out' at *train coming from*.

Should an attempt be made to withdraw a token at the receiving end for a train to proceed in the opposite direction, the line will be energised with an out-of-phase current, because the sending end commutator has reversed, thus operating the receiving end lock relay in the out-of-phase direction, and not energising the commutator lock.

Once a token has been withdrawn at the sending end, should a further attempt be made to withdraw a token, the line will be energised with the same in-phase polarity as previously, but because the sending end commutator has been reversed, the lock relay will be energised in the out-of-phase direction, thus failing to energise the commutator lock.

When the train arrives at Z, the token is inserted into the instrument and turned clockwise through 180 degrees, thus reversing the commutator. The signalman sends the train out of section bell signal, holding in on the last plunge for a few moments. This transmits a pole-changed current which is now in phase.

At Y, the galvanometer deflects. The in-phase current energises the indicator A coil, thus operating the needle to *normal*. The galvanometer restores, at which the signalman gives a plunge for a few moments, thus transmitting an in-phase current. At Z, the galvanometer deflects, the in-phase current energises indicator A coil and operates the needle to *normal*.

Thus it can be seen that the commutator at both ends of the section must be reversed, initially by the withdrawal of the token at the sending end, and then by the insertion of the same token at the receiving end, before the system can be considered as in phase for a further token to be issued, unless the line is pole-changed by the commutator of an intermediate instrument.

To ensure that the home and distant signals, and the opposite direction starting signal, are normal before a release can be given for a move to be made towards the signalbox, these signals are interlinked in the initial release circuit. If the instruments are connected by a direct line circuit, the bell can be operated by means of a circuit which bypasses these controls; a resistor is then included which prevents sufficient current being transmitted to operate the lock relay, but enough to operate the bell relay. If the connection between instruments is by means of a multiplex system, this is not possible and a telephone must be provided for passing operating messages.

A passing contact on the commutator is made as the token is withdrawn which operates the KTSR relay. This releases the starting signal for one movement only, being reset by the

SINGLE LINE SIGNALLING

Fig. 3.2 Key token main section (signalman) instrument

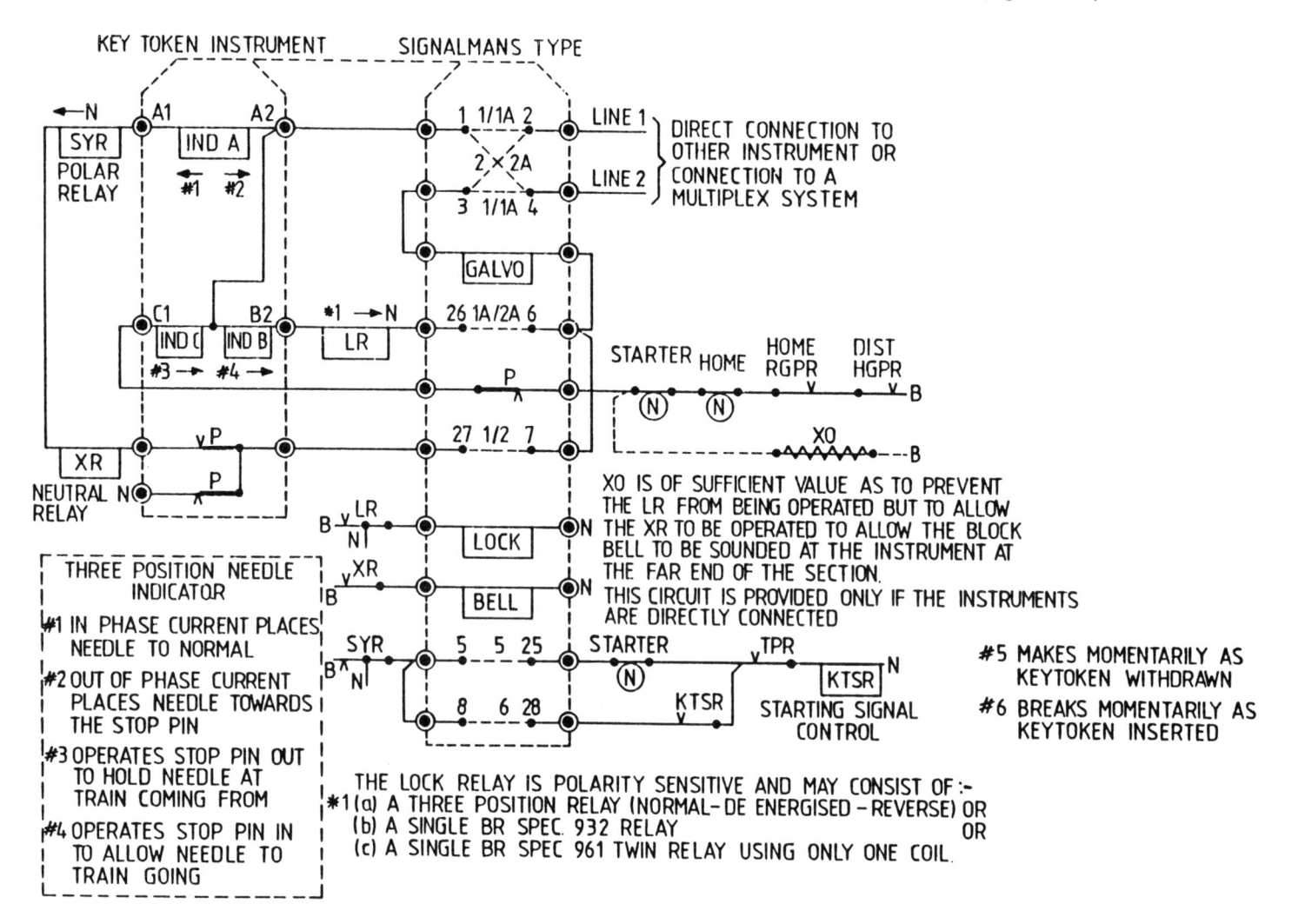

train entering the section. If the movement does not proceed through the section and the token is replaced into the instrument from which it was withdrawn, the KTSR relay is released by commutator contacts which break as the token is returned to the instrument. If the token is withdrawn for engineering purposes and is placed into the instrument at the far end of the section, the KTSR relay is reset by the operation of the SYR relay. This is a polarised relay which operates on the receipt of an in-phase current received during the out of section procedure, breaking a contact in the KTSR circuit.

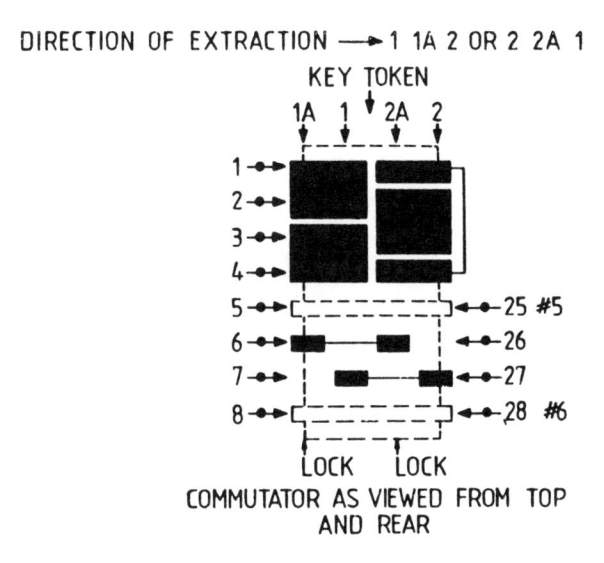

Fig. 3.3 Key token instrument commutator arrangement

SINGLE LINE SIGNALLING

Fig. 3.4 Key token automatic operator instrument

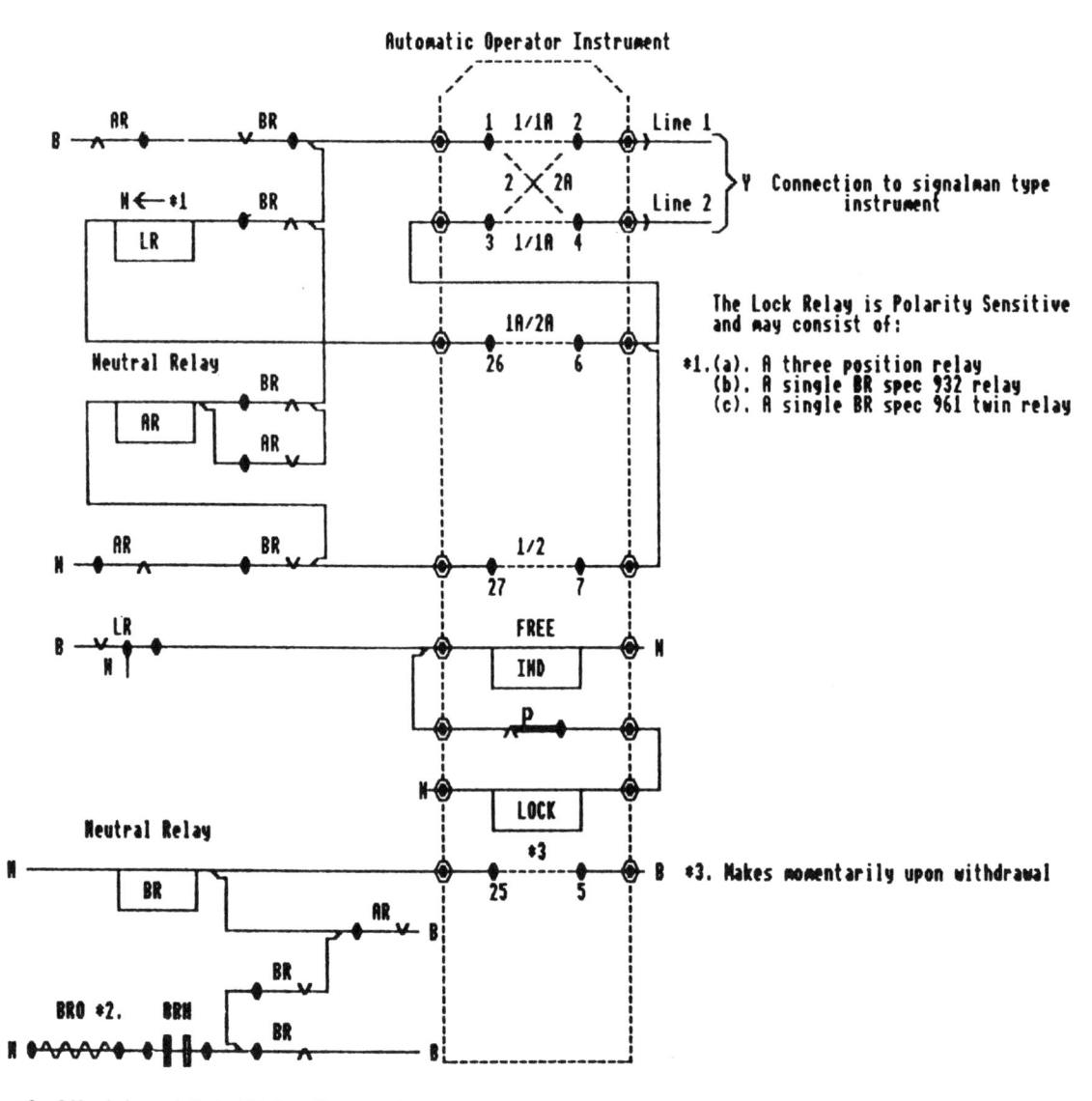

Operation of an Automatic Operator Remote Instrument **(Fig. 3.4)**

The instrument which is at the remote end of a branch line, simulates the action of a signalman, and is used in conjunction with a standard signalman's instrument, complete with an indicator, at the controlling signalbox. To obtain a token, the signalman depresses the plunger, which energises the line monitoring relay AR at the remote instrument. This operates the transmission relay BR which transmits a current back to the main instrument for a period of 10 s after the signalman has released the plunger. Providing the system is normal the current will be in phase, allowing the withdrawal of a token, which reverses the commutator and causes the incoming current to place the indicator to *train going to*. When the train arrives at the remote instrument, the train crew insert the token, informing the signalman by telephone. The indicator can now be placed to *normal* by the signalman pressing the plunger which will again cause the automatic operator to give an in-phase pulse. When a train requires to proceed towards the signalbox, the train crew contact the signalman and request permission to withdraw a token. If this is granted, the signalman presses the plunger and when the train crew turn the token through 90 degrees anticlockwise, the incoming current is diverted to the lock relay (LR), which, assuming that the current is in phase, causes the operation of the commutator lock. The token can now be withdrawn, thus reversing the commutator. This is fitted with a passing contact which momentarily operates the relay BR, thus transmitting an out-of-phase pulse for 10 s to the main instrument, placing the indicator to *train coming from*. Once the train arrives at the signalbox and the token is returned to the main instrument, the indicator is placed to *normal* by the signalman causing the automatic operator once again to send an in-phase current.

SINGLE LINE SIGNALLING

Operation of an Intermediate Instrument
(Fig. 3.5)
This instrument would be placed at an outlying siding, at which a train would be required to shut inside. The commutator is in the line between the two main instruments. When a train arrives at the siding in possession of a token, transmissions between the main instruments are out of phase. Once shut inside, the train crew inform the signalman and place the token into the intermediate instrument, reversing the commutator, and thus causing further transmissions to be in phase. The signalman is then able to restore the indicators to *normal* by carrying out the usual train out of section procedure. Further trains may then pass. For a train to leave the siding the train crew contact the signalmen, who co-operate by depressing their plungers, which providing that the section is normal causes the transmission of in-phase currents. The train crew is instructed to turn the token 90 degrees anticlockwise, which allows the operation of the commutator relays to interrupt the line and diverts the current to the two lock relays. Providing the phases of the incoming currents are the same (ie in-phase even or in-phase odd), the lock relays will both operate in the same direction, thus completing the circuit to the commutator lock. The token can now be withdrawn, reversing the commutator. This disconnects the lock relays and once again reconnects the through lines which have been reversed through the commutator, so that any further transmissions between the main instruments will then be out of phase. The through lines must not be reconnected until the commutator is in such a position that the token can be withdrawn. Once the train clears the single line section, the indicators are restored to *normal* by carrying out the normal train out of section procedure.

Mechanically-tuned Twin Reed Communication between Instruments using Public Telecommunications Lines

GENERAL DESCRIPTION
This system was developed in order to overcome the high cost of the replacement of railway-owned overhead linewires with cable. The line polarity, which is the vital information to be passed between the instruments, is translated into three FDM channels in each direction. In all phases of the instruments, two channels out of three are in use, thus enabling the transmission of three distinct conditions.

TRANSMISSION EQUIPMENT
The equipment uses the so-called Reed/ETS FDM (electric token signalling/frequency division multiplex) transmission system. The transmitters and receivers, and associated equipment of this system are fail-safe, making use of twin mechanically-tuned reeds. The system may work in either the duplex mode (bothway transmission on one pair of lines) or the simplex mode (two pairs of lines). A separation of seven channels is used to avoid adjacent channel interference.

TRANSMISSION MEDIUM
The public telephone system telecommunications lines are loaded physical pairs direct to the nearest convenient telephone exchange, with either PCM channels between exchanges, (no FDM group carriers are employed) or physical pairs. A barrier unit is fitted to separate the railway and public telecommunications circuit.

SINGLE LINE SIGNALLING 73

Fig. 3.5 Key token intermediate instrument

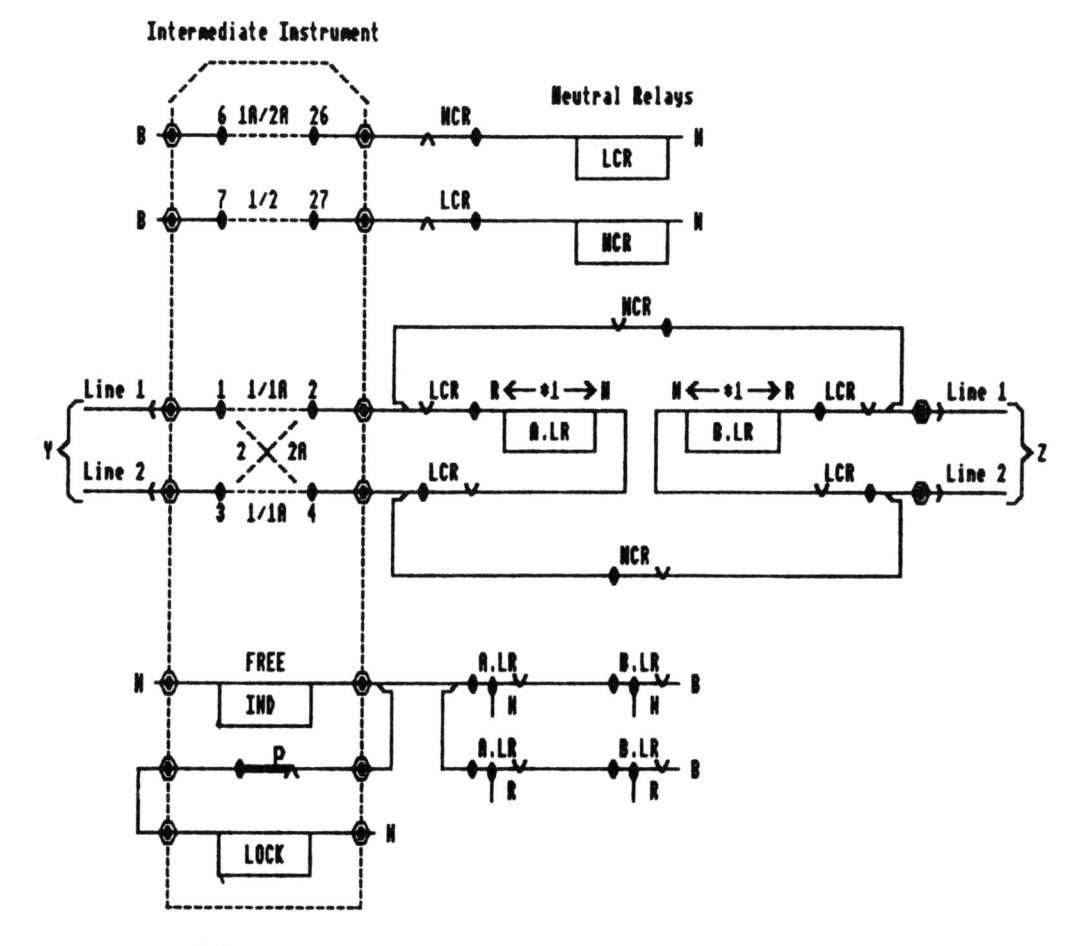

74   SINGLE LINE SIGNALLING

Fig. 3.6   Key token system — Reed-type transmission

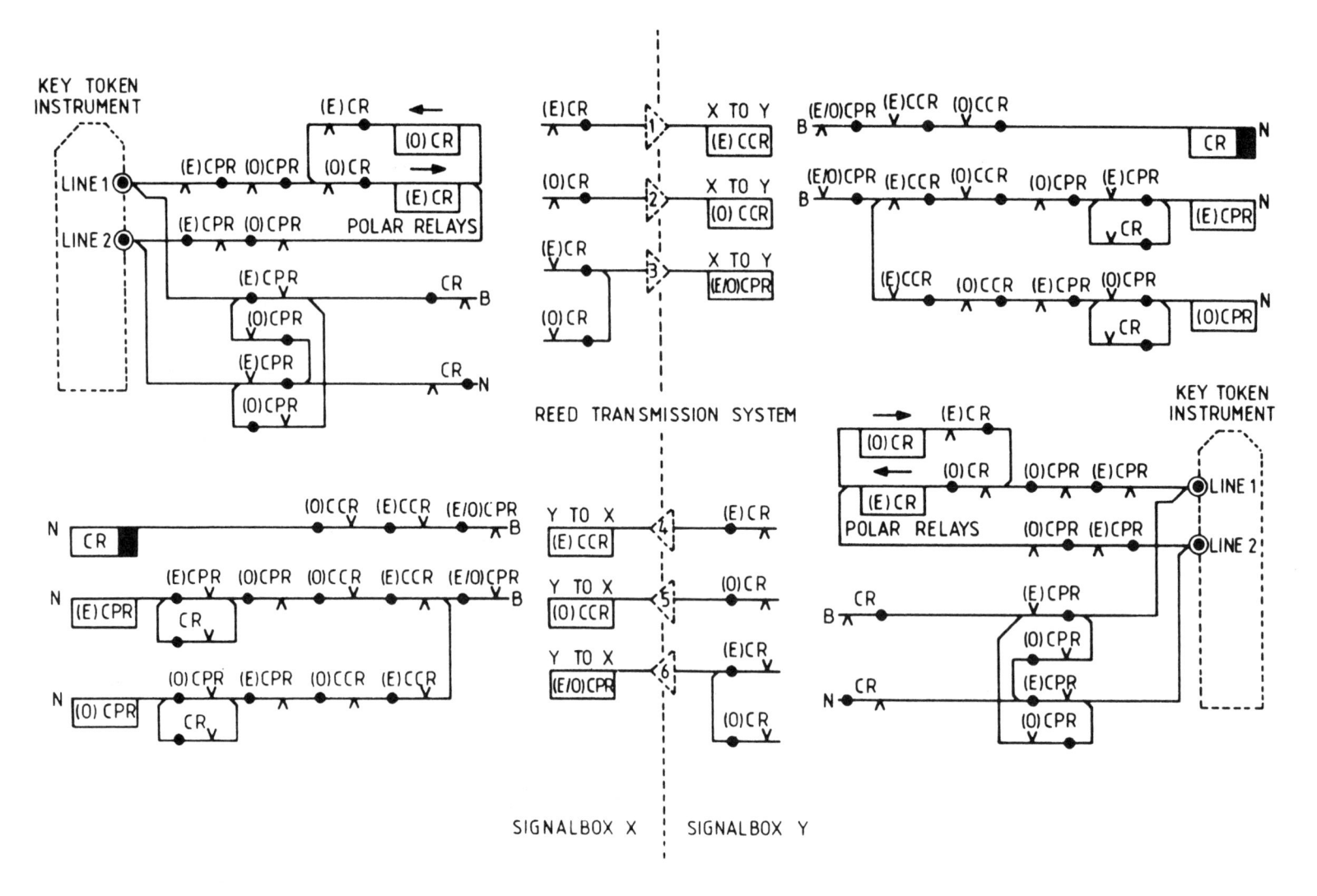

MODE OF OPERATION
(Figs. 3.6 and 3.7)

Consider a train to travel from signalbox Y (sending end) to signalbox X (receiving end).

To allow the signalman to withdraw a token at Y, it is necessary for the signalman at X to operate and hold in the bell plunger on the token instrument. When the bell plunger at X is normal, channels 1 and 2 are transmitting, thus operating at Y both polarity receiving relays (reed following relays) (E)CCR and (O)CCR, and maintaining the operation of the checking relay CR. Similarly when the bell plunger at Y is normal, channels 4 and 5 are transmitting to X also operating both polarity receiving relays, and maintaining the checking relay CR.

Depression of the bell plunger at X will operate either the (E)CR or (O)CR depending upon the position of the token instrument commutator. If the commutator is in even phase, there will be a positive feed on line terminal L1, and a negative feed on line terminal L2. Because the (E) or (O) CR relays are polarised, only (E)CR will operate. This will stop the transmission of channel 1, maintain the transmission of channel 2, and cause the transmission of channel 3. Thus at Y, polarity receiving relays (E)CCR will de-energise and (O)CCR will remain operated. The change of state relay (E/O) CPR will operate.

The operation of the change of state relay causes the checking relay CR to de-energise, but being slow to release, this remains operated for a sufficient period to allow the polarity registration relay (E) CPR to operate. This relay sticks out the CR contact, thus ensuring that the change in the state of the channels will only operate the polarity registration relays if all changes have taken place simultaneously.

The operation of the polarity registration relay (E)CPR, with the checking relay CR de-energised, will cause a feed to be sent to the line terminals of the token instrument with positive to line 1 and negative to line 2. This is the same polarity as that at the line terminals of the token instrument at X, and thus simulates a physical line circuit. It should be noted that back contacts of the polarity registration relays are included in the circuit for the polarity transmission relays to prevent an incorrect feed.

At Y, the bell will ring, and if the commutator is in even phase, it will be possible to withdraw a token.

The previous description of the detailed operation of token instruments connected by a direct lineside circuit are applicable, with the exception that the galvanometer will not operate as described. This is because an interruption of the line at one end of the system, by the reversal of the commutator, will not be seen at the other end of the section as an interruption to the line current. The signalmen must therefore remain in contact by telephone to co-ordinate the operation of the instruments to ensure that the indicator will operate correctly. Because of the operation of two relays and the start-up time of the reed equipment, it is necessary for all bell signals to be slow and deliberate. Also it is not possible to operate the bell when the interlinking circuit is not intact, as the circuit is unable to detect current increments.

This system can be used for automatic operator instruments and intermediate instruments. If used with intermediate instruments, then two full sets of relays are required to detect the line from both of the main instruments.

SECURITY

Although the system is connected to the public telephone network, it is not considered likely that any frequencies present or which may be generated, will cause a false operation of two reed channels simultaneously, coincident with both signalmen attempting to operate the instruments erroneously.

Should both ends of the section be within the area of the same telephone exchange, then public telephone physical pairs are acceptable, and no multiplexing system is necessary. The only danger would be if a telephone technician accidentally reversed the line, at the same time as both signalmen attempted to operate the instruments irregularly.

76 SINGLE LINE SIGNALLING

Fig. 3.7 Key token system — transmission over public telecommunication lines

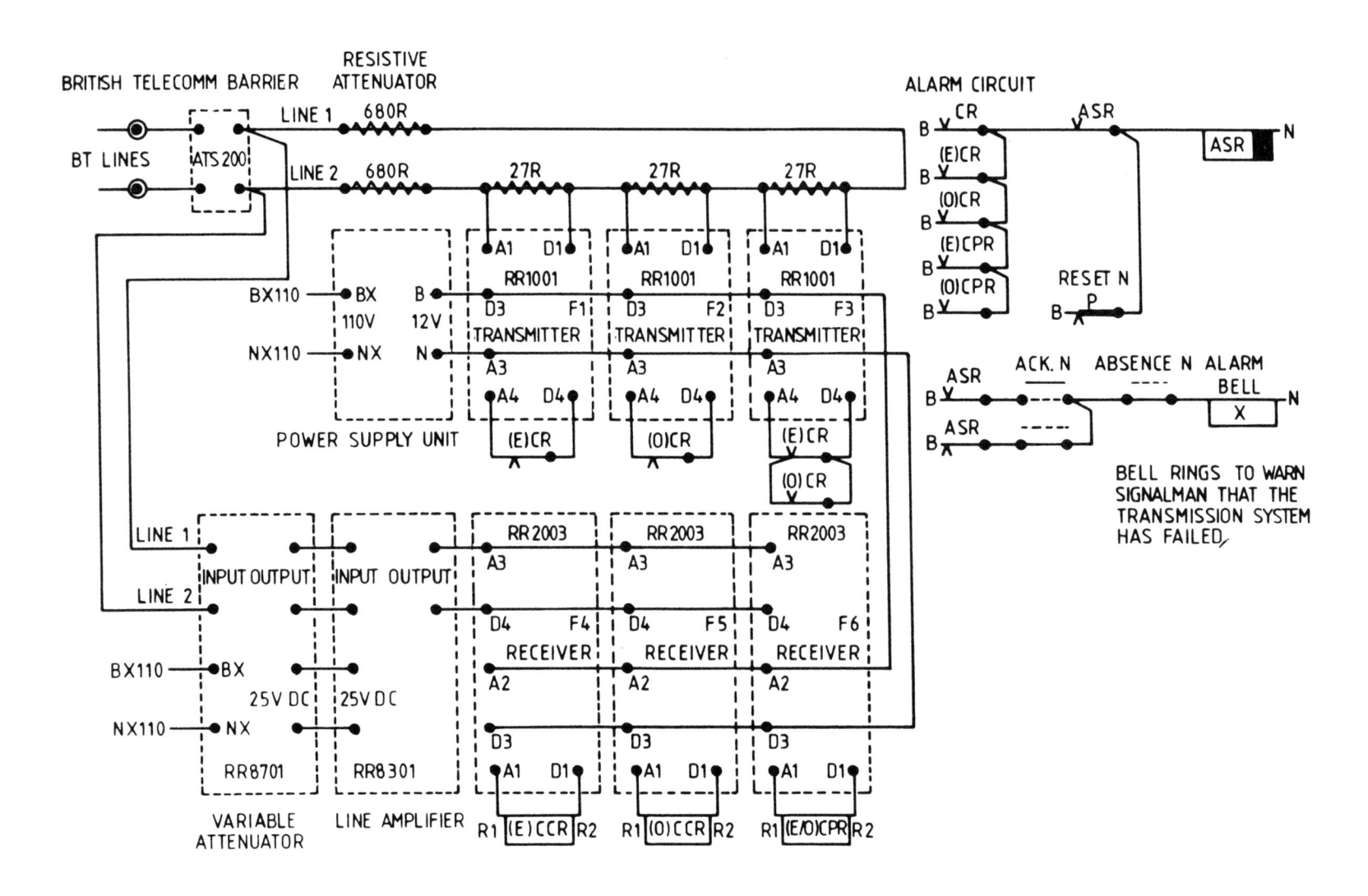

An alarm circuit is provided to warn the signalman if an acceptable pair of frequencies is not being received and thus that the transmission system has failed.

Simplified Infrastructure

In recent years there has been much pressure to reduce the cost of operating lightly trafficked lines. This can be achieved either by simplifying the infrastructure, where the saving is on the lower cost of maintenance of the permanent way and the signalling, or by the reduction of the numbers of staff required to operate such railways. Simplification of the infrastructure is usually achieved by singling the line on a former double track railway, and by the automation, or remote operation, of level crossings. Modest savings can be obtained by these methods. However, major savings can be made if the signalling system does not require a lineside cable, and if the points can be operated either from the ground or automatically.

The elements of such a system are as follows.

LEVEL CROSSINGS
The level crossings are converted to automatic open crossings, locally monitored (AOCL), or to automatic half barriers locally monitored (ABCL). A full description of the method of working of these forms of crossing is given in Chapter 6.

RUNNING CONNECTIONS
Loop points are converted to train-operated hydro-pneumatic spring points. These are described in detail in Chapter 7.

SIDING CONNECTIONS
All siding connections are provided with ground frames. A simple connection would require a frame of two levers, one of which would be the release and facing point lock. An Annett's lock is provided on this lever, which requires either a section key in the case of key token working, or a radio controlled key in the case of radio electronic token block working. Reversing this lever unlocks the points, enabling the point lever to be operated. The facing point lock would be cut without a reverse notch in the lock stretcher, thus being only locked for running moves.

FIXED SIGNALLING
(**Figs. 3.8** and **3.9**)
The driver of a train approaching a typical passing loop would pass the following fixed signals and features:

(a) Distant Signal
A reflectorised distant signal board, placed at service braking distance from the point indicator (home) signal. This distant signal is fitted with an AWS permanent magnet inductor, placed 185 m to the rear.

(b) Point Indicator (Home) Signal
A single yellow aspect signal mounted on a post, the lamp having a plain yellow lens with a reflector. The lamp is continually alight as long as the loop points remain correctly set for the loop, and the pneumatic pressure is maintained at a safe level. This yellow aspect authorises the driver to enter the platform and come to a stand at the starting signal stop board. (With RETB, the signalman is able to determine that the previous move has cleared the passing loop.) If used with NSTR, the yellow aspect authorises the driver to drive as far as the line is clear. (With NSTR, the signalman is unable to determine that the previous move has cleared the passing loop.)

An alternative form of point indicator is a position light signal mounted on a post, only having an off aspect, which shows as long as the loop points remain correctly set for the loop, and the pneumatic pressure is maintained at a safe level. This off aspect authorises the driver to pass the signal, but to be prepared to come to a stand at any obstruction before reaching the starting signal stop board.

78 SINGLE LINE SIGNALLING

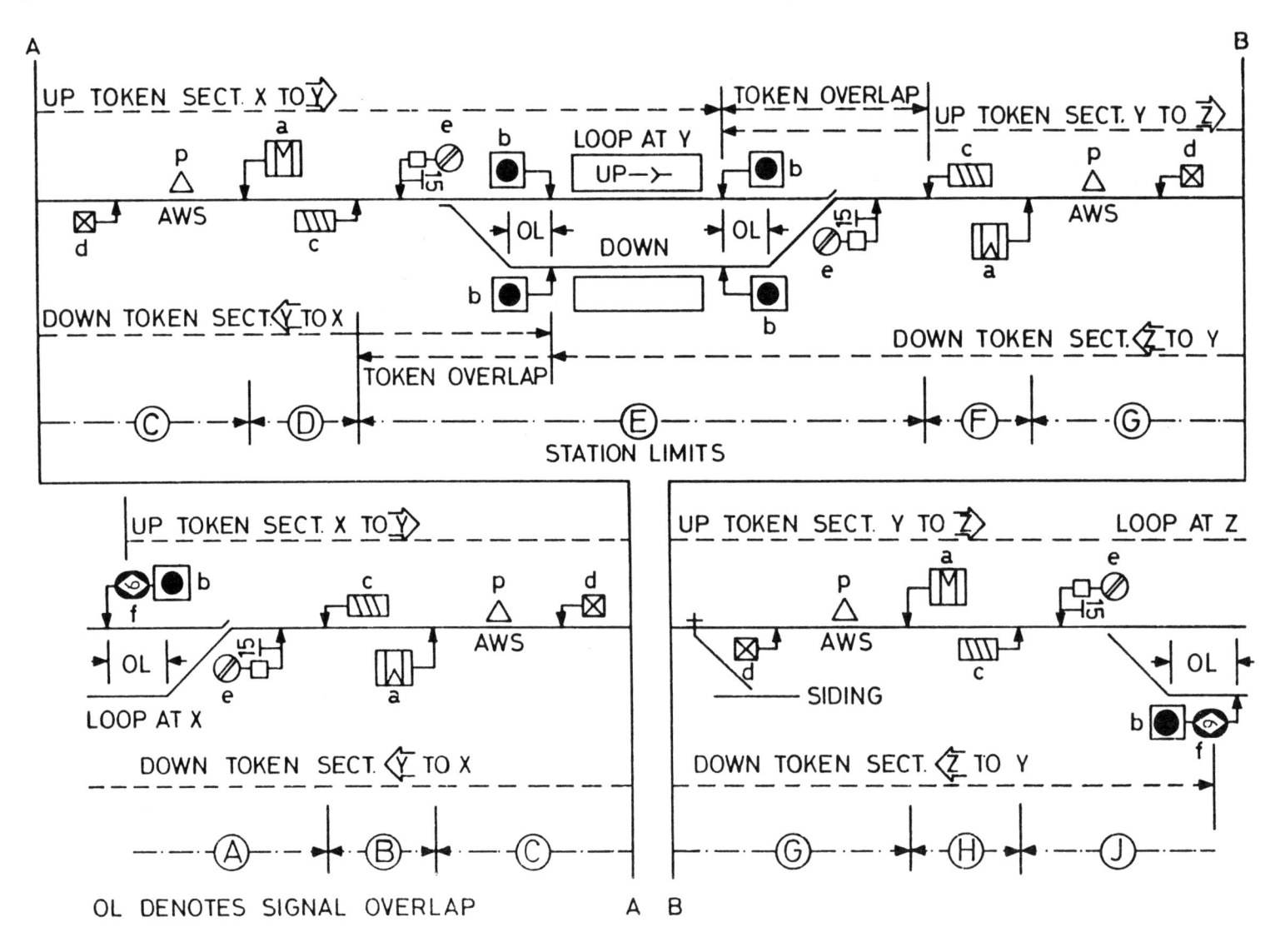

Fig. 3.8 Simplified infrastructure: radio electronic token block system
[*for details of items a–f, see* **Fig. 3.9**]

SINGLE LINE SIGNALLING 79

Fig. 3.9 Simplified infrastructure: fixed signals

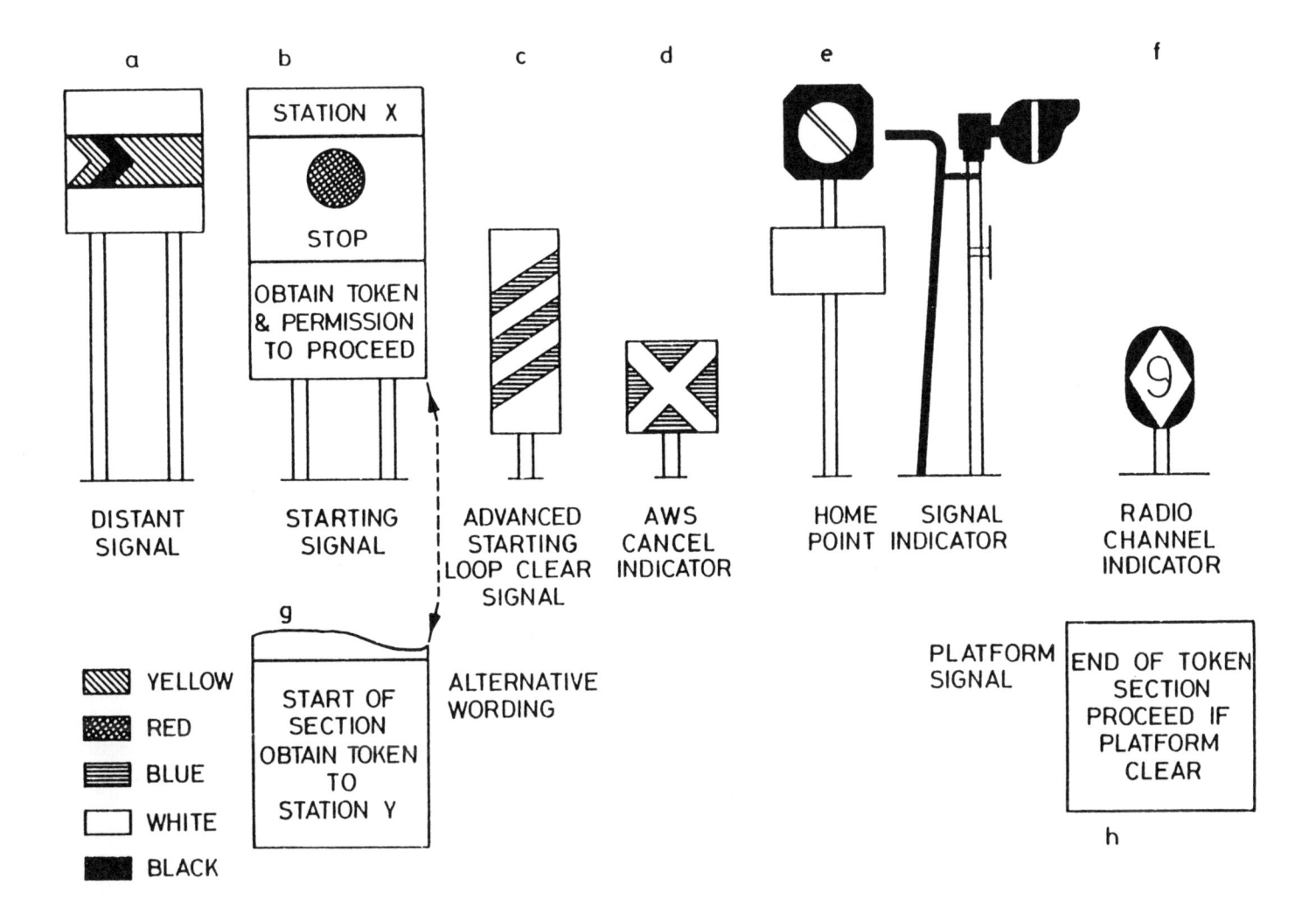

A reflectorised notice board mounted below the signal shows *point indicator*. In the event of the signal lamp not being alight,* the driver of the train must leave the cab and clamp the points for the desired position. This signal is placed 51 m from the facing points.

(c) Starting Signal Notice Board
A reflectorised red 'spot' with the name of the passing loop above, and with the legend *Stop, obtain token and permission to proceed*. The signal is placed in such a position that there is a signal overlap of at least 51 m before reaching the fouling point of the loop points, to allow a train to approach the passing loop in the opposite direction. The maximum speed in the loop is 24 km/h.

(d) Train Clear of Passing Loop Indicator
An upright white rectangular reflectorised board, with three light blue stripes. This is the marker to remind a driver to inform the controlling signalbox that the train has passed clear of the loop. It indicates the token overlap (not to be confused with the starting signal overlap), which must be clear before a succeeding train may approach the passing loop. This indicator is only required in RETB installations.

(e) AWS Cancelling Indicator
On passing the distant signal for the opposite direction, the driver will pass over its AWS inductor. 185 m in advance of this point is the cancelling indicator, a square reflectorised blue board with a diagonal white cross. This is to remind the driver that the AWS indication which he has just received is not applicable to that direction of movement.

---

*At the time of writing a decision is awaited as to whether the yellow aspect will be supplemented with a flashing red aspect which will show when the loop points are not set. This would be to bring the signal into line with the similar AOCR and ABCL signal. (See Chapter 6.)

(f) Radio Channel Indicator
A reflectorised black lozenge-shaped board, with a white upright diamond, upon which is a number indicating which radio channel should be selected by the driver. (RETB installation only.)

OTHER FEATURES OF SIMPLIFIED INFRASTRUCTURE
Passing loops and double line sections in RETB areas are normally signalled for moves in both directions.

Radio Electronic Token Block Working

INTRODUCTION
(Fig. 3.10)
Radio electronic token block (RETB) is used as a method of controlling movements on rural railways, where, combined with simplified infrastructure as already described, it has enabled dramatic reductions in operating costs to be achieved. The essence of the system is the secure transmission of uncorrupted data by means of a radio system. The controlling signalbox is equipped with an electronic interlocking which has custody of the tokens and is responsible for their safe management. The tokens are held in the form of data which can be transmitted to the motive power unit as an addressed telegram representing an 'electronic token'. Transmission is effected by co-operation between the driver and the signalman, working over a secure radio system, which also provides a speech link between the two operators. All data and voice transmissions are recorded on tape at a secure place for later analysis if necessary. The radio electronic token instrument in the driving cab, when in possession of a token received from the interlocking, displays in alpha-numeric form, the geographical section which the train may occupy between passing loops. The displayed token can be transmitted from the instrument back to the interlocking, again by co-operative action between the driver and the signalman.

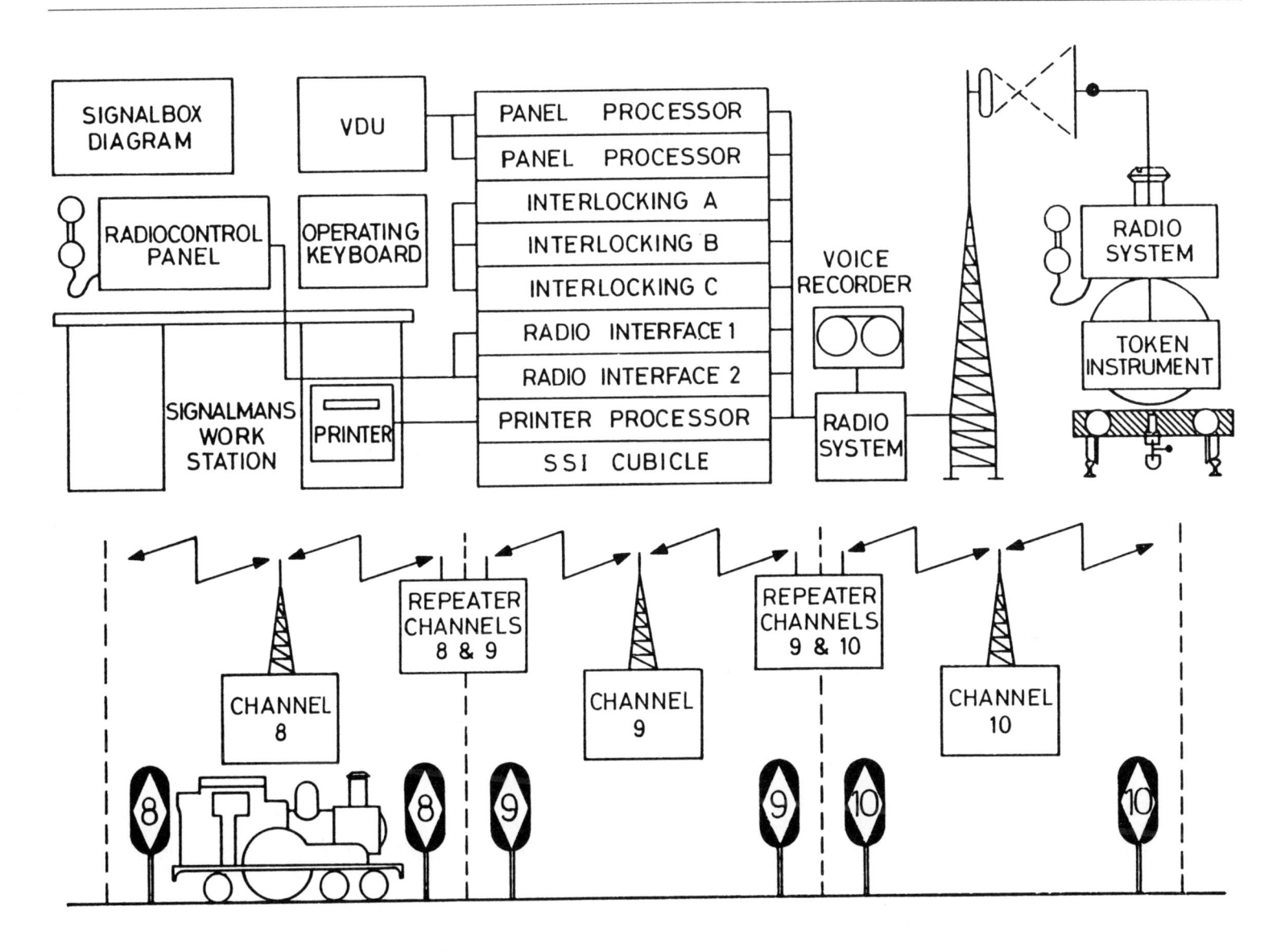

Fig. 3.10 Radio electronic token block: diagram of system

82 SINGLE LINE SIGNALLING

Fig. 3.11 Radio electronic token block system: driver's and signalman's equipment

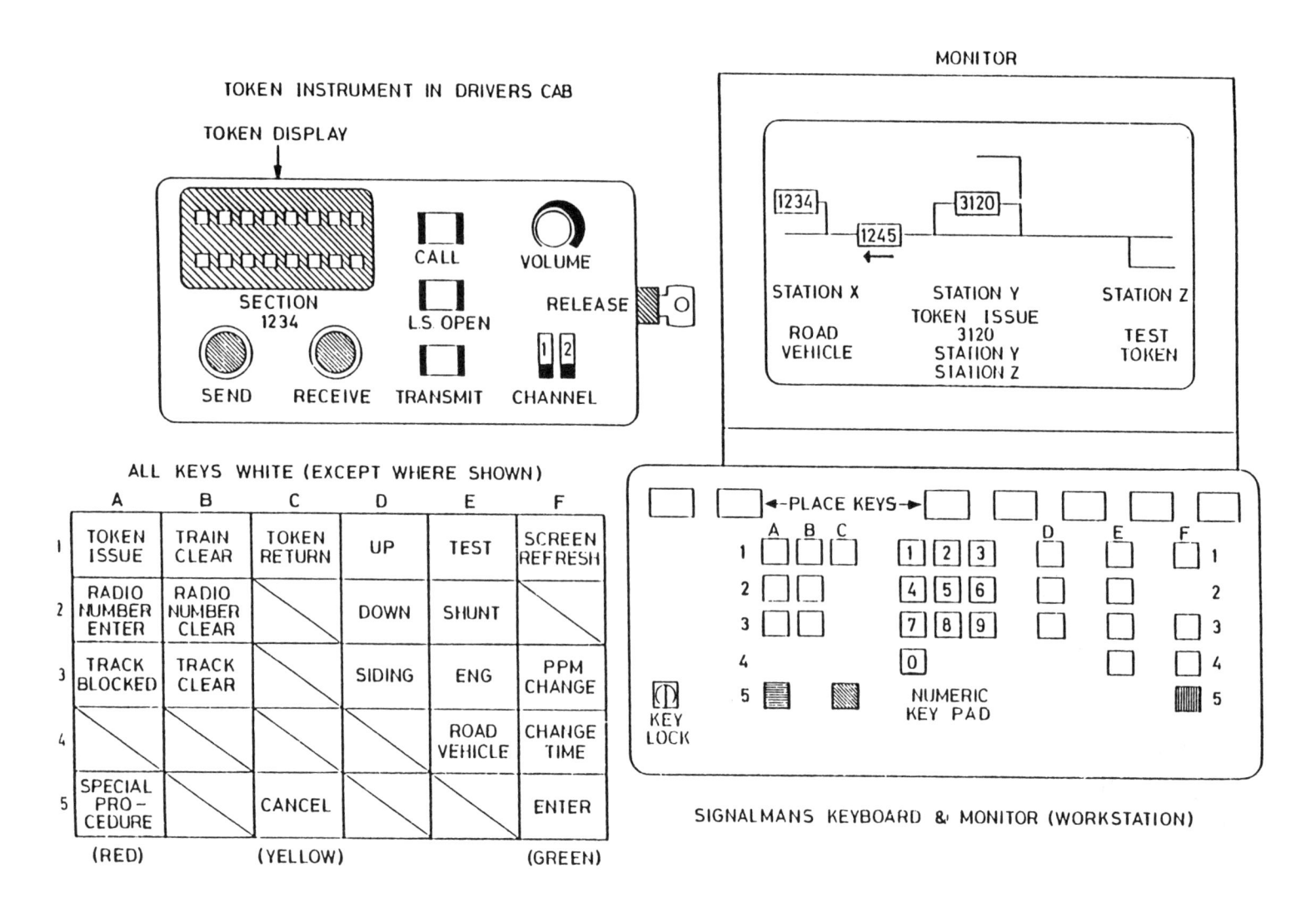

DETAILED DESCRIPTION OF THE SYSTEM
(Fig. 3.11)

The signalman has a control desk or workstation having a keyboard to control the BR standard SSI system with which the installation is equipped. Also provided is a VDU on which is shown a diagram or 'map' of the area controlled. Each train logged into the system is shown on the display by means of its unique radio identification number, superimposed upon the display in the position at which the train was last reported, with accompanying arrows to indicate the direction of the movement. The successful issue and receipt of tokens, and the status of the token (movement, shunting or engineering) is also displayed.

THE TRACTION UNIT

A traction unit can be a locomotive, a multiple unit, or an engineering on-track machine or road vehicle. It is fitted with a radio electronic token instrument which is engraved with its own unique radio number. Locomotives and multiple units are fitted with a separate instrument in each driving cab.

DEFINITION OF LIMITS

A token section (see **Fig. 3.8**) is defined as being the running line from the starting signal notice board at the commencement of the single line section, to the starting signal notice board of the next passing loop in the same direction of travel.

The token overlap is defined as being the running line from the starting signal notice board at the commencement of the single line section to the clear-of-loop marker board in the same direction of travel. (This is not the signal overlap.)

The station limits are defined as being the section of running lines, including both loops, between the clear-of-loop marker boards at each end of the passing loop.

TYPES OF TOKEN

Main Section Token

Possession of a main section token authorises a movement of the train from the starting signal notice board at the entrance to the section, through the section to the starting signal notice board at the entrance to the next section. Issue of the token is dependent upon the shunting, engineering, and opposite direction main section token, being within the custody of the interlocking. Main section tokens are unidirectional, and therefore separate up and down direction main section tokens exist for the same section.

Shunting Token

Possession of a shunting token authorises the movement of the train only between the loop-clear marker boards. Issue of the token is dependent upon the main section tokens applying to moves towards the loop from the sections at each side of the loop, to be within the custody of the interlocking. It is permissible to shunt with a move being made away from the passing loop.

Engineering Token

This authorises an engineer to take possession of the section of line between the distant signals of adjacent passing stations. Issue of the token is dependent upon the main section tokens for that section, in both directions, being within the custody of the interlocking. The token can be issued to the driver of a train, or to a portable electronic token instrument carried in an engineer's road vehicle. (This would be used, eg, if the engineer wished to place a trolley on the line.)

Test Token

This is issued for test purposes only to prove the transmission equipment effective and cannot be used as an authority for the movement of a train. It is normal practice to issue and receive a test token before a train commences from its terminal towards the single line.

Special Tokens

Any token can be issued with a suffix to remind a driver that the move has to conform to a special requirement, for example to stop and operate a level crossing or a set of points, or to arrive in a passing loop in other than the normal direction.

GENERAL

The interlocking records trains in the form of the individual and unique radio number attached to the train-borne token instrument. This number cannot be changed. The interlocking not only manages and records the issue and receipt of tokens, but also checks upon the logical sequence of events. Radio numbers may normally only be entered into the system at recognised stations and tokens may normally only be exchanged at recognised token exchange stations. Radio numbers are automatically deleted from the system as the train is recognised as leaving the limits of the system. They can only be deleted at other places by the use of special procedures.

There is no ground equipment for positive identification of trains; thus the integrity of the system is dependent upon the controlling signalbox operator inserting the radio number at the correct station when the train enters the system. If an error is made after receiving the token in the interlocking by attempting to issue the next token for a section which does not geographically adjoin the previous one, or applies to the incorrect direction, the interlocking will recognise this situation and prevent the issue of the token. In the event of the failure of an individual train-borne token instrument, it is permissible by the adoption of a special procedure, for the token instrument in the other cab to be used for the movement to continue. In the event of the complete failure of the electronic token system, it is necessary to revert to staff and ticket operation.

No-Signalman Remote Key Token Working: Trainman-Operated

GENERAL DESCRIPTION
(Fig. 3.12)

The system is intended for use on railways which have simplified infrastructure. Each passing loop is equipped with point indicators and notice boards, and the loop points are fitted with train-operated hydro-pneumatic mechanisms. The passing loops are not bidirectionally signalled, although starting signal notice boards are provided for wrong direction moves from the loop. The only difference from RETB infrastructure is that the clear-of-loop (station limits) boards are not provided. This is because the driver does not have continuous communication with the supervising signalbox and is therefore unable to inform the signalman that the move has vacated the passing loop. Thus the supervising signalman, before allowing a further token to be issued for a succeeding train to approach that loop, is unable to tell if the previous train has started away into the next section. To overcome this problem, the driver must be reminded at the loop points, to drive into the loop ready to stop short of any train which may still be standing in that loop. A single yellow aspect point indicator may be used (with special instructions), or the point indicator may consist of an elevated position light signal, each with a reflectorised notice (*Points indicator*), followed by a reflectorised notice board (*End of token section. Proceed if line clear*). The starting signal notice boards also define which token must be obtained before proceeding into the next single line section. The supervising signalbox is provided with telephone communication with each token station, and supervises the train crew on the issue of tokens. The signalman has a diagram of the line on which he can record the movement of trains by means of reminder appliances. All telephone conversations into and out of the supervising signalbox are tape recorded for investigation in the event of an incident.

SINGLE LINE SIGNALLING   85

Fig. 3.12 Simplified infrastructure: no-signalman remote key token system (trainman operated)
[*for details of items a–g, see* **Fig. 3.9**]

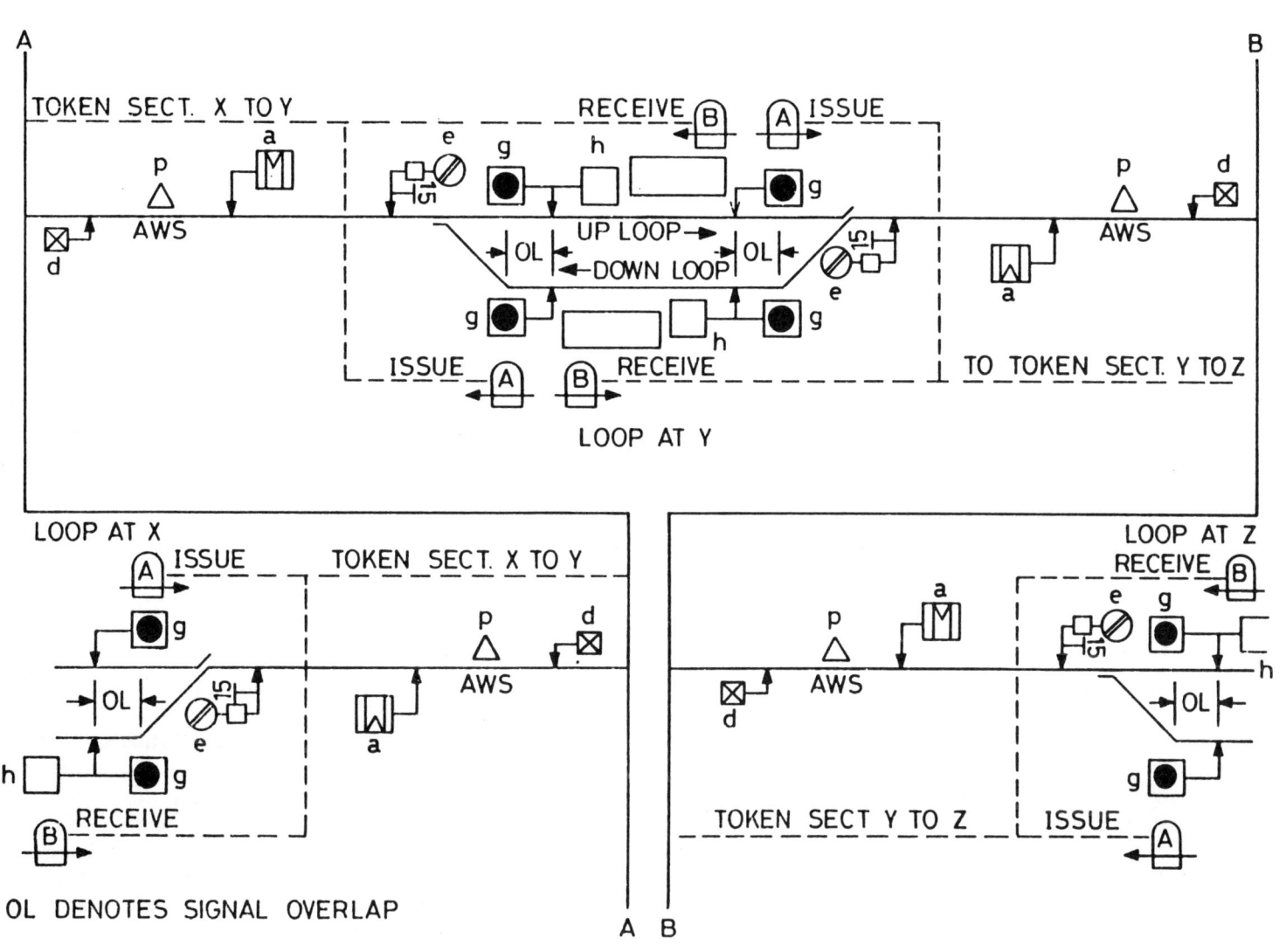

Each crossing station is provided with two identical 'no-signalman remote' token instruments for each section (so known because the signalman is only able to instruct the trainmen remotely on the operation of the instruments and cannot prevent the withdrawal of a particular token if it is safe to do so). The (A) or 'issuing' token instrument is placed adjacent to the starting signal notice board, and is the instrument from which a token is normally withdrawn. The (B) or 'receiving' instrument is placed adjacent to the issuing instrument for the next section, and is the instrument into which the token is normally surrendered after passing through the section. Adjacent sections have tokens of different configuration and colour and the tokens are engraved with the section designations. Because separate instruments are used at the intermediate passing places for the issuing and receipt of tokens, it is necessary to balance the instruments at regular intervals, and a balancing magazine for this purpose is provided.

The principle of operation of no-signalman remote token instruments is similar to that previously described for signalman-operated instruments. Once a token has been withdrawn from the issuing instrument, no further token may be obtained from the instruments at either end of the section, until the token has been surrendered into the receiving instrument after passing through the section, or returned to the instrument from which it was withdrawn. By exception a token may be withdrawn from a receiving instrument.

For lines with no lineside cable route, communication between the instruments uses a fail-safe frequency division multiplex (FDM) system, working over public telecommunications lines, similar to that already described. Three channels are required in each direction, but unlike the system described above, the transmissions only take place when the withdrawal of a token is required. Two channels transmit from the sending end to the receiving end to register a request for the release of a token, and information of the phase of the receiving end instrument is returned to the sending end by the transmission of two-out-of-three channels. This phase information is compared with that of the sending instrument, and if they are the same, the release is established and a token may be withdrawn. The circuits are so arranged that if the third frequency transmits erroneously, the release will not be given. Communication with the controlling signalbox is by telephone over the public telecommunications network.

If a reliable lineside cable route exists, this may be used to provide a direct physical line circuit. In this instance, the phase of the instruments at both ends of the section is normally transmitted to the line. When it is required to obtain a release for the withdrawal of a token, the operation of the plunger on the sending instrument causes that instrument to cease this transmission, and to compare the received phase information with that of the sending instrument. If the instruments are in the same phase, then the release is established and a token may be withdrawn. Communication with the controlling signalbox may either be by lineside telephones or by telephone over the public telecommunications network.

This system is particularly suited for use with intermediate instruments either for traffic purposes or for civil engineering maintenance purposes. (Motorised maintenance trollies are still in use on some less accessible branch lines.)

MODE OF OPERATION FOR A SECTION USING AN FDM SYSTEM
(Figs. 3.13 and 3.14)
Consider an up train from station X (the sending end) to station Y (the receiving end).

The driver of the train waiting at the starting signal notice board at X, contacts the supervising signalbox and requests permission to withdraw a token for the section to station Y from the (A) token instrument. If the single line section is unoccupied, and it is otherwise safe for the train to proceed through the section, the signalman will grant permission for a token to be withdrawn.

The driver places a token in the keyway. At this stage, one of the sending end (E)CR or (O)CR relays will be operated, detecting the commutator positions on both the (A) and (B) instruments. The driver then presses the plunger. This causes the operation of the (A)NSR, checking the status of the system, and in turn operating both the (E)(REQ)R and the (O)(REQ)R in the request mode.

At Y, providing that the (A) and (B) issuing and receiving token instruments are standing in phase (ie either the (E)CR or (O)CR is energised), operation of both the (E)(REQ)PR and the (O)(REQ)PR in request mode will in turn operate the UP(REQ)PR, causing the UP.YR to be energised at the sending end, where either the (E)(REQ)R or the (O)(REQ)R are operated in release mode according to the phase of the instruments. This will cause either the (E)(REQ)PR or the (O)(REQ)PR to be energised at Y, thus indicating at the sending end, the phase of the receiving end instruments.

At X, the energisation of the UP YR and the (E)(REQ)PR or the (O)(REQ)PR in release mode, causes the operation of the (E)BCR or the (O)BCR respectively. If the (E)BCR and the (E)CR, or the (O)BCR and the (O)CR are operated, then the instruments at both ends of the section are in phase, and therefore the circuit to the indicator lamp will now be completed. Upon seeing the illumination of the indicator, the driver turns the token 90 degrees anticlockwise, which interrupts the (E)CR or (O)CR circuit, but the relays will remain operated, being maintained by the (E)BCR or (O)BCR and stick contacts. The lock economiser contacts will close and the commutator lock will be energised. This enables the driver to withdraw the token from the instrument. The commutator will be reversed causing the operation of the opposite (E)CR or (O)CR. The driver informs the signalbox that the token has been withdrawn, and requests permission to pass the starting signal notice board and proceed through the single line section to station Y.

An attempt to withdraw a further token at station X for a succeeding train, will result in the same sequence of relay operation. However, the operation of the sending end (E)BCR or (O)BCR will be out of correspondence with the (E)CR or (O)CR. Any attempt to withdraw a token at station Y for an opposing direction movement to station X will result in the same sequence of relay operation (but from the opposite end to that described); however, the (E)(REQ)PR or (O)(REQ)PR, in release mode, received from X, will operate the (E)BCR or (O)BCR, which does not correspond with the (E)CR or (O)CR. Thus it will not be possible to operate the commutator lock to withdraw a token for either an opposing move or a succeeding move.

On arrival in the loop at station Y, the driver places the token into the keyway of the (B) receiving instrument and turns the key 180 degrees in the clockwise direction, restoring the token into the magazine. Thus the commutator is reversed, operating the opposite (E)CR or (O)CR. The system is now back in phase, and any further attempt to withdraw a token at either end of the section will be successful. The driver informs the signalman that the train is standing complete within the passing loop, and that the token system has been normalised.

88 SINGLE LINE SIGNALLING

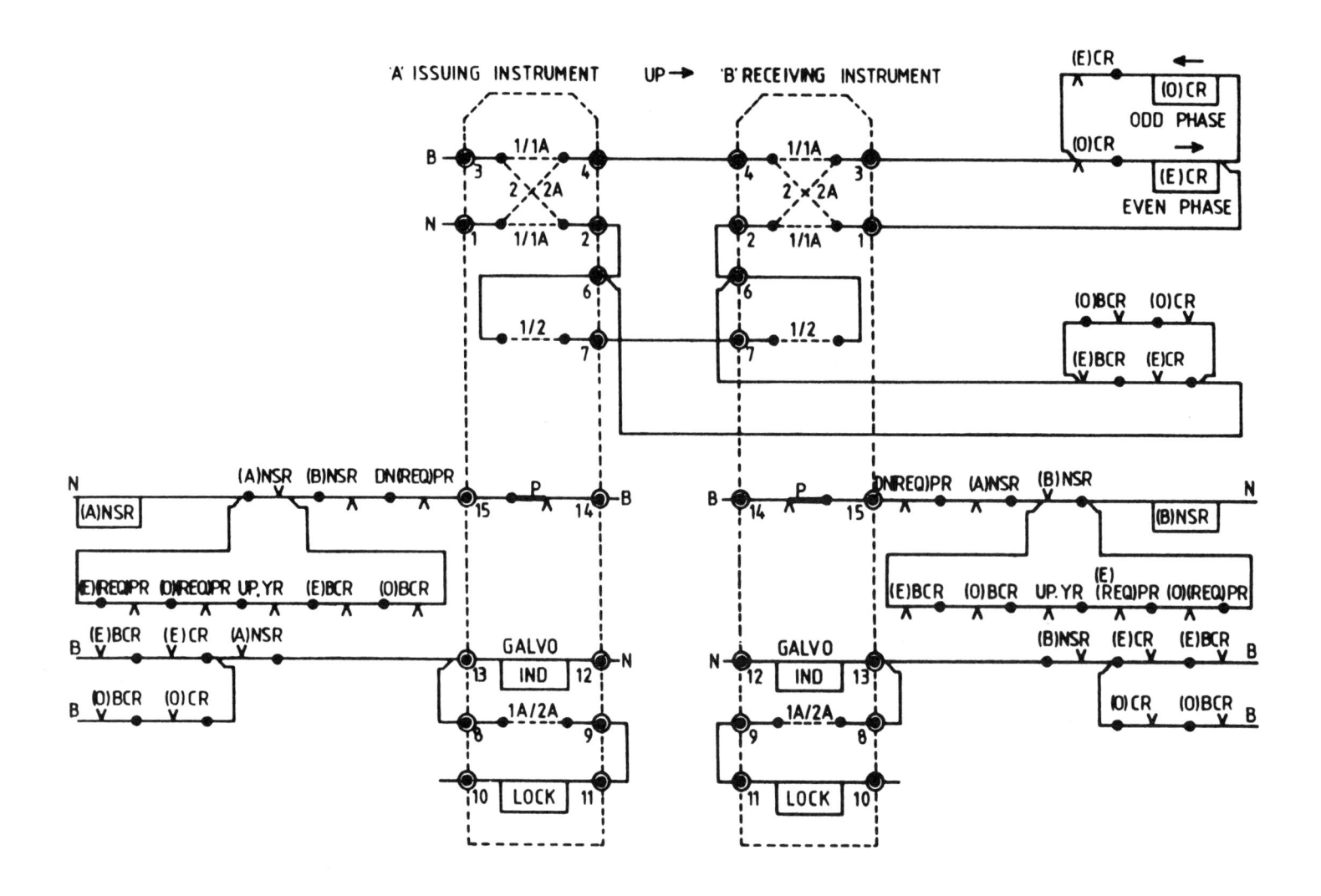

Fig. 3.13 No-signalman remote key token system (trainman operated)

SINGLE LINE SIGNALLING 89

Fig. 3.14 No-signalman remote token system: Reed-type transmission

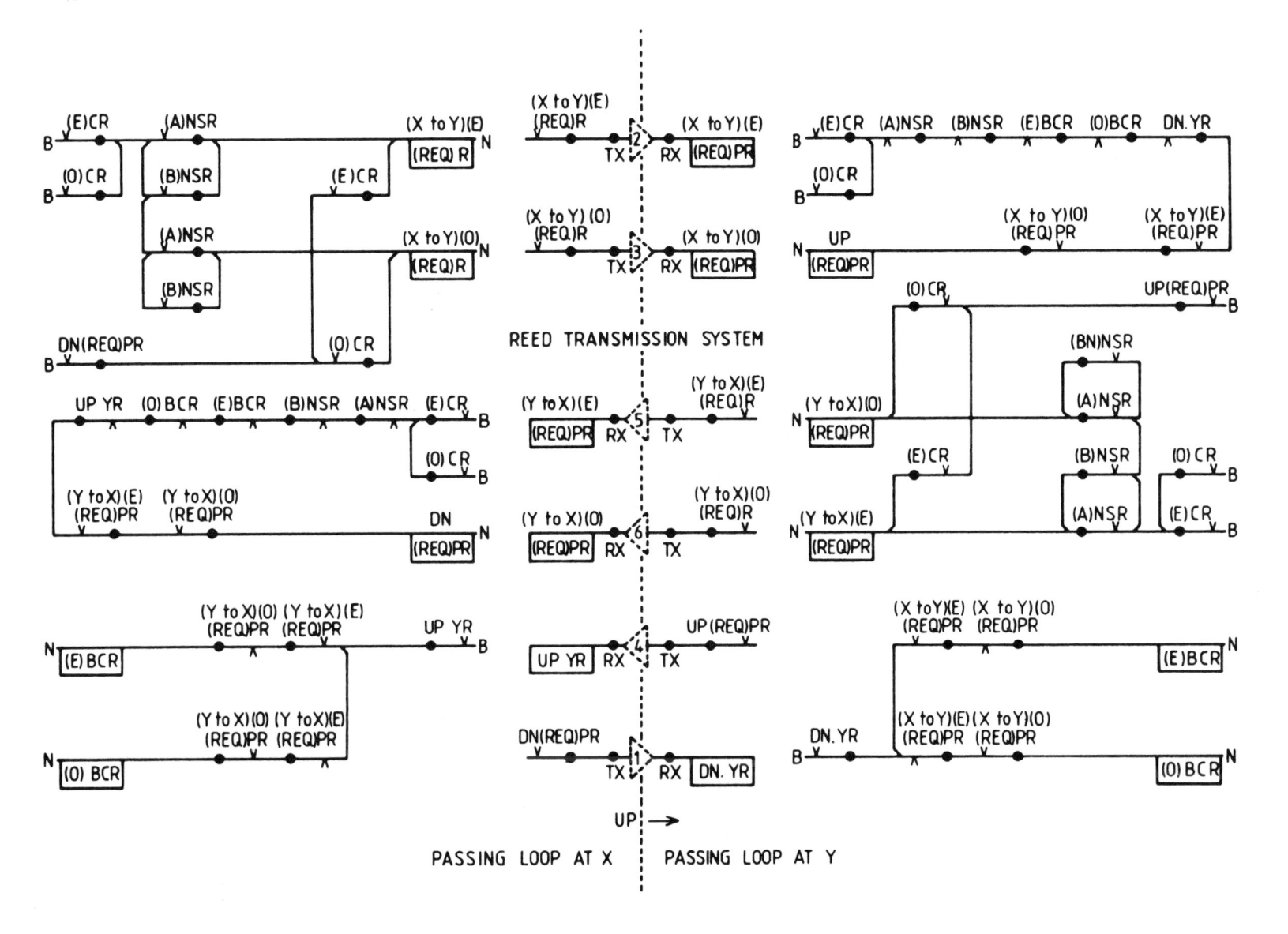

90 SINGLE LINE SIGNALLING

Fig. 3.15 Key token system: no-signalman remote instrument for use with physical circuits

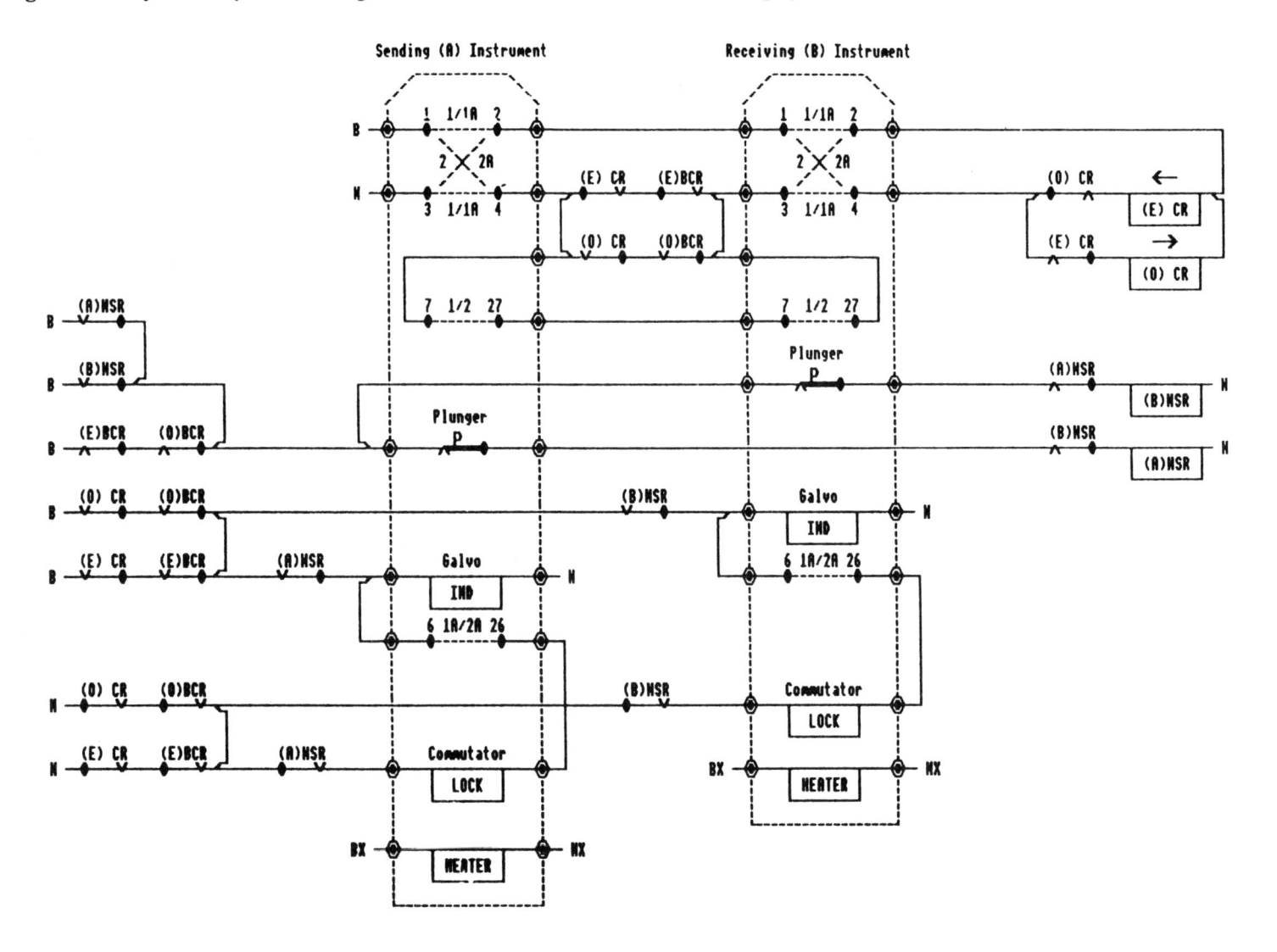

MODE OF OPERATION FOR A SECTION USING INSTRUMENTS CONNECTED BY A PHYSICAL LINESIDE CIRCUIT
(Figs. 3.15 and 3.16)

Again consider an up train from station X (the sending end) to station Y (the receiving end).

As far as the signalman and driver are concerned, the procedure is identical to that just described for the section having an FDM system. At Y, the driver places a token in the keyway. At this stage, one of the sending end (E)CR or (O)CR relays will be operated detecting the commutator positions on both the (A) and (B) instruments. The driver then presses the plunger. This causes the operation of the (A)NSR which disconnects the outgoing phase information from Y, and connects the (E)BCR and (O)BCR to the line, one of which operates dependent upon the received polarity indicating the phase of the instrument at Y.

At X, providing that the (E)CR is operated with the (E)BCR, or the (O)CR is operated with the (O)BCR (ie the instruments are in phase), the circuit to the indicator will be completed. Upon seeing the indicator show *free*, the driver turns the token 90 degrees anticlockwise which interrupts the (E)CR or (O)CR circuit; however, the relays will remain operated, being maintained by the (E)BCR or (O)BCR and stick contacts. The lock economiser contacts will close and the commutator lock will be energised. This enables the driver to withdraw the token from the instrument. The commutator will be reversed causing the operation of the opposite (E)CR or (O)CR. The driver informs the signalbox that the token has been withdrawn and requests permission to pass the starting signal notice board and proceed through the single line section to station Y.

An attempt to withdraw a further token at X for a succeeding train, will cause the same (E)BCR or (O)BCR to operate, but because the commutator was reversed by the withdrawal of the token, the opposite (E)CR or (O)CR will be energised, and the phase comparison will therefore prevent the indicator from showing *free* or the token from being released. Also the polarity of the outgoing phase information from station X will be reversed, therefore any attempt to withdraw a token at station Y for an opposing direction movement to station X will again result in the phase comparison preventing the release of a token. Thus it will not be possible to operate the commutator lock to withdraw a token for either an opposing move or a succeeding move.

On arrival in the loop at station Y, the driver places the token into the keyway of the (B) receiving instrument and turns the key 180 degrees in the clockwise direction, restoring the token into the magazine. Thus the commutator is reversed, operating the opposite (E)CR or (O)CR placing the system back into phase.

NOTE ON LINE CIRCUITS

If, due to the length of the line circuit, repeating relays are required, then it is not possible for the phase information normally to stand to line. In this case it is necessary to have a further bidirectional line circuit which is energised by the operation of the sending end plunger and which operates a relay at the receiving end instrument, causing the phase information to be transmitted back to the sending end, thus allowing any in line repeating relays to operate in the correct direction. In practice, it has been found that with BR 9XX series polarised relays, a repeater station needs to be provided if the line length exceeds 10 miles.

Fig. 3.16 No-signalman remote key token system (trainman operated): transmission over physical circuits

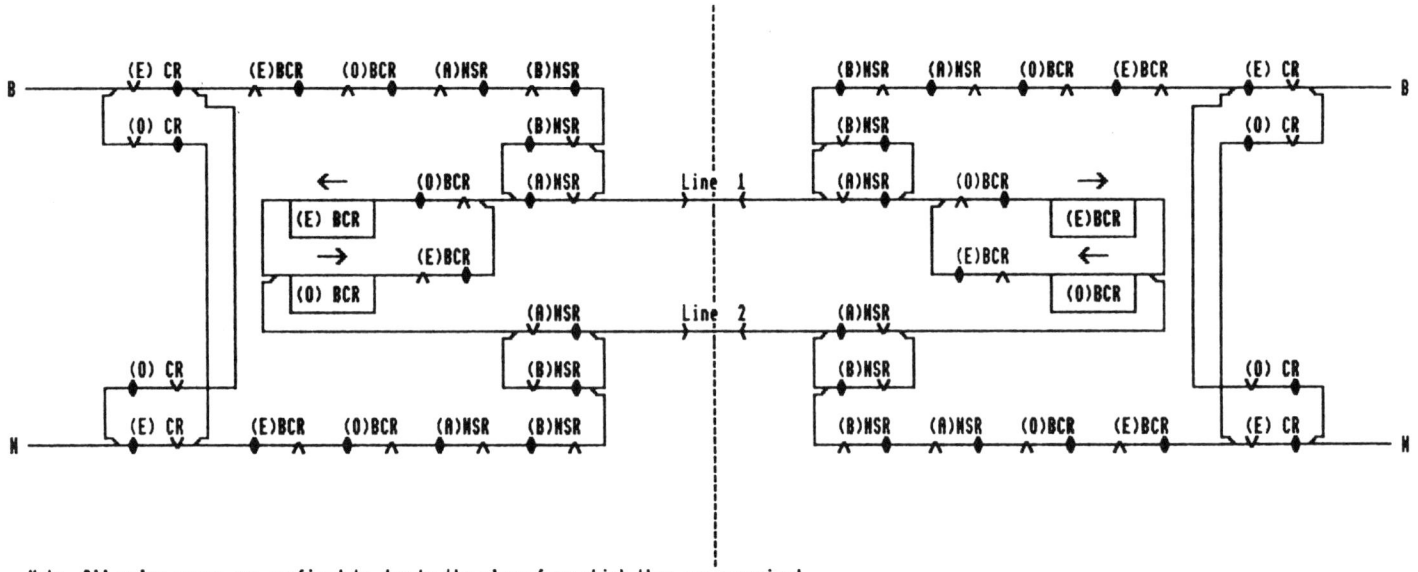

British Railways Tokenless Block System

GENERAL DESCRIPTION
(Fig. 3.17)
This system was developed for operation by staff in a porter/signalman situation, in that the signalmen at each end of a single line section do not need to co-operate in order to pass a train. Thus it is normal practice for no block bell to be provided, and it is theoretically possible to work entirely by the timetable, with the operators only requiring to communicate with each other to cover out-of-course running. On lines fitted with this system, it is usual to provide an annunciator to warn of the approach of a train, so that the operator has sufficient time to set the required route, operate any level crossings and clear the distant signal. To control moves on to the single line section, a starting signal is provided which is interlocked with the tokenless block. Home and distant signals are provided which are interlinked with the block, and must therefore be normal before a train can be accepted. Track circuits are provided between the home and starting signals, as is a berth track circuit to the home signal. The arrival of a train is registered by a treadle at the home signal, which allows the train

SINGLE LINE SIGNALLING   93

Fig. 3.17 British Railways tokenless block system

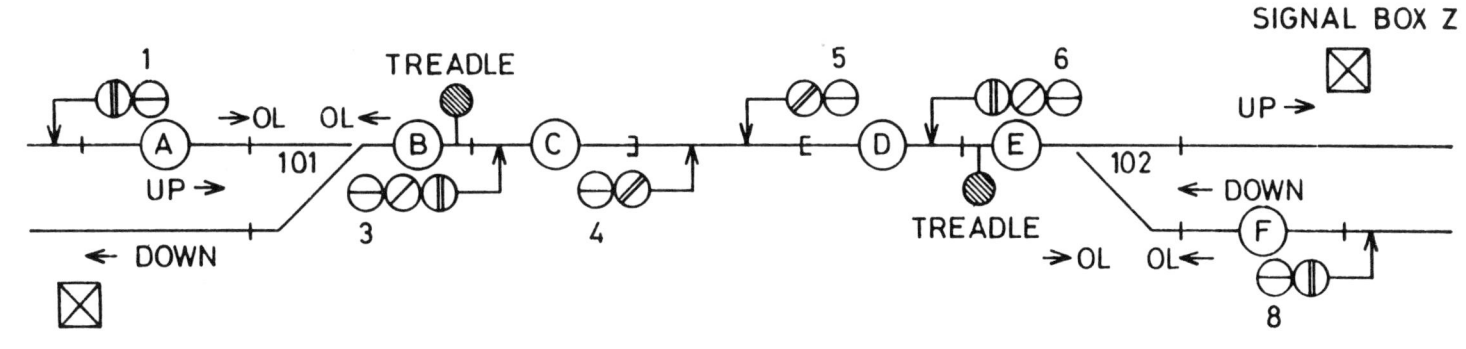

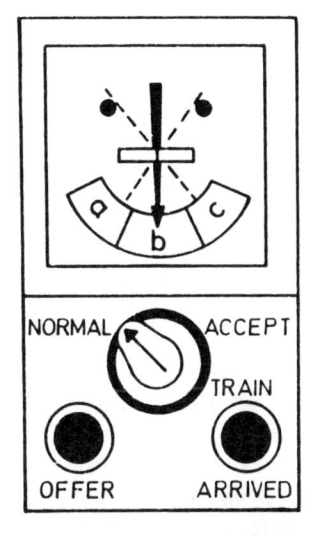

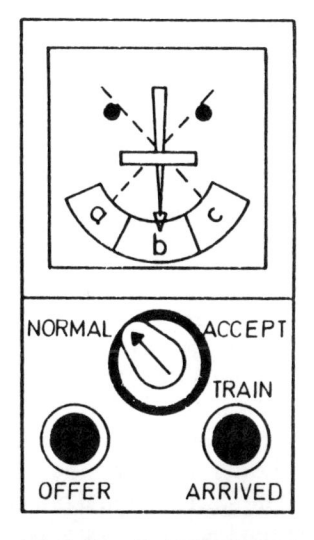

out of section procedure to be carried out. The signalbox at each end of the single line section is equipped with a tokenless block instrument, which has a three-position needle indicator. This, when pointing to the left, indicates *normal*, with no train in section or accepted. When the needle is in the centre position, it indicates *train in section*, and when the needle is to the right, *train accepted*. The indicator applies to trains travelling in either direction through the section. There is a two-position commutator, which is turned to the left for *normal*, and to the right for *accept*. There are two plungers, *offer* and *train arrived*.

MODE OF OPERATION
(Fig. 3.18)
Consider an up train to travel from signalbox Y (the sending end) to signalbox Z (the receiving end).

The signalman at Z, who is expecting to receive the train, providing that he can accept it, places the commutator to *accept*. (Both commutators may normally stand at *accept*.)

In order to offer the train, the signalman at Y places the commutator to *normal*, and operates the *offer* plunger. This transmits an 'offer' to the receiving end for an up train.

Providing that the up distant and home signals are normal, receipt of the offer causes the UP.AR to operate, thus placing the block indicator to *train accepted*. The UP.AR, providing that the down starting signal is normal, transmits an 'acceptance' to the sending end.

If the starting signal at Y is normal, the received acceptance operates the UP.BCR, thus placing the instrument to *train accepted*. The UP.BCR also maintains the transmission of the initial offer, enabling the signalman to release the *offer* plunger. The starting signal may now be cleared for one train only.

When the train proceeds into the section, occupation of the starting signal overlap track circuit interrupts the transmission of the 'section clear' to the receiving end, thus registering that the section is now occupied. This track circuit also places the block indicator to *train in section*.

At the receiving end, release of the DN.BKC.TPR (section occupied), places the instrument to *train in section* and ceases the transmission of the opposite 'section clear' to the sending end. Section occupied also causes the release of the UP.AR, ceasing the transmission of the acceptance to the sending end.

At the sending end, release of the UP.BKC.TPR (section occupied), ensures that the down 'section clear' will not retransmit to the receiving end once the train has passed clear of any track circuits. The release of the UP.BCR ceases the transmission of the initial offer.

With the train in the section, no line circuits are energised. A further attempt by the sending signalman to offer a succeeding up train, will result in the transmission of an 'offer', but this will be prevented from operating the receiving end UP.AR, by the DN.BKC.TPR being de-energised, and similarly for the opposite direction. No moves in either direction may therefore be made until the section clear relays, Dir.BKC.TPR, re-operate.

After travelling through the section, the action of the train occupying and clearing the berth and overlap track circuits of the home signal, and operating the treadle at Z, energises the UP.QSR. The signalman, after ensuring that the train is complete, places the commutator to *normal* and operates the *train arrived* plunger. Providing that the move has passed clear of the overlap to the home signal, this plunger causes the transmission of an up 'section clear' to the sending end UP.BKC.TPR. This relay operates, placing the block indicator to *normal*, and causing the transmission of the down 'section clear' to the receiving end DN.BKC.TPR, which also operates, placing the block indicator to *normal*. This relay also maintains the transmission of the section clear, enabling the signalman to release the *train arrived* plunger. The DN.BKC.TPR resets the UP.QSR.

The section is now normalised for the next movement.

If the movement is cancelled, and has not entered the section by de-energising the BKC.TPR, the signalman at the sending end, after having returned the starting signal to

SINGLE LINE SIGNALLING 95

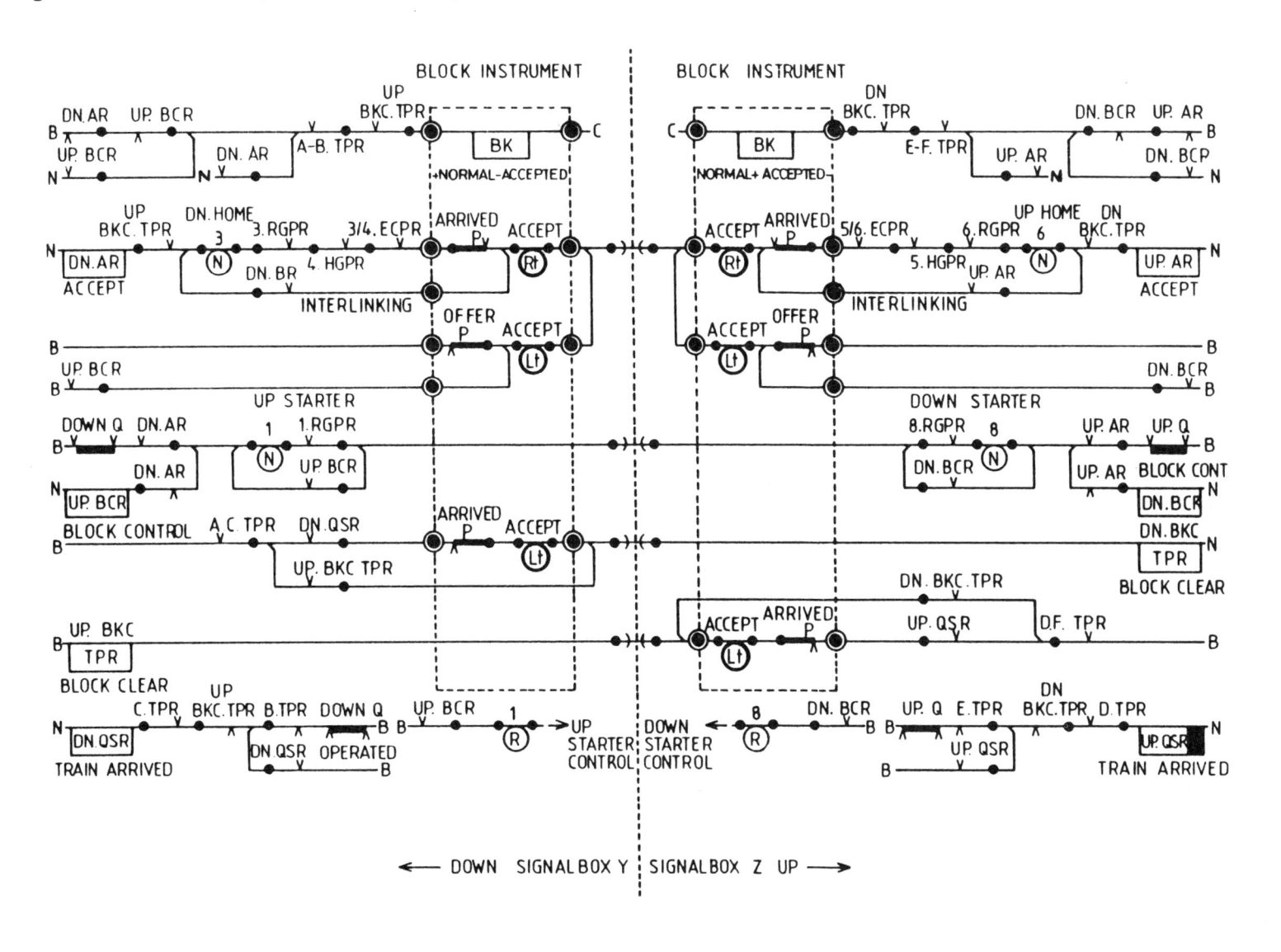

Fig. 3.18 British Railways tokenless block system: transmission over physical line circuits

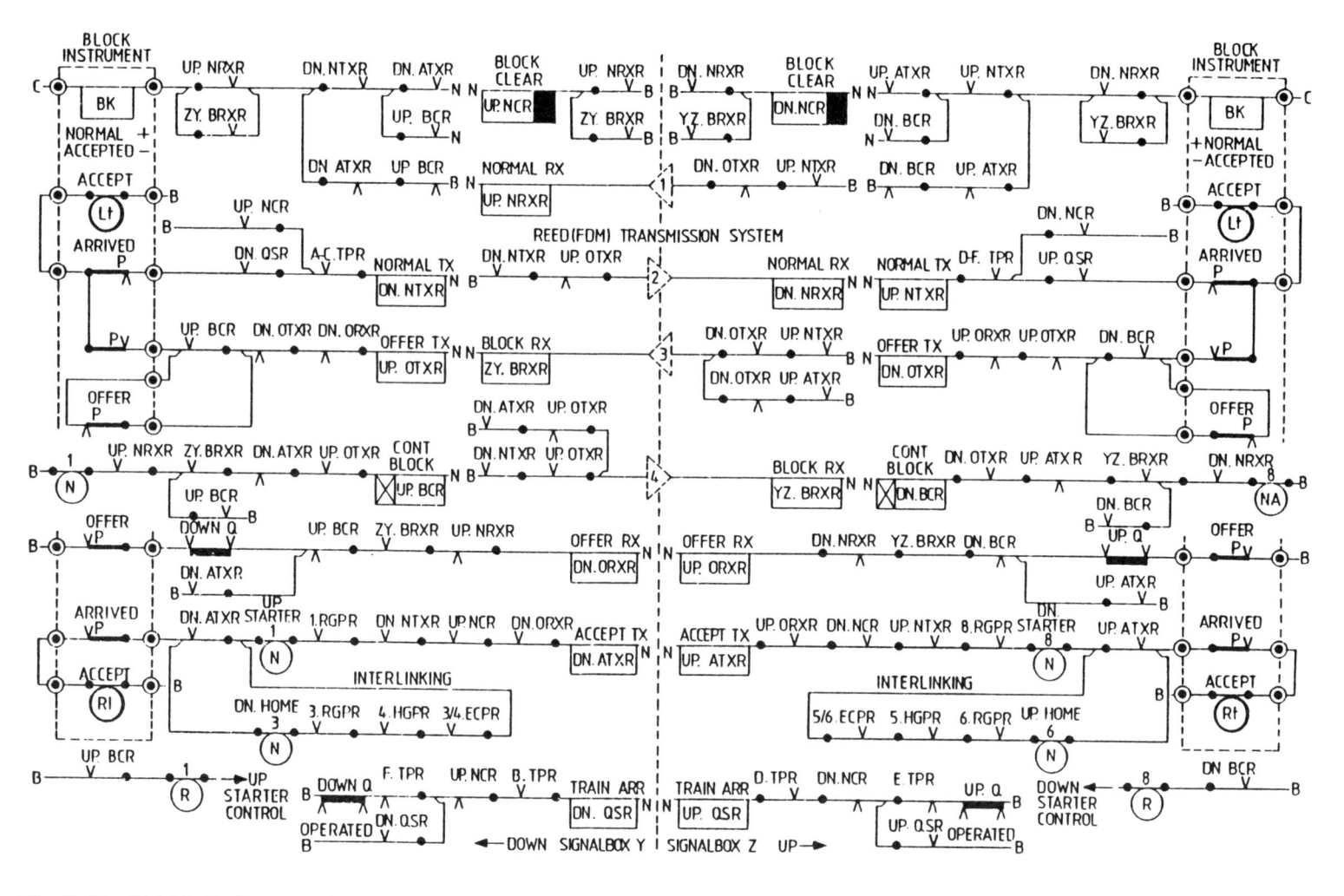

Fig. 3.19 British Railways tokenless block system: Reed-type transmission

normal, requests that the block commutator be placed to *normal*. This interrupts the acceptance circuit, causing both block indicators to revert to *normal*.

If the train returns to the sending end, the signalmen co-operate by replacing all signals to danger and restoring the commutators to *normal*. The signalman at the initial sending end must then carry out the train out of section procedure.

In the event of an emergency, the receiving end may withdraw the acceptance by placing the commutator to *normal*. De-energisation of the BCR replaces the starting signal. (If this is a mechanically operated semaphore, an alarm is provided to indicate that the release has been withdrawn.) The block indicators at both signalboxes will revert to *normal*. If the withdrawal of the acceptance results in the move running past the starting signal at danger, the BKC.TPR relays will be released, placing both block indicators to *train in section*. The train must then draw clear of the section, and the block be normalised by carrying out the train out of section procedure.

The disadvantage of this system is that it requires four insulated line circuits, which can be a major constraint over long distances. This can be overcome by multiplexing the circuits using a fail-safe transmission system.

British Railways Tokenless Block with Reed Transmission

GENERAL DESCRIPTION

The general description of the system and its appearance to the signalman is identical to that using physical lineside circuits (see **Fig. 3.17**).

TRANSMISSION SYSTEM

The equipment uses the Reed fail-safe transmission system which can be operated in the simplex (4 wire) mode, but is more usually in the minimum line duplex (2 wire) mode. Two channels are required in each direction, these being up block normal, down block normal, YZ block operate and ZY block operate. With the section clear, both block normal channels are transmitting.

MODE OF OPERATION
(Fig. 3.19)

Consider an up train to pass through the section from signalbox Y (the sending end), to signalbox Z (the receiving end).

As in the previously described system, the signalman at Z, who is expecting to receive the train, places the commutator to *accept*.

To offer the train, the signalman at Y turns the commutator to *normal*, and operates the *offer* plunger. This operates the UP.OTXR, causing the transmission of the 'block operate' channel, which is identified as being in the offer mode by the 'down block normal' channel ceasing to transmit to the receiving end.

At Z, the block operate channel energises the YZ.BRXR. The cessation of the block normal channel DN.NRXR, identifies the block operate channel as being an 'offer', thus causing the operation of the UP.ORXR. With the commutator at *accept*, and providing that the up home and distant signals and the down starting signal are normal, the UP.ATXR operates, placing the indicator to *train accepted*, and causing the transmission of the block operate channel to the sending end. Here it is identified as being in accept mode by the up block normal channel continuing to transmit.

At Y, the block operate channel energises the ZY.BRXR relay. The block normal channel maintains the UP.NRXR, which identifies the block operate channel as being in accept mode, thus causing the operation of the UP.BCR. This places the block indicator to *train accepted*, and maintains the initial offer, enabling the signalman to release the *offer* plunger. The up starting signal may now be cleared for one train only.

Occupation of the up starting signal overlap track circuit by the train entering the section, de-energises the DN.NTXR.

This ceases transmission of both the down block normal and the down block operate channels to the receiving end.

At Z, the cessation of the down block normal and down block operate (offer) channels releases both the DN.NRXR and YZ.BRXR, in turn releasing the DN.NCR, which registers a train as having entered the section. This also releases the UP.NTXR and places the block indicator to *train in section*. The DN.NCR releases the UP.ATXR, thus ceasing the transmission of both channels to the sending end.

At Y, cessation of the up block normal and block operate (accept) channels releases both the UP.NRXR and ZY.BRXR, in turn releasing the UP.NCR. The ZY.BRXR places the indicator to *train in section*.

With the train in section, no channels are activated. A further attempt by the sending signalman to send a succeeding train will result in the operation of the UP.OTXR, but this will be prevented from transmitting the offer to the receiving end by the DN.NTXR being de-energised. Similarly, an attempt by the receiving signalman to offer an opposing down train will result in the operation of the DN.OTXR, but this will be prevented from transmitting the offer to the sending end by the UP.NTXR being de-energised.

After travelling through the section, the action of the train occupying and clearing the berth and overlap track circuits of the up home signal at Z, and operating the treadle, energises the UP.QSR, which prepares for the normalisation of the block. The signalman, after ensuring that the train is complete, places the commutator to *normal* and presses the *train arrived* plunger. This operates the UP.NTXR, causing the transmission of the up block normal channel to the sending end.

At Y, the up block normal channel operates the UP.NRXR, placing the indicator to *normal* and operating the UP.NCR, which in turn operates the DN.NTXR and thus causes the transmission of the down block normal channel to the receiving end.

At the receiving end, Z, the down block normal channel operates the DN.NRXR, placing the indicator to *normal* and operating the DN.NCR. This maintains the UP.NTXR enabling the signalman to release the *train arrived* plunger.

The section is now normalised for the next movement.

Should the train be cancelled, or leave the section at the end which it entered, the manipulation of the block is identical to that shown for a system with physical line circuits.

CHAPTER FOUR

Immunisation and Earthing of Signalling Systems

Introduction

Modern signalling equipment is based on the use of electrical and electronic circuits, the design of which relies upon their segregation by insulation to eliminate electrical interference between each other and from external sources.

Railway signalling equipment must co-exist with a variety of other electrical systems. In some cases, these may share common electrical conductors. It is possible that these conductors may lie on the ground, close to each other. In some cases, although they will be insulated from other electrical circuits, the insulation may be breached under abnormal circumstances. Their circuits may be influenced by induced electrical interference from the neighbouring circuits of other signalling equipment, or power supply systems.

Electric traction faults, or the general environment in the form of lightning strikes, may cause interference to signalling equipment. This may be of a destructive nature from which the equipment must be protected.

Electrical systems must be constructed in a manner which ensures the safety of staff working on them. This in itself may produce circuit configurations which hinder the mode of operation required by the signalling equipment.

Electrical interference is capable of causing various types of failure to signalling equipment, and these must be categorised. To obtain successful and efficient protective measures, all sources and mechanisms of interference must be taken account of, and their effects must be fully understood. Methods of achieving protection from, or tolerance of, the different types of electrical interference have been devised, and successfully applied.

Categorisation of Interference to Signalling Equipment

Electrical interference may cause signalling equipment to fail to fulfil its required functions. These may be categorised as follows:

- To operate without suffering wrong-side failures.
- To operate without suffering an unacceptable level of right-side failures.
- Not to be destroyed except in extreme conditions.
- To operate without endangering staff working on the equipment.

The primary design criterion of signalling equipment is to prevent wrong-side failure due to electrical interference. In this category, the worst operational situations must always be considered. If there is any doubt as to whether the worst situation has been analysed, it is necessary to leave suitable margins of safety.

The occurrence of right-side failures must be kept to an acceptable level. To achieve this, the probability of the electrical interference exceeding a level which will cause right-side failure of the equipment must be considered. There will be a cost associated with providing immunity to right-side failure, and it may not be justified to prevent failures which would occur very rarely, or transiently, and which have a minimal effect on train movements.

Equipment can sustain direct lightning strikes, or electric traction supply flashovers. However, whilst such occurrences must not be permitted to destroy large areas of equipment, there has to be a compromise between limiting the amount of destruction and the cost of protective measures.

The voltages used in signalling systems (apart from power distribution) do not constitute a danger to trained staff working on such systems. However, the interference mechanisms of other electrical systems may produce different voltages on signalling equipment and suitable design is required to ensure that these voltages do not exceed safety levels.

Mechanisms of Interference from External Electrical Systems

The dominant mechanism of interference from a particular traction supply system will depend on whether alternating or direct current is being used. The former produces a large and fluctuating electromagnetic field which causes inductive and electrostatic interference in its locality. Direct current systems (although having a small alternating current component) produce minimal inductive fields, but will propagate conductive interference over a large area.

The running rails are used for traction return purposes, but may have other electrical circuits connected to them. They run parallel to lineside conductors, and are in contact with the ground.

By virtue of their location and usage, the running rails may have a voltage generated along and across them, or they may be raised to a voltage with respect to earth (see **Fig. 4.1**). This voltage may be applied to signalling equipment which is attached to the rails, or comes into contact with them.

When used for electric traction return, the running rails are joined together to propagate the traction current return, and are connected to other metal constructions for safety purposes. These connections constitute electrical paths through other rails, or via the ground, which can permit electrical energy to pass from one track circuit into another, giving false energisation. Such connections may also provide multiple paths for the track circuit energy, which are not all shunted by the presence of vehicles.

The lineside circuits of the signalling system use conductors running parallel to each other and to the track. These are

Fig. 4.1 Mechanism for generating conductive interference

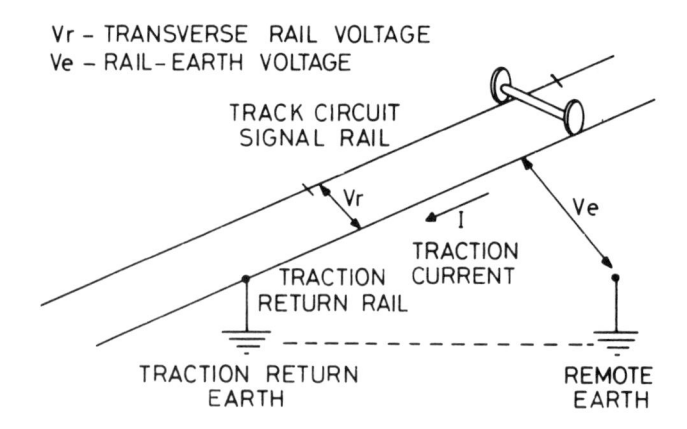

V_r – TRANSVERSE RAIL VOLTAGE
V_e – RAIL-EARTH VOLTAGE

prone to induced alternating current interference from electric traction currents or other signalling circuits.

If two conductors of the same circuit run close together, they will both experience induced voltages of equal magnitude, giving no resultant voltage in the circuit. However, if there are earth faults at either end of the circuit, then the induced voltages do not cancel, but drive a current through the earth circuit (see **Fig. 4.2**).

Electrical circuits are insulated from each other. However, the insulation may fail, or be damaged, which can permit signalling conductors to come into contact with each other, with the earth, or with the conductors of some other electrical system. The consequences of such false connections must be analysed.

In some cases, electrical circuits may become disconnected without discernible effect on their own operation, but with a reaction on the signalling equipment. Such situations include

Fig. 4.2 Mechanism of effects of induced voltages

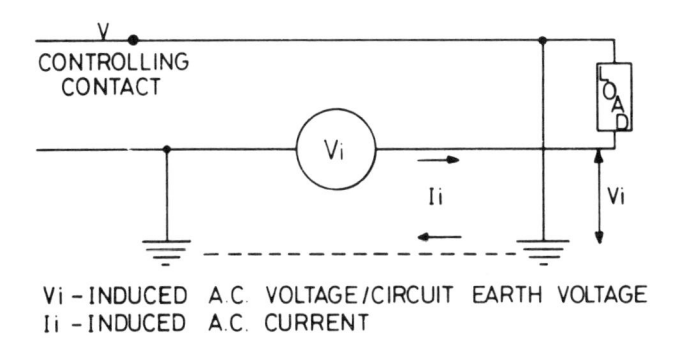

```
       V
  ─────●────────────────────●─────
  CONTROLLING               ┌─┐
   CONTACT                  │L│
                            │O│
  ──────────────┬───────────│A│──
                │    Vi     │D│
                │  ( Vi )   └─┘
                │           
                │    Ii →    ↓ Vi
         ═══════╧════════════╧════
                   ─ ─ ─ ─
```

Vi – INDUCED A.C. VOLTAGE/CIRCUIT EARTH VOLTAGE
Ii – INDUCED A.C. CURRENT

broken rails, disconnected impedance bonds, broken traction bonds, and substations being switched off.

The potential for these interference mechanisms to become operative must be examined for each signalling and traction situation.

Principles of Immunisation of Signalling Equipment

When signalling equipment is required to operate in areas suffering electrical interference, it must be constructed in such a way as to be immune to this interference. This can be achieved either by operating the signalling equipment from a type of electrical energy which is totally different from the interference or which is readily distinguishable from the interference, or by designing the signalling equipment so as to tolerate a workable amount of the interference. If the source of the interference is likely to change its character under failure conditions, then the expected characteristic must be monitored, and controlled.

The simplest method of immunisation is to operate signalling equipment from DC power, in the presence of AC interference, and vice versa.

In areas where there is AC interference, relay circuits should use DC relays having a high inductance, or an AC flux shunt, which will operate satisfactorily in the presence of heavy AC interference.

In areas with DC traction interference, the relay circuits should be operated with AC relays. These are generally constructed on the principle of an induction motor, and will not respond to DC.

In certain areas, both AC and DC interference will be present. In this case, the signalling equipment must be operated by a type of energy which can be discriminated from the interference.

This is generally achieved by the use of an AC frequency (or multiple frequencies) which cannot be generated by the interference sources, but may be detected in the presence of the interference. The main source of AC interference is the 50 Hz industrial mains frequency and its harmonics. This has led to the frequencies of 16.7, 33.3, 75, 83.3 and 125 Hz to find favour. However, because these are close to the interference frequency of 50 Hz, it is difficult to discriminate between them by means of filters. This has been overcome by the use of phase-sensitive relays relying on the presence of a reference phase, which is kept 'clean' from interference by screening protection against earth faults.

Another frequency band which has been used is in the range from 360–385 Hz. These frequencies cannot be produced from harmonics of the 50 Hz mains within its specified frequency variation. A special electromechanical Reed filter can provide a bandwidth of 0.5 Hz with a band spacing of 1.5 Hz at these frequencies, and a very high quality filter is thus available to permit the discrimination of a pure frequency in the presence of interference.

At higher frequencies, harmonics of the 50 Hz mains can completely fill the frequency spectrum. Where higher frequencies in the range 1,500–3,000 Hz are required for the operation of jointless track circuits, safety is ensured by the use of two frequencies, separated by 35 Hz. This means that both frequencies cannot be produced from the mains harmonics at the same time.

Heavy traction interference requires special measures for protective immunisation. However, there are other sources of interference which are not so severe. These may include self-interference of the equipment, due to its mode of application, and mutual interference between adjacent equipment. A heavy interference source may propagate widely, although at remote locations it will have been attenuated into less severe interference. Other interference sources may have an inherently relatively small magnitude.

In these cases, it is possible to design and apply the signalling equipment so that it may tolerate a degree of false energisation. Before such an application is made, the source of the interference must be analysed with great confidence, such that its maximum value under worst case conditions may be determined. Even then, a permissible inband interference level of 35% is customary. This level may be raised to 50% in the presence of undetected equipment failures, and to 100% with detected equipment failures. This analysis and design modification may look costly, but it saves expenditure on the more complex designs referred to in the previous section.

The interference from the normal operation of traction supplies and traction control systems may be characterised and protected against. When these systems suffer failure, the interference may change its character (frequency and magnitude), and the immunising measures may not protect against such changes. This may be countered by monitoring the interference either to ensure that it retains its correct character, or for the presence of dangerous frequencies.

Sources of Electrical Interference from Electric Traction Systems

The heaviest, and most widespread, electrical interference to signalling equipment is generated by electric traction systems. Electric traction current flows in the rails, and hence primarily affects track circuits. Other circuits not attached to the rails will suffer induction from AC traction currents if they are parallel for long distances, and the AC currents are large.

The traction current flowing in the rails is the sum of the current from the sub-station as determined by the impedance of the train, and the current driven by the train voltage perturbation into the supply system and other trains (see **Fig. 4.3**).

DC ELECTRIC TRACTION SYSTEMS

The DC electric supply system is derived from rectified AC mains supply. As such, any non-linearities in the rectification will produce AC components. For a typical 3-phase mains supply, with 12-pole rectification, apart from the main ripple frequency of 600 Hz, there will be indeterminate levels of components at 300, 200, 150, 100, and 50 Hz. A full disconnection of one rectifier arm will result in a 15% component at 50 Hz (see **Fig. 4.3**).

AC ELECTRIC TRACTION SYSTEMS

The AC electric supply system uses 50 Hz. This can have significant distortion, which may produce harmonics of the basic frequency. The nominal 50 Hz may vary between 48.5 and 50.5 Hz creating a band which is spread at the higher harmonics, until at 1,250 Hz there is no gap between the harmonics of 48.5 Hz and 50.5 Hz.

The AC currents will induce a voltage in parallel circuits, the magnitude of which will depend on the spatial location of the traction and signalling conductors, and the magnitude of the traction current.

Apart from this, the traction currents flowing in the running rails will cause conductive interference.

Fig. 4.3 Generation of traction interference currents

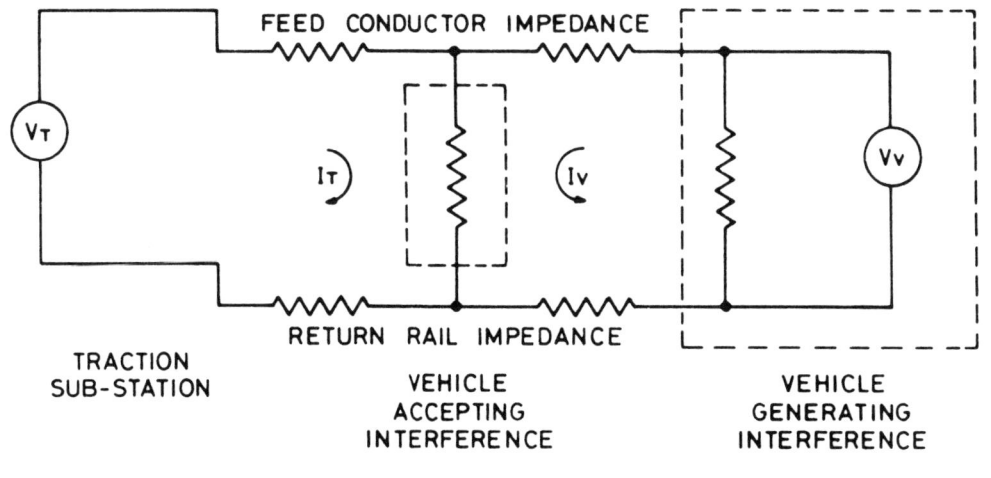

```
                    FEED CONDUCTOR IMPEDANCE
TRACTION           VEHICLE              VEHICLE
SUB-STATION        ACCEPTING            GENERATING
                   INTERFERENCE         INTERFERENCE
                    RETURN RAIL IMPEDANCE
```

I_T – CURRENT IN THE RAILS/VEHICLES GENERATED BY SUB-STATION
I_V – CURRENT IN THE RAILS/VEHICLES GENERATED BY VEHICLES

DUAL ELECTRIC TRACTION SYSTEMS
Combined AC and DC electric traction systems combine the effects of both individual systems.

An area of interest occurs when the two types of system meet in a linking section of dual electrification. Because the two traction return systems are connected together, the return currents from the dual electrified section may propogate into the other electrification system. If possible, the two systems should be separated by using isolating transformers in the AC system. If this cannot be done, then an analysis must be carried out to ascertain the magnitude of the traction return current propagation into the other electrification system.

SEMICONDUCTOR ELECTRIC TRACTION CONTROL SYSTEMS
Recent advances in semiconductor design have permitted their usage to make traction control systems more efficient.

In AC traction units, the use of thyristor drives generates components of the 50 Hz supply frequency. However, these may be present anyway due to supply waveform distortion.

In DC traction units, chopper control has been introduced. This uses semiconductor switches, working at a frequency of several hundred Hz, which generate current perturbations at the chopping frequency. Some traction units use interlaced and alternate chopping, which may give rise to derivatives of the nominal chopping frequency. Each design of vehicle must be analysed for this situation.

Both AC and DC units may be fitted with 3-phase variable frequency control equipment. This may generate any frequency. These vehicles will operate in multiple, and the resultant interference may be determined on a statistical basis.

It is possible for intermodulation to occur between interference sources of different frequencies. However, the significant non-linearities in the supply system required for this to occur, have not been found in practice.

These control systems generally require an input filter to condition the line supply. This produces an impedance to the supply system, and its effect must be examined.

Operation of Signalling Equipment in Electric Traction Areas — Track Circuits

When considering the operation of signalling equipment in electrified areas, the equipment may be placed into three categories:

- Track circuits.
- Equipment connected by tail cables.
- Lineside circuitry.

These categories of equipment will be considered in the various systems of electrification.

DC ELECTRIFIED AREAS

The track circuits must not suffer energisation from DC energy. This means that they must be operated by AC.

The track circuits must not be falsely energised from the adjacent track circuit if an insulated rail joint fails. With DC track circuits this is achieved by polarity stagger. This is not so easy with AC track circuits, but may be achieved either by using different AC frequencies on adjacent track circuits (and as far away as necessary), or by using phase-sensitive track relays, and providing phase stagger at insulated rail joints.

A further problem with AC track circuits is the lateral induction between adjacent tracks. This only occurs at audio frequencies, but must be prevented by ensuring that there is sufficient lateral distance between parallel tracks using the same frequency. At frequencies up to 500 Hz an intervening track provides sufficient distance, whereas at 2,000 Hz, three intervening tracks are needed. This again requires further frequencies to permit operation of parallel tracks.

Phase-sensitive track circuits operating at 50, 75, and 83.3 Hz have been used on BR, while London Underground has used 33.3, and 125 Hz. At these frequencies there is no problem with mutual induction, but phase stagger must be used at the joints.

BR has also used Reed (360–390 Hz) track circuits, whose filters are proof against DC and any rectification harmonics. A total of ten different frequencies are provided, which makes it possible to use four frequencies in rotation on each track to prevent any parallelism between track circuits of the same frequency, and to keep two spare frequencies.

Jointless track circuits in the 1,500–3,000 Hz range may also be used. These require two frequencies per track, and at least three tracks between parallel tracks using the same frequency. To achieve this, a total of eight frequencies is provided.

DC traction systems supplying heavy traffic require the use of both running rails for traction return. These rails must be electrically connected in parallel to provide for the DC current return, but must be electrically isolated, as far as the AC track circuits are concerned.

This is achieved by the use of impedance bonds, which permit longitudinal passage of DC current, but block the lateral passage of AC current.

AC ELECTRIFIED AREAS

The classic method of track circuiting in an AC electrified area is to use DC operated track circuits. The design of these must take into account the traction loading and type of supply system, so that the maximum interference voltages may be determined. While equipment immunisation may be improved to sustain any viable interference voltage, the length of the track circuit may be limited by staff safety restrictions.

DUAL ELECTRIFIED AREAS

The presence of both AC and DC traction reduces the options for equipment that can operate safely and reliably. Track circuits are required which operate at frequencies in the windows of the 50 Hz harmonic spectrum, or with encoded multiple frequencies. In the past, phase-sensitive track circuits at 75 and 83.3 Hz have been used, but Reed frequency equipment, or jointless track circuits are installed at present.

In general, double rail return systems are required to minimise corrosion to the earthed structures needed for the safe operation of the AC electrification. This introduces the requirement for impedance bonds, with their associated disadvantages of cost and reliability.

ELECTRIC TRACTION RETURN CIRCUITRY AND TRACK CIRCUIT APPLICATION

To assist in the operation of the electric traction supply system, the running rails used for traction return are cross-bonded in parallel. In AC electrified areas, the overhead line masts are also connected to the traction return system to achieve staff safety requirements. This creates secondary paths for the track circuit current to pass from the feed to the relay along a parallel track, or via the ground. As a first result, this means that a broken rail in the traction return system will not cause the track circuit to de-energise. This feature, although not a required function of track circuits, will be achieved by double rail track circuits.

The second problem is that the trainshunt of a vehicle is not applied to the secondary paths, and the energy passing along them may keep the track relay energised when there is a train occupying the track circuit. If the track circuit is DC or low frequency AC, there is no significant impedance in the rails, and the trainshunt will not be attenuated wherever it is applied.

However, with audio frequency track circuits, the rails have a significant impedance. A trainshunt will not shunt the secondary paths, and will only be applied to the feed and relay ends in series with the rail impedance. Worse still, a broken rail will cause the trainshunt to be applied to only one end, and that in series with the rail impedance (see **Fig. 4.4**). To overcome this problem, the track circuit must be commissioned such that it will be de-energised either by a disconnection in the genuine feed path in the track circuit, or by the presence of a shunt anywhere in the track circuit. This may limit its operational length, or tolerance of ballast resistance variation.

Where double rail track circuits are used, the traction bonding is carried out with impedance bonds. In normal operation, the impedance bond is balanced, and there is no resultant voltage to drive the track circuit current out of the genuine track circuit and into a secondary path. The system may become unbalanced if there is a broken rail or a disconnection of one side of an impedance bond. If there are two impedance bonds in the same track circuit, the unbalance will drive a proportion of the current along the secondary paths, where it will not be shunted.

These problems are only significant with audio frequency track circuits, and result in the track circuit sustaining residual energisation levels when it is shunted by vehicles. The magnitude of the residual energisation may be analysed, and commissioning methods which ensure satisfactory operation must be implemented.

Fig. 4.4 Track circuit current paths through the traction return system

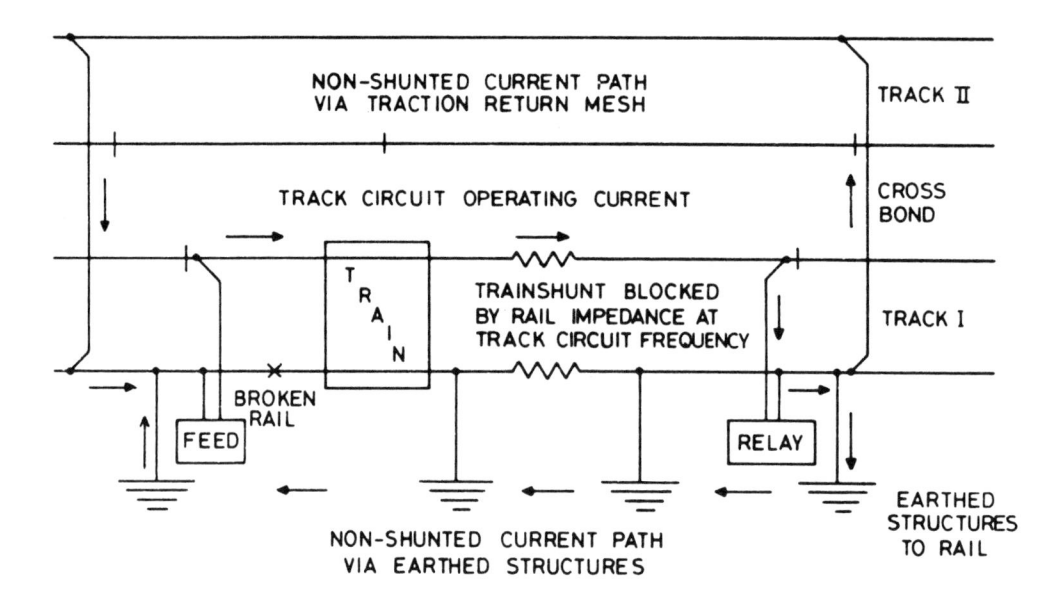

Operation of Signalling Equipment in Electric Traction Areas — Tail Cables

Tail cables in electrified areas can sustain false energisation from two sources:

- Direct contact, due to insulation damage.
- Induction from parallel AC circuits.

Because the tail cables lie on the surface of the track, they are very prone to mechanical damage from a variety of sources. Most tail cables run locally for relatively short distances, and hence do not suffer large induced voltages. The power required by trackside equipment generally means that a relatively high operating voltage is used and the circuit can tolerate a degree of induced voltage without prejudicing the operation of the circuit.

Tail cables run to three main types of equipment:

- Track circuits.
- Point machines.
- Signal lamps.

By virtue of their operating principle, track circuits will operate satisfactorily in this particular electrical environment. However, their tail cables will be in direct contact with the traction return rails, so that a loss of insulation to the other equipment wiring can permit the traction return voltage to propagate into the signalling system wiring and overload the cables.

Unless the method of construction of the location wiring ensures that the insulation will remain intact, the track circuit tail cables should incorporate fuses to protect any lineside circuits with which they may come into contact (see **Fig. 4.5**).

The voltages developed between traction return rails and earth are large enough to compromise the operation of point machines. False connections between tail cables running to point machines and the traction return system must be regarded as a viable failure mode.

Electrically operated point machines may be constructed with AC or DC motors. Point machines with AC motors may be operated satisfactorily in DC electrified areas, as the amount of AC components on the DC traction would not be enough to cause false operation.

Point machines with DC motors may either have enough intrinsic inductance (permanent field magnet design) to prevent any movement from AC currents, or may require the addition of a choke to improve their operation when used in an AC electrified area.

Another method is to install a local point controller very close to the point machine, and drive it with an immunised relay.

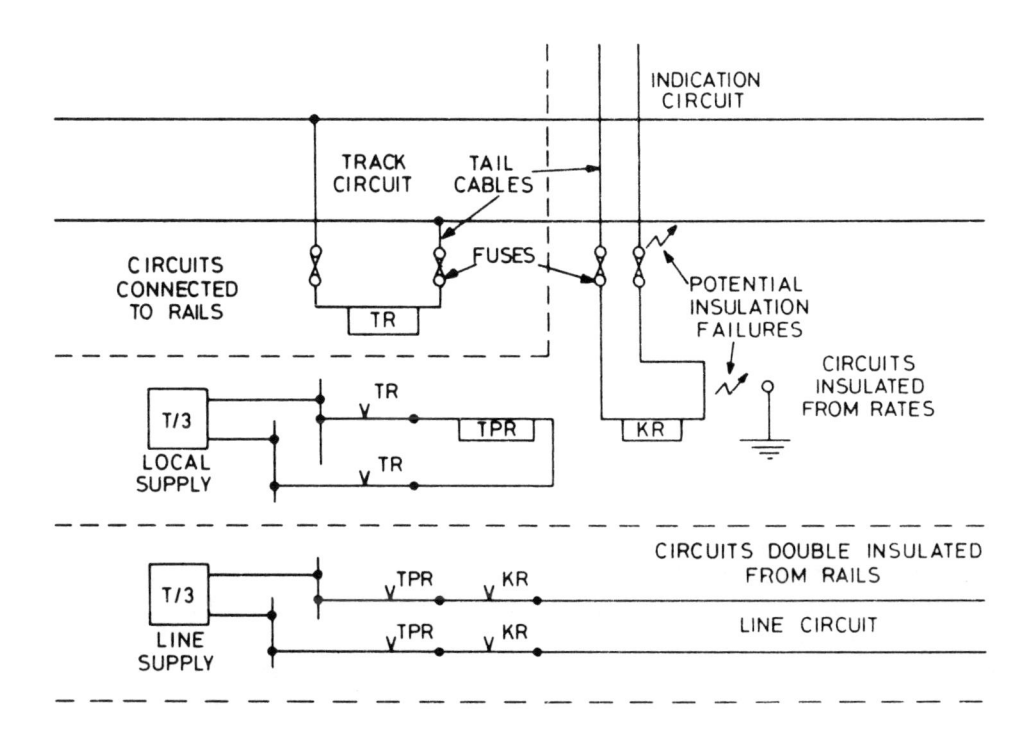

Fig. 4.5 Arrangement of tail cables: local and line circuits

For a dual electrified area, it would be difficult to provide a frequency selective point machine. However, it is possible to use an AC point motor with separate circuits to the armature, and field windings. This means that multiple insulation failures are required to cause false operation. Individual insulation failures can be detected by earth leakage detectors, or periodic insulation testing, before such multiple faults can occur.

Electro-hydraulic point machines use valves and motors. These may be operated by AC or DC and have the intrinsic advantage of using separate valve and motor circuits.

Point indication circuits run over the trackbed and into equipment which may come into contact with the traction return system. They require operation in the same manner as track circuits.

It is necessary to ensure that the power supply used for point detection is not used for lineside circuitry, as there would be a direct connection capable of propagating traction currents.

Signal lamps have no immunity to false energisation. Modern lamp circuits are AC operated, with a local signal head transformer. As such, they do not experience problems with DC traction interference. However, they may sustain false energisation from direct contact, or induction, in AC electrified areas.

The signal head circuitry may come into contact with the signal structure, which will be connected to the traction return system, and can sustain a voltage to earth. This may be overcome by double-cutting the feed circuits.

The tail cables will run parallel to the track and are hence prone to the effects of induction. Because there is no way of preventing this induced interference, it must be tolerated. This results in the operational length of signal lamp feed circuits being restricted to a length over which the service traction current cannot induce sufficient voltage to illuminate the lamp irregularly.

A false energisation of 20% can be tolerated without danger of illumination. With a typical theoretical induction coefficient of 9 volts/100 amps/km, and a maximum service traction current of 1,250 amps, the maximum permissible feed length would be 200 m.

Operation of Signalling Equipment in Electric Traction Areas — Lineside Circuitry

Lineside circuits can suffer interference from electric traction currents if they come into direct contact, or if induction occurs. Whilst it is quite possible for tail cables to come into contact with traction return conductors, or the earth, the lineside conductors should not suffer from this problem.

It is therefore acceptable to operate DC line circuits on DC electrified routes, as long as all the circuits from the busbar run entirely in a lineside cable. Care must be taken during the design of lineside locations that separate rectifiers are used for lineside circuitry and any tail cables.

If it is not expedient to design the circuitry in this way, then the lineside circuits must be operated with an AC supply. One method is to use AC relays, sometimes phase-sensitive, which will permit multiple indications over the same circuit.

Another method is to use an AC feed over the line circuit with an isolating transformer and rectifier feeding a DC relay at the far end. The section of the circuit which may sustain traction faults is operated with AC, and will hence be protected.

Although many AC circuits may operate over the same lineside multicore cable, there is insignificant mutual induction between them. This is because the currents are small, the operating voltage is high, and the currents are not able to induce an unacceptable voltage.

In AC electrified areas, the dominant form of interference is induction from the traction current. The passage of the traction current along the rails and the traction return conductors will induce a voltage in parallel conductors. If the two conductors of a signalling circuit are close together (as in a lineside multicore cable), they would experience nearly identical induced voltages, and there would be no resultant voltage applied to the controlled apparatus.

However, if earth faults occur at different locations on the two conductors, a fault circuit is produced. In this circuit there is a resultant induced voltage, which drives current through the apparatus and the earth, and which may falsely energise the apparatus (see **Fig. 4.2**). To counter this situation the lineside circuits may be operated by DC relays, designed to be immune from AC currents (similar to track relays). To ensure adequate immunity, the maximum traction currents and length of circuit must be determined.

If the route is dual electrified, then it is necessary to use methods as required for DC electrification. If the design arrangements required for the use of DC relays cannot be met, then AC line circuits must be used. However, because of the presence of 50 Hz induction, it will be necessary to use relays operated at a different frequency.

These relays must have discriminating filters, or be phase-sensitive. Such relays must be supplied from a 'clean' reference phase. If the traction current is induced in both the phase circuits, a false energisation can result. This is prevented by running the reference phase in a screened pair, the screening being used to prevent earth faults on this pair.

At this stage, another factor becomes important. The induced voltage in the circuit must not exceed the permitted safety level for signalling circuits of 110 volts. For a given maximum traction current, this requirement will limit the length of the line circuit. For a typical return system, the theoretical induction will be 9 volts per 100 amps of traction current per km. For a 600 amps service traction load, the line circuits will have to be limited to 2 km.

For DC circuits, any circuit longer than 2 km will have to be terminated, and repeated for onward transmission. In this instance, because the circuits are controlled by contacts, there is no question of enhanced energisation from rectifiers which experience an impressed AC voltage. If a busbar is used to feed in both directions, the 'exposed' length is extended in both directions, and the total length between the extremities of feeding must not exceed the limiting distance.

In some situations, Reed equipment, detailed previously, is used for frequency division multiplex remote control systems. When used on AC electrified lines, only the frequencies which are clear of the 50 Hz harmonics spectrum may be used. Such circuits may well exceed 2 km in length, and will require repeating. The repeating is carried out by installing isolating transformers to curtail the build-up of longitudinal voltage under fault conditions. These transformers must not permit the transmission of the induced 50 Hz voltage and achieve this by being constructed to saturate at 50 Hz and to attenuate this frequency while passing on the 360–390 Hz Reed frequencies.

Signalling power distribution cables will also sustain induced voltages. Because they are medium voltage (above 110 volts), staff safety is ensured by other methods. However, induced voltages cannot be permitted to increase indefinitely, and to limit these, the distribution mains are fitted with isolating transformers at 1.5 km from the supply source and every 3 km afterwards, unless the line is electrified with booster return, which limits the induction to an acceptable level even for very long circuits. Any other longitudinal conductors, such as pneumatic mains, will sustain induced longitudinal voltages, and must be sectioned into insulated lengths at the same distance as the line circuits.

Some signalling functions will require transmission over circuits of telecommunications standard. Because these systems operate with signals at relatively low amplitude, they are more prone to interference. Hence, they must be run in cables designed to mitigate this interference by their internal construction and screening. Since these qualify as telecommunications circuits, they are governed by the CCITT standards which limit the induced voltage to 60 volts.

Electrical Interference from other Electrical Equipment

RAIL CONNECTED EQUIPMENT

There are various examples of railway electrical equipment being directly attached to the rails. Electric point heaters and strain gauges are the best known examples. They are insulated from the rails, but this insulation may be damaged, and thus permit the electrical energy powering them to be transferred to the rails and track circuits.

The simple solution is to operate track circuits on such sections with equipment which is immune to interference from the rail-connected equipment. If this is difficult to achieve (eg because the track circuits already need to be immune to traction interference), then examination of the particular application may enable suitable arrangements to be devised.

For electric point heaters, it is possible to supply each rail from a separate secondary winding of the supply transformer, and allow the use of AC track circuits. For strain gauges supplied with DC, fusing of the connecting leads permits DC track circuits to be operated.

TRAIN-BORNE ELECTRICAL SYSTEMS

Almost all modern designs of rolling stock have electrical circuits passing from one vehicle to another. These circuits are normally earthed to the vehicle chassis at some point for circuit protection purposes. If another earth occurs between the return system and a vehicle chassis, then the running rails will become a secondary return path. The train will then become a voltage source along its length, and this voltage may affect track circuits. The same situation may occur in stations which provide shore electrical supplies to feed power to vehicles when they are stationary and are not connected to traction sources.

Because such rolling stock may travel over various types of track circuits, it is not feasible to guard against interference by installing a particular type of track circuit.

Analysis of the mechanisms by which such voltages may affect the track circuits will permit acceptable limits to be defined. These limits may be enforced by design of the train-borne circuitry, or by the provision of residual current detectors and circuit breakers.

ELECTRIC POTENTIALS IN THE GROUND

Buried metal constructions, chiefly pipelines, are often protected by cathodic protection systems. These systems impress a DC voltage on the ground close to the pipeline, and the groundbed which feeds the system. This may be transferred to the rails. The pipeline would not be expected to be at more than 2 volts relative to earth, although the groundbed may be up to 30 volts. Sections of track which pass over a pipeline with cathodic protection, or in the locality (closer than the pipeline) of the groundbed, may sustain a significant DC voltage.

Certain chemicals, occurring either naturally or because of the dumping of waste products, cause electric cells to be formed in the ground. The voltages produced may be transferred to the rails. Again, these voltages are not expected to exceed about 2 volts.

Both these voltages may falsely energise DC track circuits. This interference is protected against by the use of AC track circuits, or DC track circuits which can tolerate the magnitude of interference produced by the cathodic protection or earth electrochemical effects.

POWER TRANSMISSION LINES

The passage of AC power transmission lines alongside or across the railway permits induction into signalling circuits. The mechanism of this induction is the same as for railway AC traction supplies. The degree of interference, and area over which it is present, may be calculated from CCITT guidelines. This interference may affect both track circuits and lineside circuits. In both cases, AC immune equipment must be used for the signalling functions.

IMMUNISATION AND EARTHING

RADIO INTERFERENCE

The circuitry associated with electronic components may act as an aerial and pick up radio frequency transmissions. This is only a significant problem if portable radio transmitters are held close to analogue amplifying circuits or very sensitive digital circuits. Protection is achieved by installing ferrite beads on sensitive vital circuits, and prohibiting the use of portable radios in close proximity to such equipment.

POWER SUPPLY TRANSIENTS

When relatively heavy loads are switched on and off, the currents flowing in their circuits change very quickly. This current variation will produce a field which may induce voltages in sensitive electronic circuits such as axle counters and remote control systems, causing maloperation. Switching inductive circuits such as point machines, and producing sparks, is a severe example of this. The blowing of a busbar fuse will also cause current and voltage surges.

The effects of such transients can be mitigated by the installation of snubbing diodes on inductive devices, the installation of transient absorbers on busbars, and the design of wiring so as to avoid parallelism between high current and voltage sensitive circuits.

Earthing of Signalling Equipment

Earthing of signalling equipment is carried out to ensure that the staff are protected from both signalling power supplies and electrification systems, and to assist in protecting equipment from lightning. However, the presence of earth faults on wiring is a definite danger to signalling circuitry, and calls for protective measures to be taken.

It is a safety requirement that staff shall not be exposed to voltages greater than 110 volts AC. This is achieved by connecting all touchable metalwork to an earth electrode. Generally, only incoming power supplies require this earthing.

Another safety requirement is that separate earthing systems must not be within touching distance. If this is not possible, then they must be connected together or be shielded from each other. This means that on AC electrification, any metalwork within 2 m of traction return conductors must be connected to, or shielded from them. Because of this, signalling equipment enclosures should be kept 2m away from the traction return conductors.

This requirement causes further problems in relay rooms where several earthing systems such as water, gas and electricity mains, are present. These must either be connected together, or they must be isolated or shielded. The latter measures are often necessary if electronic equipment which cannot tolerate a 'noisy' earth is present. Because of this, the general practice is to provide a separate signalbox earth for equipment requiring earthing inside the signalbox.

Protection of Signalling Equipment from Earth Faults

Signalling circuits with controlling contacts or protective devices, rely upon the integrity of electrical insulation to prevent the equipment being falsely energised by power sources other than the genuine supply. The electrical insulation may be breached by mechanical damage, or degradation. This situation must be protected against.

Signalling circuits operate with a floating supply, with neither pole earthed. This means that a single earth fault will not cause maloperation of the circuitry, and that contacts can be placed in both poles.

The floating signalling circuitry can suffer dangerous failures with earth faults on the wiring. It is possible for there to be resistances to earth on both sides of controlling contacts. This will permit an operating current to bypass the contacts, via the earth, and falsely energise the relay (see **Fig. 4.6**). This problem may be overcome by two methods.

Firstly, the circuitry may be double-cut, with controlling contacts in both poles. This means that four earth faults would be required to cause the contacts to be bypassed and these faults would shunt the entire circuit.

However, one of the pairs of faults could occur on a relay rack. To overcome the effect of such a failure, the relay rack should be earthed, so that it will tend to shunt the circuit (see **Fig. 4.6**).

Double-cutting will introduce the extra cost of duplicated contacts, and the preclusion of common return circuits.

Nevertheless, it does give protection against earth faults, or false energisation from a single false contact, due to mechanical damage or faulty installation.

Secondly, the circuitry may be fitted with an earth leakage detector which gives an alarm when the first earth fault is detected. This fault must be remedied before a second one is likely to occur. The question of what is a permissible earth fault current must now be addressed. Obviously it must not be able falsely to energise the equipment on the circuit, and this produces a variety of permissible currents depending on the circuit. A 50 volts 2,000 ohms relay may remain energised with 8 mA of current. A current in excess of 100 mA on a 110

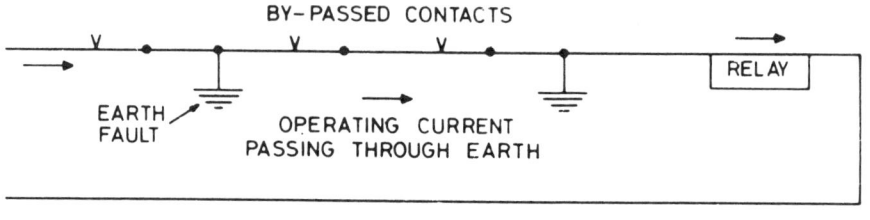

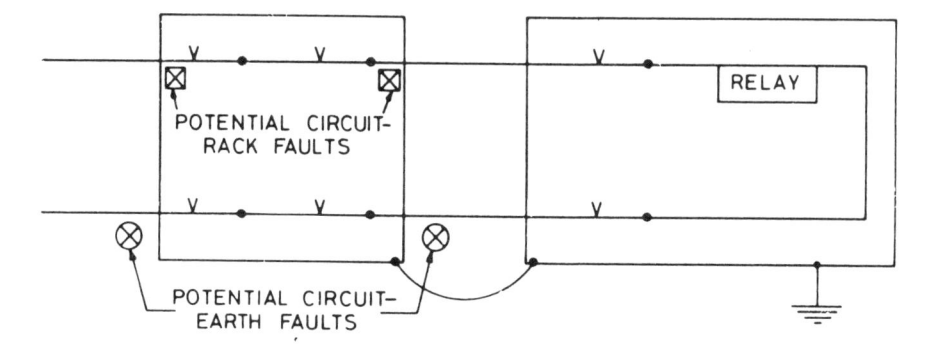

Fig. 4.6 Mechanism of earth fault finders

volt feed circuit may cause a 24 watt signal lamp to glow. A current of more than 1 amp may cause a point machine to move. Earth leakage detectors must have the capability of being adjusted to detect these values of current with AC and DC power.

The decision whether to take protective measures against earth faults, and how it should be done, will depend on the operational circumstances of the circuitry concerned. Some circuits will be prone to earth faults, while others will not.

Tail cables (which may be damaged), and cables running to electromechanical equipment such as lever frames (which may suffer chafing of insulation), will expect to suffer earth faults. Power supplies feeding more than one set of points, or point detection, can be falsely energised from multiple earth faults.

Line circuits in main cables, and internal circuits would not expect to suffer damage and consequent earth faults. Tail cables in AC electrified areas could expect to sustain solid earth faults to the traction return system. Busbars supplying many circuits would expect to suffer more earth faults than those supplying just a few circuits.

The cost of double-cutting will depend on the number of contacts involved.

The time taken to respond to earth leakage detector alarms is short for centralised interlockings, but long for lineside locations. In general it is most effective to use double-cut circuits for tail cables and lineside circuitry.

Main interlockings should use single-cut circuitry with earth leakage detectors installed on all busbars.

Internal circuitry of equipment cupboards does not merit earth leakage detection, but double-cutting should be used.

Particular situations such as lever frames, and multiple point supplies, merit the use of earth leakage detectors.

Lightning and Electrification Protection

Lightning strikes, and electric traction system short circuits, are able to deliver very high levels of power into the earth and any metallic constructions lying on its surface.

A typical lightning strike may raise the local earth voltage to 80 kV above remote earth voltage, for a period of less than 100 ms. A typical 25 kV traction system flashover may raise the local earth voltage to over 1,000 volts, for a period of 200 ms. These transient electrical currents can cause damage because of the following mechanisms.

The application of a high voltage to a point on the earth will cause a large current to flow to that point and develop a voltage gradient between it and remote earth. An insulated conductor lying in the ground, and having a distributed impedance to earth, will experience a voltage across the insulation, whose value is dependent on the voltage gradient in the ground over the length of the conductor (see **Fig. 4.7**).

If this voltage is large enough to break through the insulation, an arc will be struck to the conductor. This will connect the conductor to the local ground voltage at the site of the arc, and transfer the voltage to the far end of the conductor. If this transferred voltage breaks down the insulation at the far end, the conductor may carry a significant current between the two arcs. As far as electrical equipment operation is concerned, even the breakdown of insulation and arcing will be destructive, and should be protected against.

As would be expected, the local earth voltage due to a lightning strike decays away roughly in proportion to the inverse square of its distance from the strike. As the voltage becomes higher, its area of effect becomes much smaller.

An overhead electrified system is a special case. The multiple conductors will assist in propagating the strike voltage, but the multiple structure earths will reduce the impedance to remote earth. This means that although the strike voltage will be more widespread, it will be of a reduced magnitude.

114 IMMUNISATION AND EARTHING

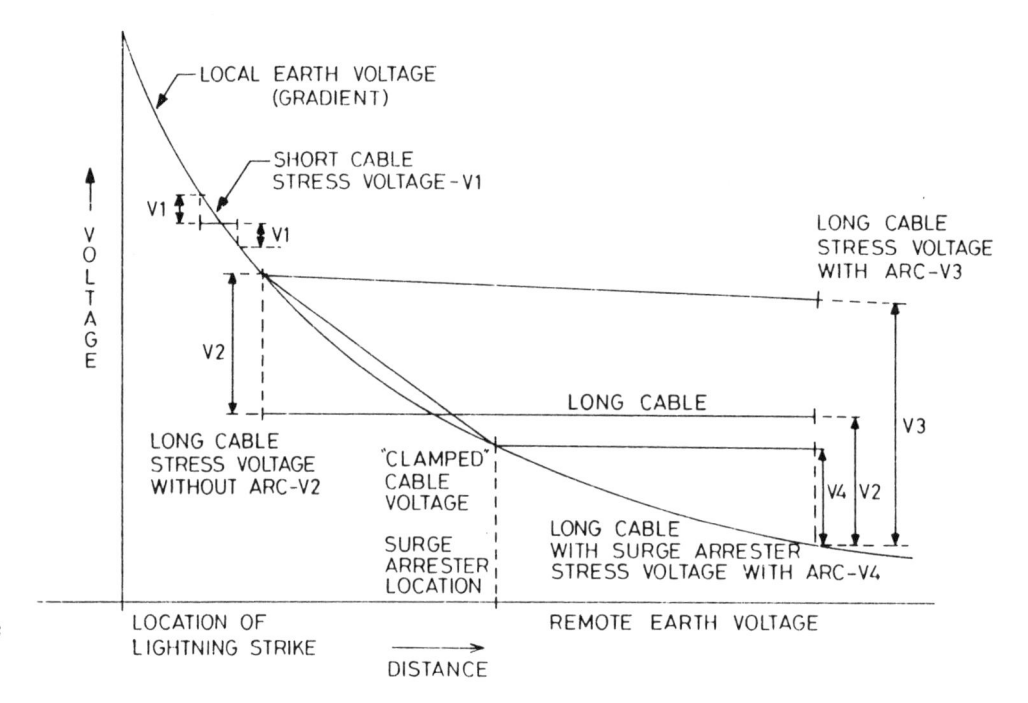

Fig. 4.7 Lightning strike voltage gradient

To take protective measures against a direct lightning strike would be very difficult. However, protective measures against a relatively remote strike can be more easily achieved. This means that protection becomes an exercise in damage limitation rather than damage exclusion.

EFFECTS ON SIGNALLING EQUIPMENT
Electrical equipment operates on the basis of insulating different circuits. If the insulation is broken down by a high voltage, a temporary arc is created. This arc may destroy a circuit, causing right-side failure, or may destroy insulation, perhaps leading to wrong-side failure. Such arcs will be created between one conductor, and the terminal frame, signalling equipment, and other conductors. The signalling equipment will be constructed with a specific insulation voltage to ensure protection against insulation breakdown.

Standard signalling cables have to withstand a voltage of 1,500 volts and metal equipment has a specified insulation voltage of 2,000 volts. Protective measures are required to keep the voltages between conductors, and other signalling equipment, below these values.

Long conductors have a distributed capacitance to earth, which will transmit rapid changes in local earth voltage. This effect will only transmit small amounts of power, but must be taken into account when sensitive equipment is connected to such conductors. Long circuits are also prone to electrical imbalance between conductors of the same circuit, giving rise to small but significant transient voltages across the terminating equipment.

PREVENTATIVE MEASURES
As previously mentioned, the voltage occurring at a direct strike is too great to be protected against. However, such voltages are very localised, and their propagation may be prevented. This is achieved by installing surge arresters at appropriate locations.

A surge arrester consists of an earth electrode with a voltage limiting device, which is connected to the propagating conductor (see **Fig. 4.8**). If the conductor rises to a voltage greater than the breakdown voltage of the limiting device, the device strikes an arc to the earth electrode, and clamps the conductor voltage to earth. This action prevents propagation of the lightning voltage into other equipment (see **Fig. 4.7**). Modern surge arresters are gas discharge tubes, capable of peak currents of 20 kA and breaking down at between 1,000 and 2,000 volts, depending on the rise time of the surge. This is a very effective protection, being able to dissipate a great deal of power while maintaining a low voltage.

In some circuits, a false connection of a conductor to earth after the surge arrester has operated, would be dangerous. For such circuits, a type of surge arrester which fails to an open circuit, has been designed.

For some electronic equipment, the surge arrester does not act fast enough. To cater for this, a semiconductor transient absorber is installed. This cannot dissipate much power, but does act quickly to limit voltages. To prevent too much power passing to it from the conductor, a low resistance may be installed in series with the conductor (see **Fig. 4.8**). The surge arrester may be protected by a fuse in the conductor circuit. However, this may be a source of unreliability, and a suitably rated fuse may be difficult to determine. Because conductors generally utilise two conductors, the surge arresters are normally constructed to protect two conductors, with two line

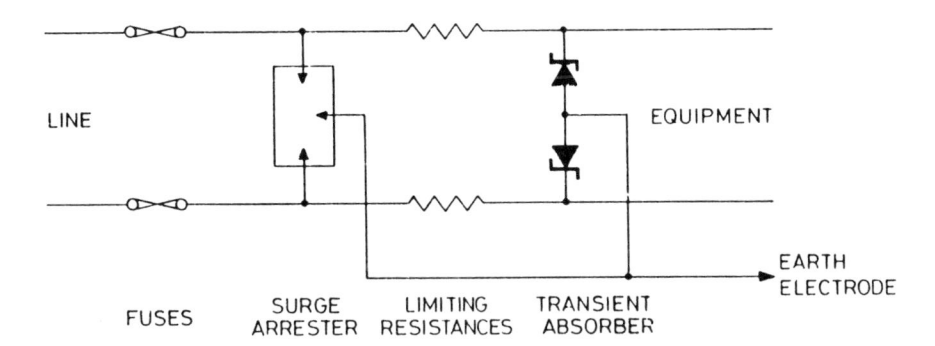

Fig. 4.8 Lightning protection for a line circuit

electrodes, and an earth electrode. A typical arrangement is shown in **Fig. 4.8**.

Another protective device, used on communications circuits, is the line isolating transformer. This transformer features a specially high voltage withstand of about 20 kV between primary and interwinding screen, which will not break down or suffer damage. A costly alternative method of main cable protection is the use of an earthed metal sheath. The metal sheath tends to short-circuit any voltage gradients in the ground. By doing so it carries a heavy current, but limits the voltage between the cable and earth.

The decision whether to provide lightning protection, and of what form, is dependent on the type of equipment connected to the circuit, the length of the circuit and the probability of lightning strikes.

Line circuits and communications circuits, because of their length, can be expected to sustain a large number of remote lightning strikes, and interference from capacitive coupling. The insulation voltage of cabling and relays is sufficient to permit operation without lightning protection in temperate zones, although tropical areas may merit surge arresters. Electronic equipment connected to such lines will require protection from capacitive interference, which is best effected by transient absorbers. In areas prone to lightning, a line isolating transformer would be worthwhile.

Tail cables running to track-connected equipment are vulnerable to local lightning strikes. Direct strikes are unlikely, and the damage incurred is not widespread. Tail cables will also suffer from traction system faults, which are not unusual. In view of this, equipment connected to tail cables in electrified areas should be designed with insulation able to withstand traction fault voltages. This is normally achievable with relay equipment, but electronic equipment would require protection, either in the form of an isolating transformer, or surge arrester and transient absorber.

Relay circuitry provides at least two insulations between tail cable circuits and line circuits. Hence, lightning effects on one are unlikely to propagate into the other system. However, electronic interlocking and data transmission systems may have tail cables and communications circuits, coming into the same equipment. In this instance, it is desirable to install lightning protection on all circuits to prevent the propagation of lightning voltages along the system.

A lightning protection system relies on a suitable local earth electrode and connections. The general rules for such constructions are that lightning protection devices should be installed as close as possible to the conductor entries; the earth points of the protective devices should be connected in 'star' form to an earth busbar; cables for such connections should be adequately rated, and run separately from other cables and metalwork without sharp bends; the earth electrode should be close to the earth busbar.

CHAPTER FIVE

Train Detection

Introduction

Despite technological advances such as solid state interlockings, electronic control centres and radio block, detection of the passage of trains by the use of track circuits remains the basis of railway signalling on the majority of lines in the United Kingdom. Although modern track circuits are in general more sophisticated than their early counterparts, the basic principles of operation remain unchanged.

The first volume of the Textbook dealt with the general principles of track circuits, their layout on the track, the design of track relays, and immunisation from traction interference. It also examined a number of types of track circuit in common use for both jointed and jointless track.

This chapter considers methods of train detection which have come into common usage on main line railways in Britain since the first volume was written. Of particular interest is the axle counter, which is now regarded by British Railways as an acceptable alternative to the conventional track circuit. The chapter also refers to other train detection systems which may well become standard within the next few years.

Type TI21 Track Circuits

INTRODUCTION
TI21 track circuit equipment is designed for use in areas where there is AC or DC electrification, and it can be applied to both jointless and jointed track, with track circuits being up to approximately 1,000 m long subject to the condition of the track. It is a 'voltage-operated' type of track similar in principle to the Aster 'U' track circuit, although with some improvements. The equipment exhibits great immunity to high values of interference signals.

PRINCIPLE OF OPERATION
The track circuit operates on a frequency shift principle whereby an audio frequency carrier signal is modulated between a pair of frequencies 34 Hz apart at a shift rate of between 2–8 Hz. Eight forms of track circuit are available, each operating at a unique pair of frequencies. Each line uses at least two such types of track circuit to ensure that consecutive track circuits do not work at the same frequencies. This is a basic requirement of the equipment to ensure safe operation. By providing eight types it is possible to equip up to four parallel lines fitted with cross-bonding without risk of frequencies from one line wrongly operating a track circuit on another.

The electrical separation of the ends of consecutive track circuits is accomplished on jointless lines by electrically tuning a short length of track using two series-resonant tuning units connected across the rails and mounted adjacent to them. For main line applications the carrier frequencies are in the range 1,500–2,600 Hz, and the tuned length of track is about 20 m. For rapid transit applications and overlay track circuits, frequencies up to 10 kHz are employed, giving a correspondingly shorter tuned length.

A schematic diagram showing the major items of equipment which comprise a typical track circuit on jointless track is shown in **Fig. 5.1**.

THE TRANSMITTER
A block diagram of a transmitter is shown in **Fig. 5.2**. A multi-vibrator produces a square wave at a frequency in the range 2–8 Hz which is set by external connections. This square wave controls the frequency of a voltage-controlled oscillator, the output from which is a signal alternating between two closely spaced audio frequencies at a rate determined by the multi-vibrator frequency.

118 TRAIN DETECTION

Fig. 5.1 T121 basic track circuit block diagram

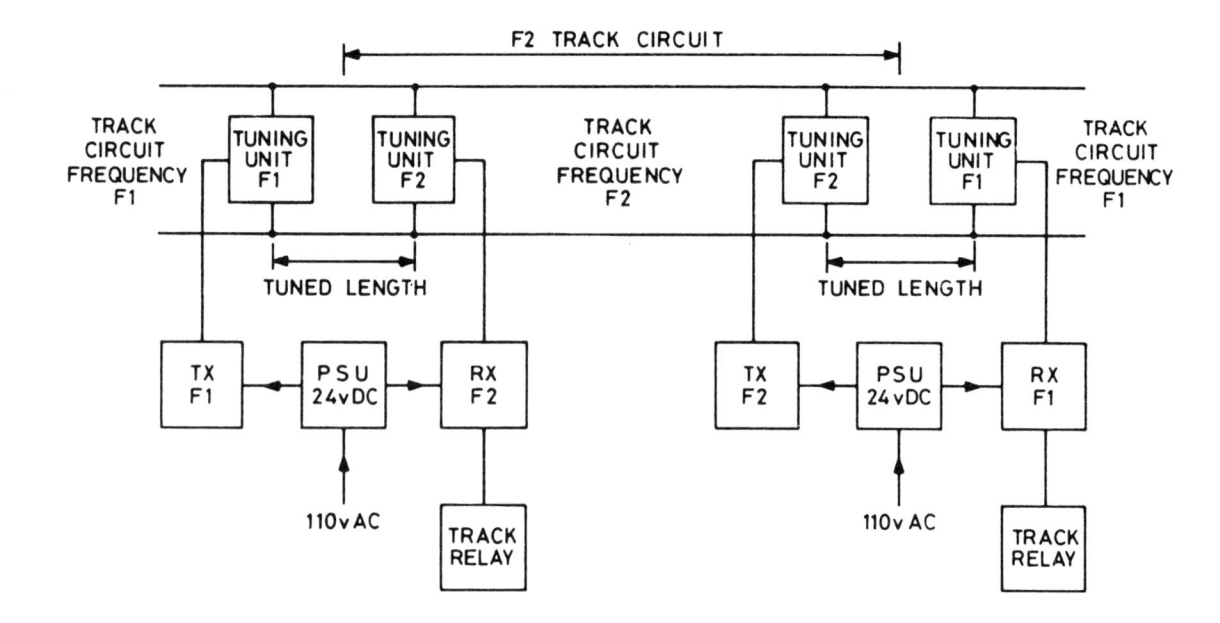

Fig. 5.2 TI21 transmitter block diagram

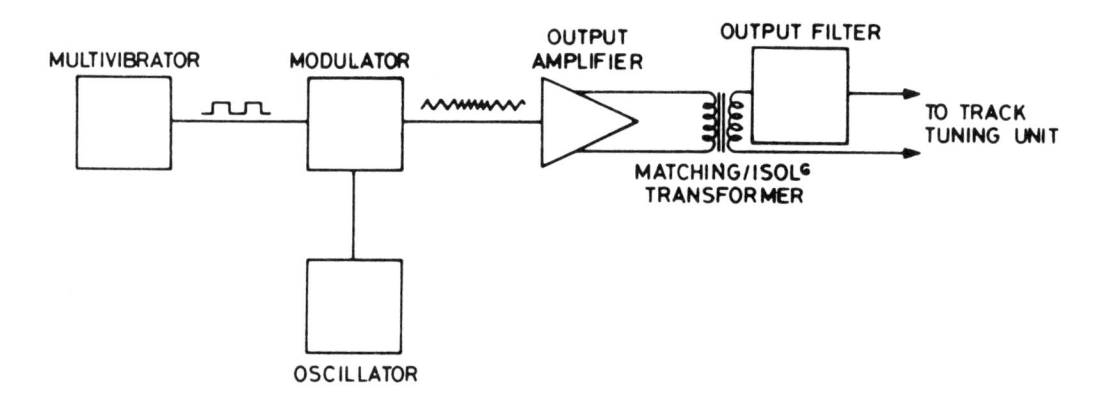

The output amplifier raises this signal to a power level suitable for transmission along the track. Special design features are included to provide an output impedance which ensures a suitable shunt value for the track circuit as a whole.

The transformer provides a means of matching the amplifier output to the load, and finally the filter serves to isolate the unit from unwanted AC and DC currents on the track. The output from the filter is fed to the track tuning unit at the feed end of the track circuit.

THE RECEIVER

The receiver is shown in block diagram form in **Fig. 5.3**. The signal from the track tuning unit at the relay end of the track circuit is fed to an input transformer which is used both to isolate the receiver from the track and to set the receiver gain. The latter is selected by suitable tappings so as to achieve the desired drop shunt. Protection from the effects of common mode interference is provided by means of an interwinding earth screen.

Fig. 5.3 TI21 receiver block diagram

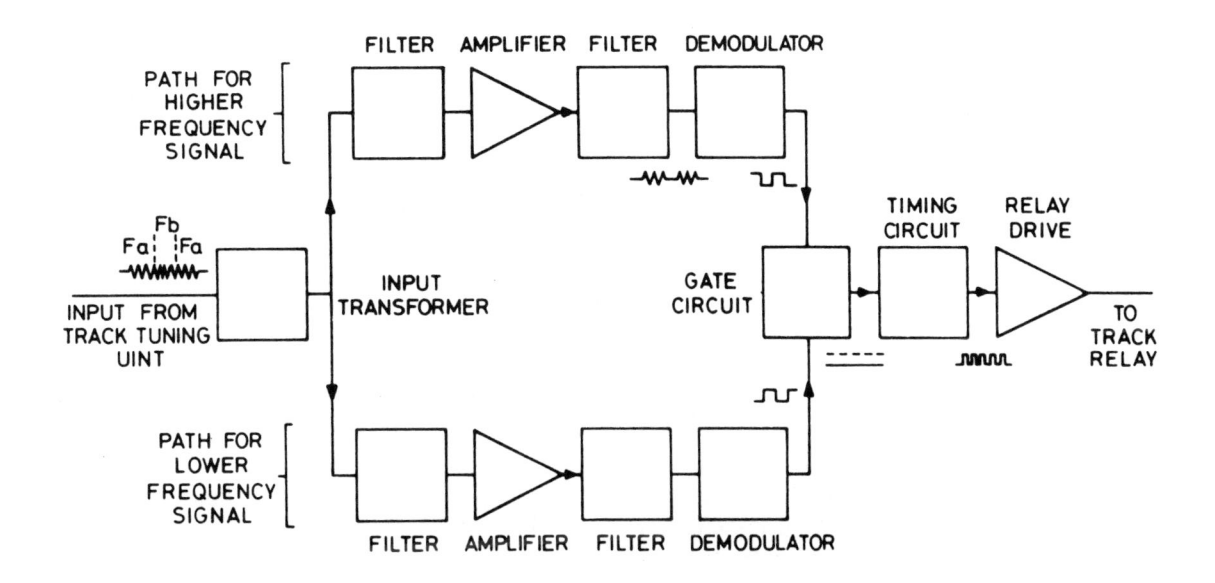

The signal is then fed to two filters. One is tuned to pass only the higher of the two frequencies generated by the transmitter, and the other only passes the lower frequency. Each of the resulting two signals is then amplified, further filtered and demodulated. If the receiver is receiving its correct frequencies, the demodulator outputs will be two square waves in anti-phase.

The demodulated signals are combined in a special gate circuit, and the resulting signal is then presented to the input of a timing circuit. This timing circuit produces an output in response to an input only after a predetermined time delay so as to ensure that even at the slowest rate of transmitter frequency variation (2 Hz) both low and high frequency signals are being received before an output is produced. This timer also features a very fast reset time so that if either input is removed, even for a very short period, the timing cycle will be restarted.

The output from the timer is taken to a relay drive circuit which in turn operates the track relay. Standard BRB miniature relays to BR Specification 930A may be directly operated by the receiver. Slow to energise repeat relays are not normally necessary due to the inherent slow response of the receiver.

TYPICAL INSTALLATIONS

Fig. 5.4 depicts a typical tuned area installation corresponding to the basic configuration shown in **Fig. 5.1**, with the transmitter of one track circuit butting up to the receiver of the next. Repeating this form of installation along the track provides a full set of end-fed track circuits.

Other configurations are also possible, for instance with two transmitters or receivers butting up to each other. Special arrangements are used for centre-fed track circuits (where a single transmitter feeds two track circuits), for track circuits adjoining non-track circuited lines, and for track circuits ending at an insulated joint. In the latter case, a special end termination unit is used, connected across the rails close to the joints. This single unit presents the same effective impedance to the transmitter as a pair of ordinary tuning units in a conventional configuration.

By utilising alternative connections on the track tuning units the equipment can also be used in a low power mode to ensure that on very short track circuits the equipment is not over-energised.

Impedance bonds can be provided for connection across the running rails outside the tuned areas in order to equalise the traction return current. These bonds are tuned for resonance at the track circuit frequency by means of a separate tuning unit, thus minimising the additional loading on the transmitter. The bonds also feature a centre-tap connection for use in AC electrified areas to balance traction currents between adjacent lines and to return the current to the supply substation. Other arrangements are available for low voltage DC electrification systems.

TI21 track circuits can be used through switch and crossing work.

INFORMATION TRANSMISSION VIA THE TRACK CIRCUIT

Switching the transmitter signal between two frequencies is provided not only to improve security against unwanted interference and the risk of a wrong-side failure, but also to enable additional information to pass along the track and be collected at the trackside or by a train-borne receiver. The track circuit could be used, for example, to transmit signal aspect and speed restriction information to the train driver. There is provision for as many as six unique pieces of information to be transmitted on a one-out-of-six basis, each using a different rate of frequency switching, in the range 2–8 Hz.

For applications where this facility is not required, a version of the track circuit which has its switching rate fixed at 4.8 Hz is employed.

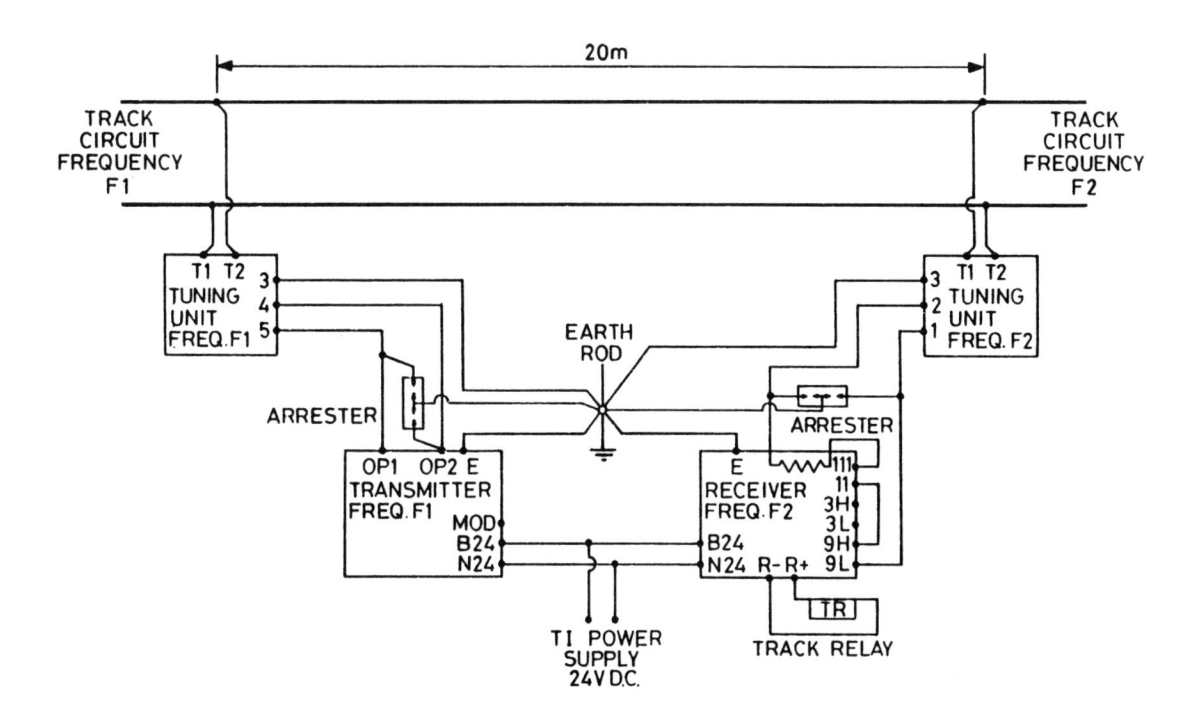

Fig. 5.4 Tuned area with one transmitter and one receiver — wiring schematic for normal power mode

Axle Counters

INTRODUCTION

Ever since the track circuit was invented, railways have become more and more dependent upon it to detect whether a section of line is clear for points to be moved or for a train to travel over it. However, there are places where the track circuit is not the best means of achieving this, and even some instances where the use of a track circuit is not possible at all.

An alternative option in such cases is a solution based on the principles of absolute block working, namely that a section of line may be considered to be clear when any train which has entered the section has also been shown to have left it. Such a solution would have to use a sophisticated system and, unless trains can be assumed never to divide in section, or tail lamps (whether conventional or electronic) can be reliably detected by trackside equipment, the only practical means of proving that an entire train has left a track section is to show that as many axles have come out of the section as entered it. This is the principle of the axle counter.

A few early systems by various manufacturers were installed in the United Kingdom, but although usage

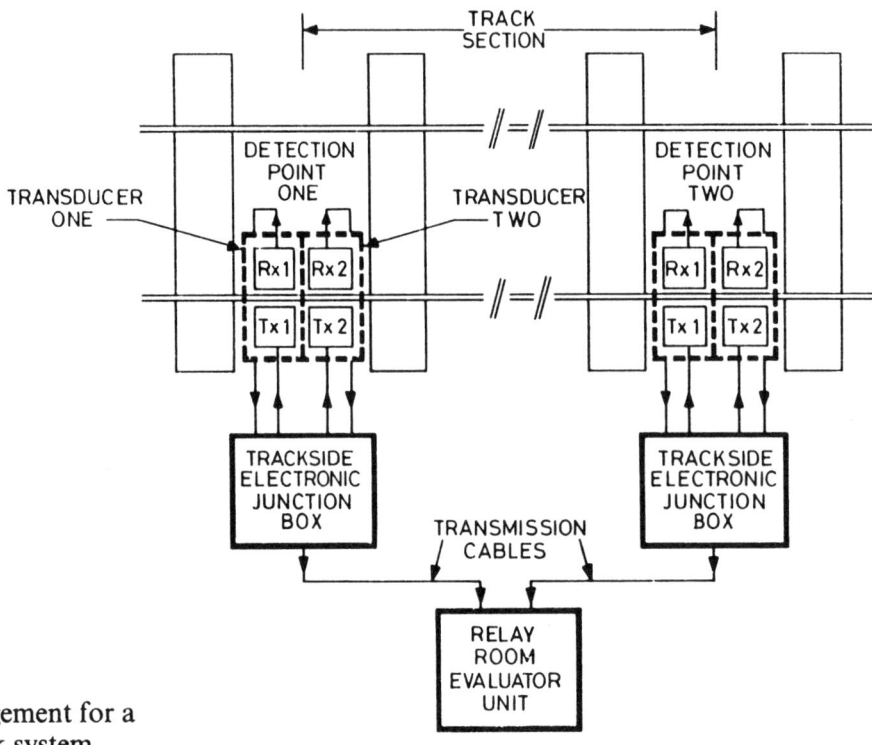

Fig. 5.5 General arrangement for a typical axle counter track system

increased in other countries, it was not until the late 1970s that British Railways started to consider the axle counter as a standard item of equipment. Trials were undertaken with a system offered by a German manufacturer. Those trials were successful and, apart from two places where older equipment remains, the SEL AzL70 axle counter is currently the only one used by BR. The description which follows is therefore based on this system.

SYSTEM OVERVIEW

A typical axle counter track section consists of an evaluator unit (frequently, but not always, housed in a relay room), with one set of wheel detection equipment at each position on the track layout where trains may enter or leave the section. The detection equipment comprises a trackside electronic junction box and two rail-mounted transducers which detect the presence of wheel flanges. Axles can be counted in or out by each detection location according to the order in which the wheels are detected by the transducers.

The evaluator unit takes the outputs from the trackside detection equipment and determines whether the track section is occupied or clear by counting the signals corresponding to the wheels as they enter and leave the section.

Fig. 5.5 shows the general arrangement of a typical axle counter track section. Until the mid-1980s the two transducers which comprise a detection point were mounted one on each rail, but on the latest version (shown in the diagram) both are mounted side by side on the same rail.

TRACKSIDE DETECTION EQUIPMENT

As **Fig. 5.5** shows, each transducer consists of one transmit (Tx) coil on the outside of the rail and one receive (Rx) coil on the inside. The two transmit coils of each detection point are fed with different frequencies (approximately 29 kHz and 30 kHz). The resultant fields couple around the rail with the two receive coils, into which voltages are induced.

If the wheel is more than 200 mm from the centre line of the transducer, the electromagnetic flux lines meet the winding of the receive coil at an angle from the line perpendicular to the axis of the coil, and an AC signal, in phase with the transmit voltage, is received (see **Fig. 5.6(a)**).

If the wheel flange is approximately 200 mm from the centre line of the transducer, the flux lines meet the receive coil at right angles and the received voltage is zero (see **Fig. 5.6(b)**).

When the wheel is located directly over the transducer, the flux lines meet at an angle α and thus generate an AC signal in anti-phase with the transmit voltage (see **Fig. 5.6(c)**). It is this phase change which the system uses to detect the passage of the wheels.

Fig. 5.6 Coupling between transducer, transmitter and receiver

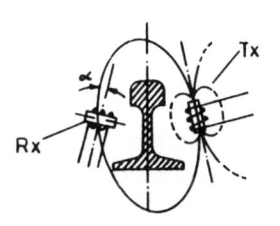

(a) WITHOUT WHEEL

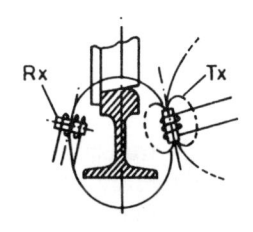

(b) WHEEL APPROXIMATELY 200mm FROM CENTRE LINE OF TRANSDUCER

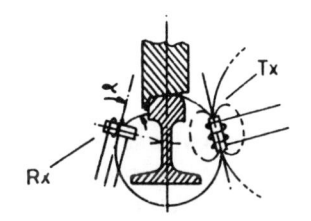

(c) WHEEL DIRECTLY OVER CENTRE LINE OF TRANSDUCER

124 TRAIN DETECTION

Fig. 5.7 Trackside electronics for an axle counter detection point

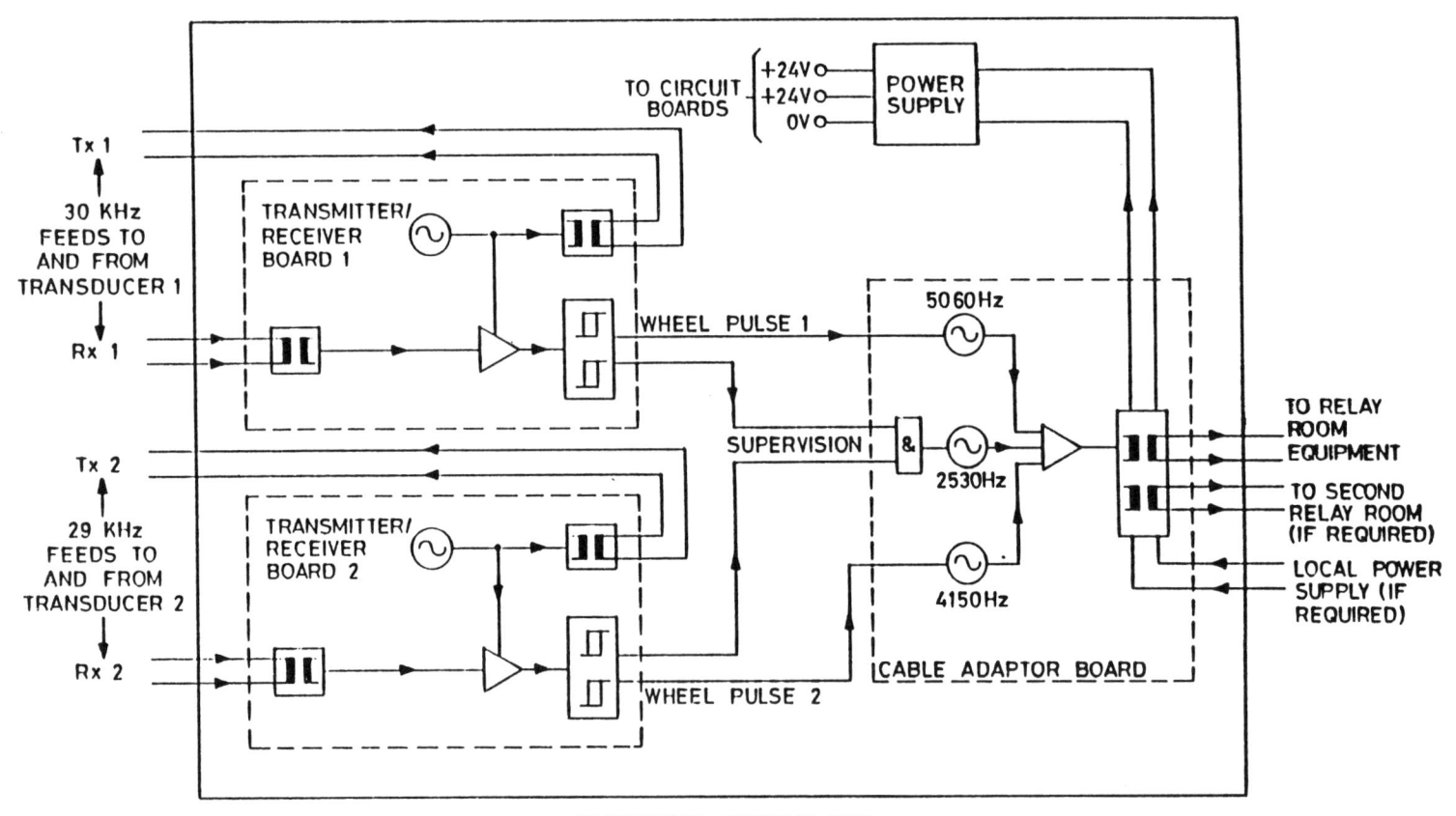

Each transmit or receive coil has a 2-core cable sealed into it for connection to the trackside electronic equipment. The equipment is housed in a special junction box mounted on a short post close to the track. The electronic components are mounted on four printed circuit boards (see **Fig. 5.7**) contained within the box.

Each transmitter/receiver board generates a frequency modulated 29 kHz or 30 kHz signal to one of the transmit coils, the actual frequency being determined by the impedance of the transmit coil itself. The signal from the corresponding receive coil is amplified and its phase compared with the transmitted signal. With no wheel present, the signals are in phase and a voltage is passed to the cable adaptor board which in turn transmits 5,060 Hz (for rail transducer 1) or 4,150 Hz (for rail transducer 2) to the evaluator. With a wheel present, the signals are out of phase, the voltage passed to the cable adaptor board drops to zero, and no signal is transmitted to the evaluator.

The cable adaptor board also transmits a 2,530 Hz signal to the evaluator when both transmitter/receiver boards indicate that they are functioning correctly. This signal is likely to be omitted from future versions of the equipment as the manufacturer considers it to have been rendered superfluous by the latest design developments.

All three frequencies transmitted to the evaluator are carried on a single cable pair (normally an unloaded telecommunications cable), which also carries the DC supply voltage for the trackside equipment in the opposite direction. Where the loop resistance is excessive, a local supply voltage must be provided. This situation occurs where the evaluator is more than 4 km from the trackside equipment (assuming that standard 0.9 mm telecommunications conductors are used). For sections of track in excess of 20 km, a speechband version of the equipment, using frequencies of 2,040 Hz and 2,520 Hz, has been developed.

Fig. 5.9 shows the principal signals generated within the detection equipment and associated electronics.

RELAY ROOM EQUIPMENT

Fig. 5.8 shows a general overview of the evaluator. After cable matching and amplification, the 5,060 Hz and 4,150 Hz inputs are used to produce detected wheel pulses as shown in **Fig. 5.9**.

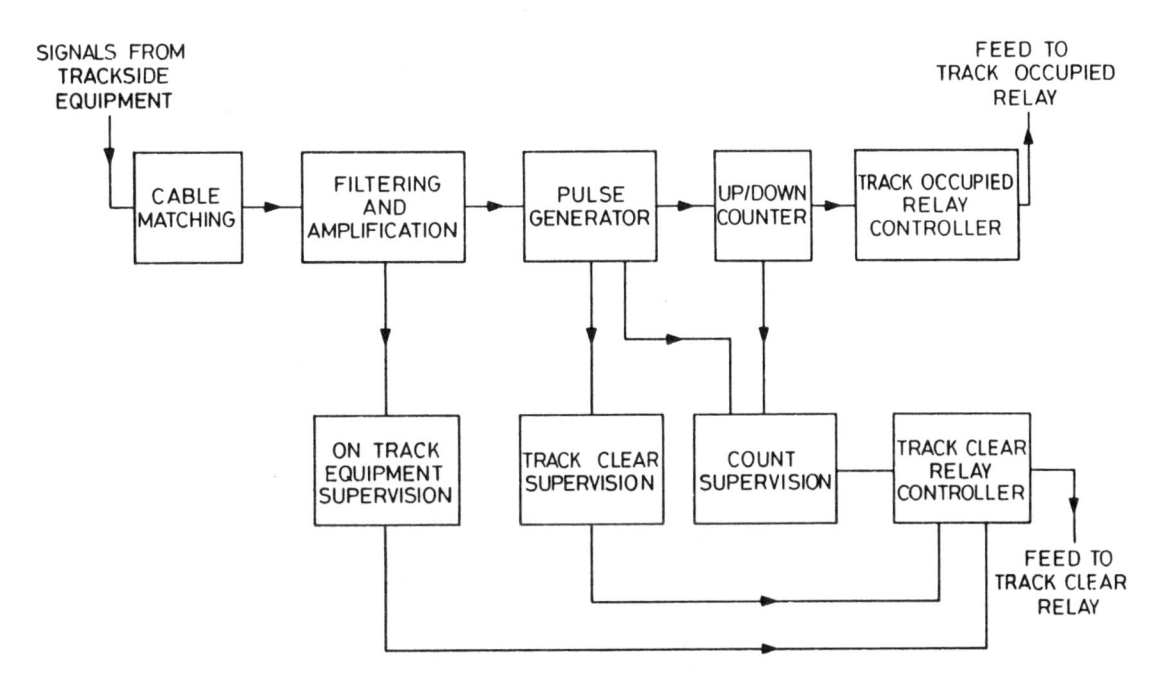

Fig. 5.8 Schematic diagram of axle counter evaluation unit

From these, 'in' and 'out' pulses are generated to count the movement of wheels into and out of the track section. An 'in' pulse is generated when a wheel arrives at transducer 2 with transducer 1 already showing the presence of the wheel. An 'out' pulse is generated when a wheel leaves transducer 2 with transducer 1 showing its presence. These pulses are used, respectively, to increment and decrement the up/down counter shown in **Fig. 5.8**.

Although the original signals from the transducers are inherently fail-safe insofar as the presence of a signal indicates the absence of a wheel, when the signals are processed to generate the 'in' and 'out' pulses in order to detect the actual passage of a train, this fail-safe feature is lost. Therefore a redundancy technique is used to achieve safety by way of a count supervision circuit.

Another supervisory circuit monitors the 2,530 Hz signal from the trackside electronic junction box, and yet another ensures that as soon as counting commences, the 'track clear' status is lost, and is not restored until the total axle count passes from one to zero.

The output from the evaluation unit operates non-safety relays with down-proving, to indicate track occupied and track clear. To drive a track repeat relay (TPR) in a manner which conforms to British safety principles, it is necessary to provide an interface relay circuit between the internal relays and the TPR. Thereafter the TPR forms part of the interlocking system in the same way as for a track circuit, and the indications given to the signalman are the same as for a normal track section.

Following a failure or other incident affecting normal operation, it is necessary to reset the counter to zero and bypass some of the supervision circuits to restore normal operation. As this negates the safety provided by the normal sequence of operation, the reset facility must only be used when the track section is known to be clear of all vehicles. The responsibility for operating the reset facility rests with the technician, although the information regarding the

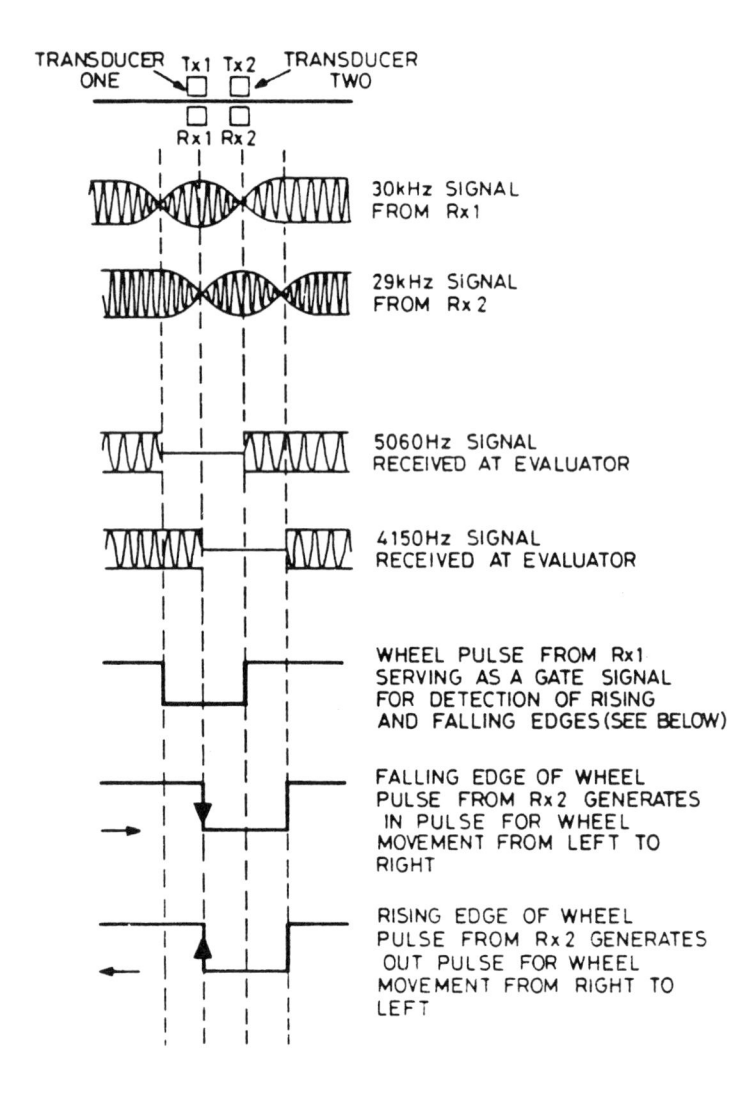

Fig. 5.9 Principal signals in the axle counter system

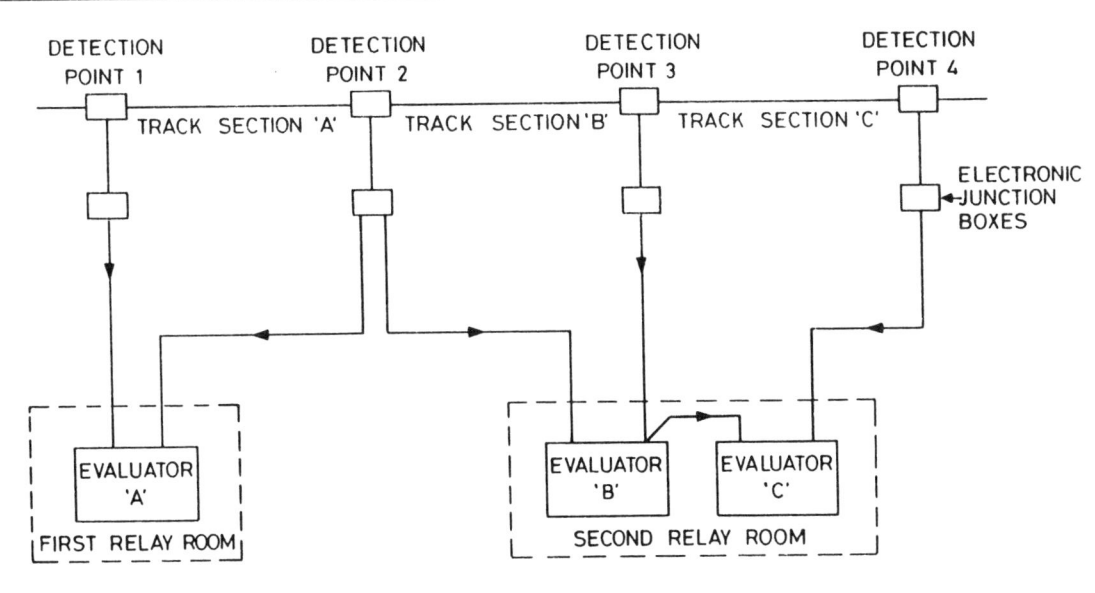

Fig. 5.10 Axle counters in a multi-section configuration

occupancy of the section comes from the signalman. In keeping with railway policy, the signalman has no facility for resetting the counter himself.

MULTI-SECTION AND DEAD-END TRACK
Where two or more axle counters are arranged consecutively on the same track, one set of detection equipment can be used to serve the evaluators of adjacent sections. This can be achieved by one of two methods, depending on whether the evaluators are adjacent or in different relay rooms. Both methods are illustrated in **Fig. 5.10**.

Detection location 2 is used by evaluators A and B. As these are in different relay rooms, separate cables are required, and a second set of output terminals on the electronic junction box is used to feed the second evaluator.

Detection location 3 is used by evaluators B and C. Because these are in the same relay room, a single cable from the detection location suffices, with local wiring between the two evaluators.

For dead-end lines, only one detection location is required. For the system to function, it is then necessary to simulate the second detection location by means of interconnections on the evaluator terminal board.

SWITCH AND CROSSING LAYOUTS
Where an axle counter section includes switches and crossings, one set of detection location equipment is required for each position where insulated joints would separate the equivalent track circuit from its neighbours or from a non-track circuited line.

By providing various additional printed circuit boards, up to eight detection locations can be monitored by a single evaluator.

RELATIVE MERITS OF TRACK CIRCUITS AND AXLE COUNTERS

The first considerations when comparing axle counters and track circuits must be safety and reliability. Exact comparisons can only be made on the basis of individual installations, but in general both systems are equally acceptable on these grounds.

Axle counters have the following advantages:

- They are able to operate with low or zero ballast resistance.
- There are no problems with rusty rails or lightweight vehicles.
- They avoid the need for multiple section track circuits and associated location cupboards and power supplies.
- There is no need for insulated joints or bonds.

They do, however, suffer from the following disadvantages:

- They are relatively expensive, except as a replacement for multi-section track circuits.
- A special power supply must be provided, usually with a battery to avoid the need to reset after brief supply interruptions.
- Cabling is required from each detection location to the evaluator.
- Care is needed to avoid incorrect operation of the reset plunger.
- There is no broken rail detection.
- Some small wheels, such as those on the skates used to remove damaged vehicles, can cause problems of miscounting.

THE USE OF AXLE COUNTERS IN THE UNITED KINGDOM

At the time of writing, there are some 40 axle counters on British Railways, with more included under several proposed schemes.

Nine installations arise through zero, or near-zero, ballast resistance — one because of reinforcing rods in an early design of concrete sleeper, two where the civil engineer is not prepared to have insulated joints within a flat crossing on 200 km/h lines, and six where rails are bolted through to bridge girders (including two on the Forth Bridge).

Most of the other installations are on long sections where the cost of providing multiple track circuits exceeds that of an axle counter. Such a decision depends not only upon the cost of the equipment, but also upon civil engineering costs for joints, maintenance costs, and whether or not a suitable cable route exists.

The remaining few axle counters have been installed in tunnels, such as the Severn Tunnel, where the maintenance costs for track circuits are high, and their reliability is unacceptably low.

Research is currently under way to link detection location equipment directly to SSI, which would carry out the evaluation, thereby removing the need for dedicated cabling and separate evaluator units.

Transponders

INTRODUCTION

Transponders are being used increasingly as part of signalling systems. They are best known in applications for passing information between track and train, but they can also be used to detect the passage of trains.

A transponder system comprises two basic components. One is the transponder itself, which is normally passive, requiring no external power supply or cabling. The other is the interrogator, which obtains coded information from the transponder. For most applications it is more cost-effective to mount the transponder on the track and the interrogator on the locomotive, although they can be mounted the other way round.

Fig. 5.11 shows a typical transponder system incorporated into a signalling system, with a radio link between the locomotives and the signalling control point.

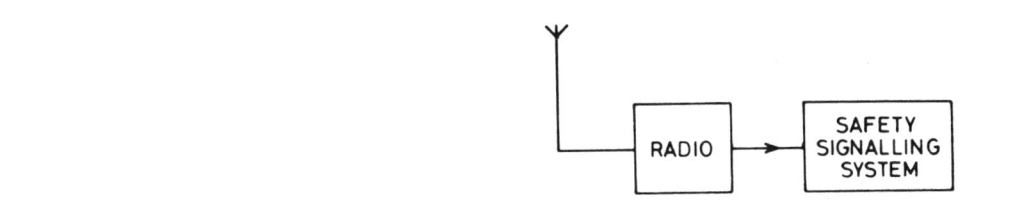

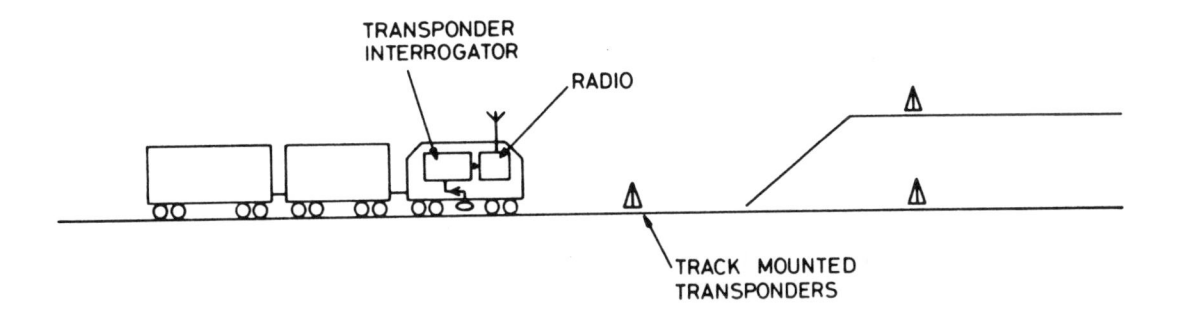

Fig. 5.11 Typical transponder system for train detection

PRINCIPLES OF OPERATION

Fig. 5.12 is a block diagram showing the track transponder and the train-borne interrogator which might be used as the basis of a train detection system.

The locomotive is fitted with an oscillator which feeds power into a coil mounted under the locomotive. The field produced by this coil provides power for the electronic circuits in each track-mounted transponder as the train passes over.

The transponder is a robust sealed unit fixed to the track. It contains a power receiving coil tuned to the frequency of the energising field from the locomotive. A rectifier converts energy from this coil into a power supply for the electronic circuits of the transponder. As a locomotive approaches the transponder, a code generator circuit in the transponder produces a digital code repeatedly until power is lost as the locomotive moves past the transponder. The digital code modulates a transmitter, which in turn sends the modulated signal to the locomotive by means of a transmitting coil in the transponder.

The locomotive is equipped with a receiver coil which detects the signal from the transmitter coil in the transponder. Circuits in the interrogator then demodulate and decode the signal to recover the digital codes and assess their validity. If accepted as being valid, the data content of the message is then passed on to other train-borne equipment.

The frequencies at which transponders are energised, and at which they reply, vary greatly from one proprietary system to another. The reply frequency may be harmonically related to the energising frequency, or the frequencies may be in different bands entirely. For instance, one such system uses inductive energisation at 69 kHz with a radio frequency response in the form of a bit message transmitted at 56 kbits/s on a 10 MHz carrier.

The interrogator must communicate with the transponder for a sufficient time to allow a locomotive passing at maximum speed to receive several complete messages. The data rates for transmission to the interrogator therefore have to be high, and the design of the coils must be such that coupling occurs over an adequate distance along the track. However, the range of communication must be limited to prevent locomotives detecting transponders on adjacent tracks.

The digital code programmed into each transponder is unique. For train detection applications, the data content of the code represents a position on the railway. A locomotive equipped with an interrogator will therefore know the identity of each transponder it passes. These identity codes, together with the identity code of the locomotive, can be passed back to the signalling control centre, for instance by radio, to enable a train location function to be performed.

In addition to the data, the code also contains checking information to enable the interrogator to carry out error detection on the received message and, if required, to correct those errors. By providing such coding security, it is possible to demonstrate that the probability of a valid code being produced unintentionally is acceptably small. The security coding must protect against failures within the transponder, the interrogator, and the track-to-train and the train-to-control centre communication links.

Transponders and interrogators can fail in such a way that the passage of a train over a transponder is not detected at all. The design of a signalling system which uses transponders for train location purposes must therefore be able to cope in a safe manner with failures of this type.

The direction of travel of a train cannot be safely deduced from the code derived from a single transponder of the type described.

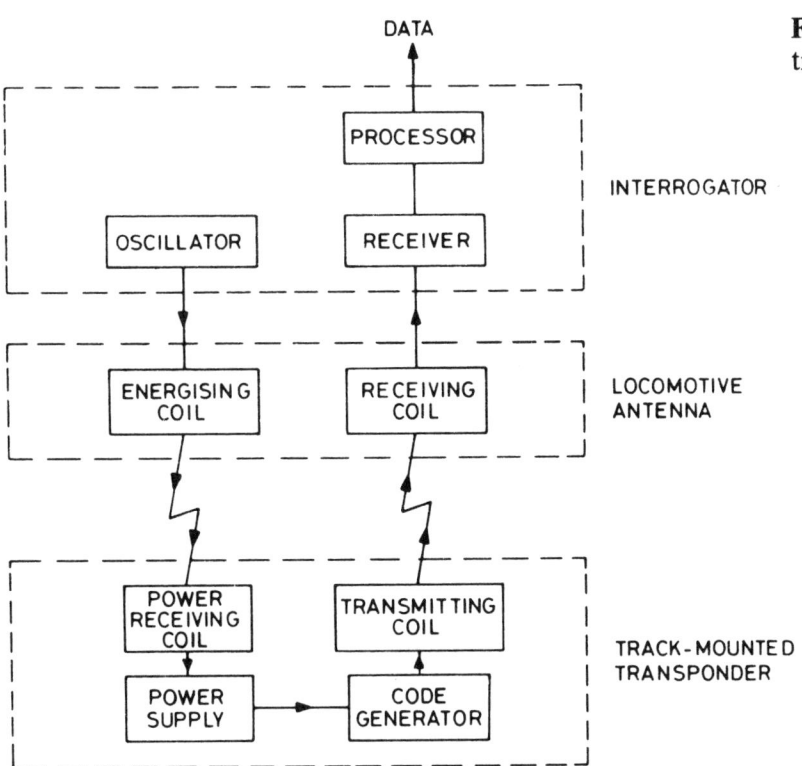

Fig. 5.12 Block diagram showing the track transponder and train-borne interrogator

THE PRACTICAL APPLICATION OF TRANSPONDERS FOR TRAIN DETECTION

Transponders are used for train detection purposes in block signalling systems, their prime purpose being to establish that a train has passed through a block section. A track-mounted transponder is located maximum train length beyond the overlap point for the home signal or stop board which marks the end of the section. When a train reports that it has passed over this transponder, it can be deduced that the section is clear, and that the next train may therefore be given authority to proceed. The safety of this method of working assumes that the train is complete and in one portion, that only one train had authority to enter the section and that no train enters a section without authority.

It is clearly vital that safety measures are incorporated into the signalling systems and/or operating arrangements to ensure that protection is afforded against any of these eventualities.

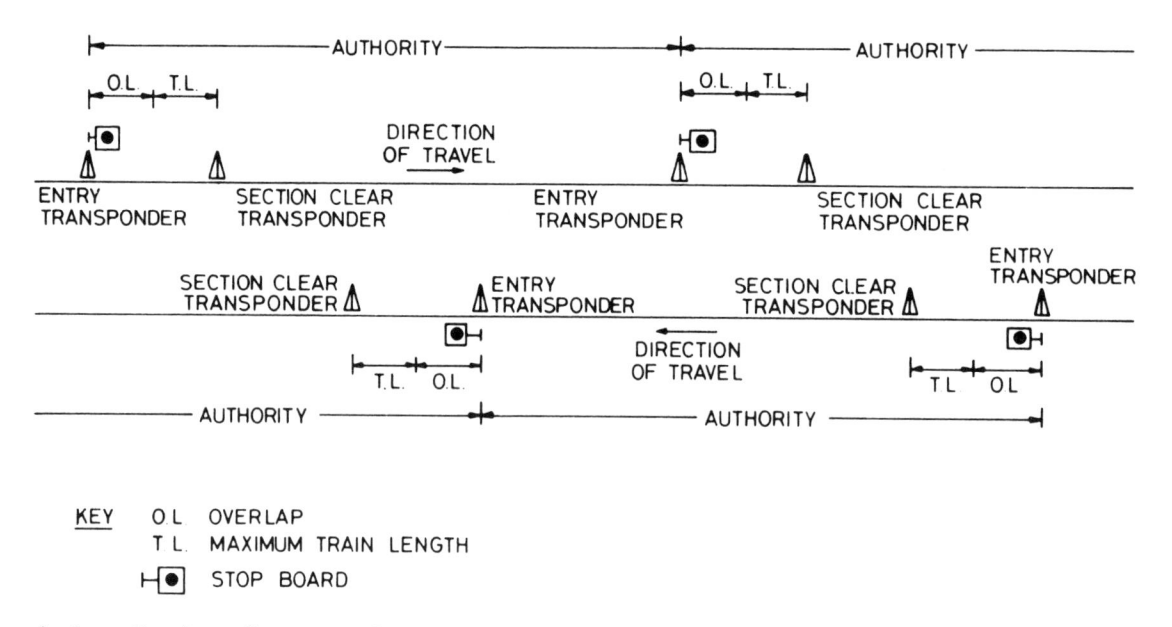

Fig. 5.13 Practical application of transponders on a simple double line section of track

Fig. 5.13 shows the application of transponders to a simple block system on double lines, with stop boards marking the entrance to each section of track. An additional transponder is shown positioned at the entrance to each block section to inform the control centre that a train has entered the section.

Fig. 5.14 shows a similar arrangement for a single line with passing loops. In the latter case, separate transponders are provided to prove that the single line section is clear and, for following trains, that the loop is clear.

Once locomotives are fitted with interrogators, it becomes an attractive proposition to use the transponders for purposes other than train detection, for instance to apply the brakes automatically in the event of a train entering a section without authority.

THE USE OF TRANSPONDERS FOR TRAIN DETECTION IN THE UNITED KINGDOM

At the time of writing, transponders have not been used as a means of train detection in the UK. Radio electronic token block (RETB) has been installed overseas using transponders for train detection, but RETB in Britain relies upon train position information being manually entered into the signalling interlocking system, on the basis of verbal radio messages from drivers.

British Railways does, however, use train-mounted transponders on coal wagons, together with track interrogators, to monitor the passage of wagons between collieries and power stations.

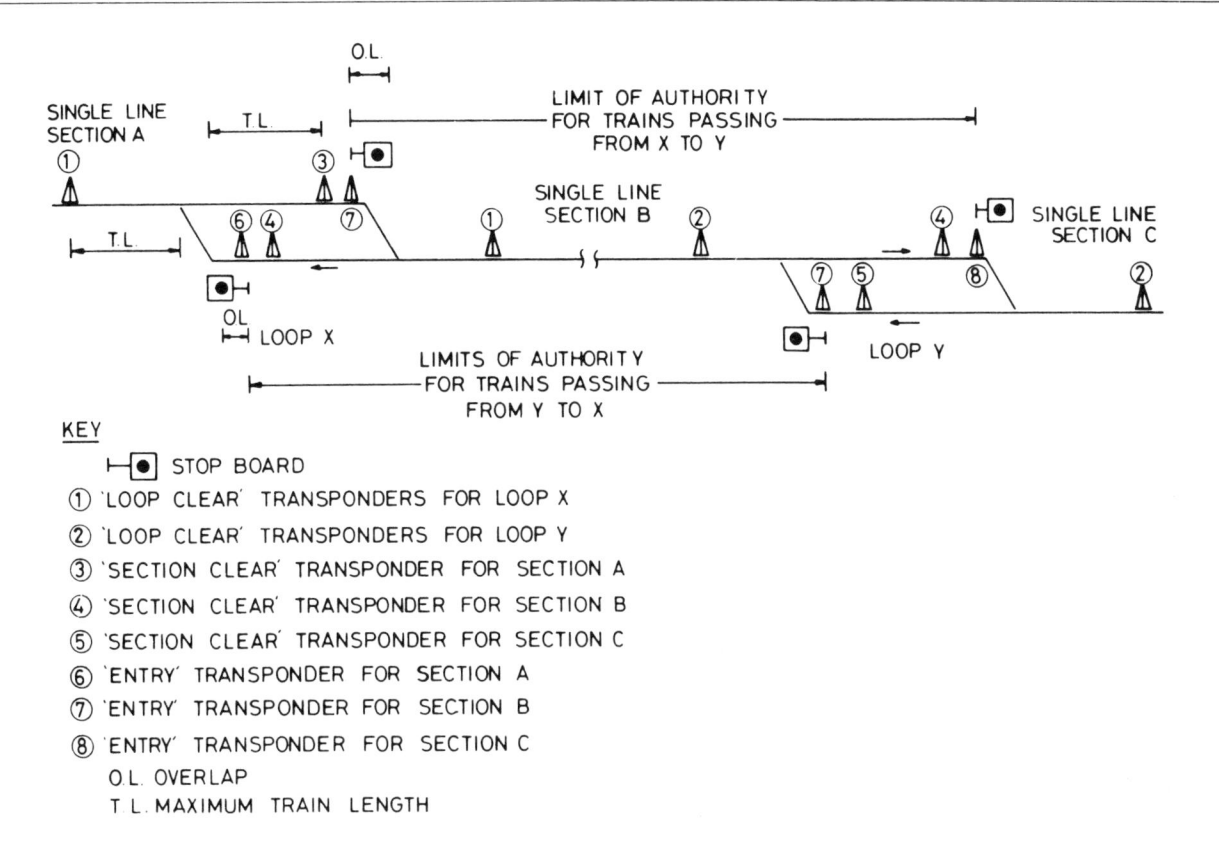

Fig. 5.14 Practical application of transponders on simple single line with passing loops

CHAPTER SIX

Level Crossings

Introduction

Although it is most desirable for the intersections between roads and railways to be in the form of bridges, either over or under the railway, economy of construction usually dictated, especially in areas where the terrain was flat, that roads and railways should cross on the level. These intersections became known as grade or level crossings. Because it is a legal requirement for most railways to be fenced, such level crossings had to be provided with a form of barrier in order to make the fencing continuous. Heavy timber gates provided this protection and were operated by an attendant. The gates were arranged such that the roadway was normally closed, the gates being opened for the passage of each road vehicle. As roads became busier, the Department of Transport allowed amendments to individual Acts concerning the railways to permit the gates to remain open for the roadway and only to be closed for the passage of trains. Usually these level crossings were protected by worked stop and distant railway signals, although in some cases only distant signals were provided. The gates were then so arranged that when they were open to road traffic, a red target fixed to the gate would be displayed along the railway to act as a stop signal. In some instances only the target was provided, there being no other signals. Even today, examples of such level crossings still exist.

In 1954, Parliament empowered the Railways Board to install lifting barriers in place of gates. These barriers were usually provided with visual and audible road signals, and were fitted with skirts to represent a gate, although on some installations on little used roads, the road signals were omitted. An attendant was still required. The first such example was at Warthill on the now closed direct line between York and Hull. In 1957, an Act was passed which permitted the installation of barriers operated automatically by the passage of trains; this was a major change as it allowed economy of operation by dispensing with the attendant. Furthermore, the principle of continuous fencing was now modified so that the barrier only closed half of the road. This allowed any trapped road vehicle to escape. These crossings were provided with audible and visual road signals, the operation of the crossing being indicated to a supervising signalbox, to where a telephone was provided for the use of the public. The first such automatic half barrier (AHB) was installed in 1961 on the now closed line at Spath between Uttoxeter and Leek. The Act also allowed for similar crossings (but without barriers) to be installed on lines where rail traffic was infrequent and the rail speeds were low. These are now known as automatic open crossings locally monitored (AOCL). Confirmation that the road signals are operating is conveyed to the driver of the train by means of a white flashing indication, lack of which is an instruction to stop short of the crossing. The first installation is believed to have been at Yafforth on the Wensleydale branch in 1963.

In January 1968 there was a serious accident on an AHB crossing at Hixon in Staffordshire, followed shortly after by an accident at Beckingham Trent Road in Nottinghamshire. This resulted in alterations to the controls of AHBs including the addition of a yellow road signal and an 'another train coming' road signal. This made the associated barrier controls very much more complex and considerably increased the cost of such installations.

In an attempt to simplify level crossing protection, a BR/Department of Transport working party was set up which reported in 1978. This recommended that AHB crossings should be modified by the removal of the another train coming road signal (see **Fig. 6.1**) and that a new type of crossing should be installed to be known as an automatic open crossing remotely monitored (AOCR) (see **Fig. 6.2**).

LEVEL CROSSINGS 135

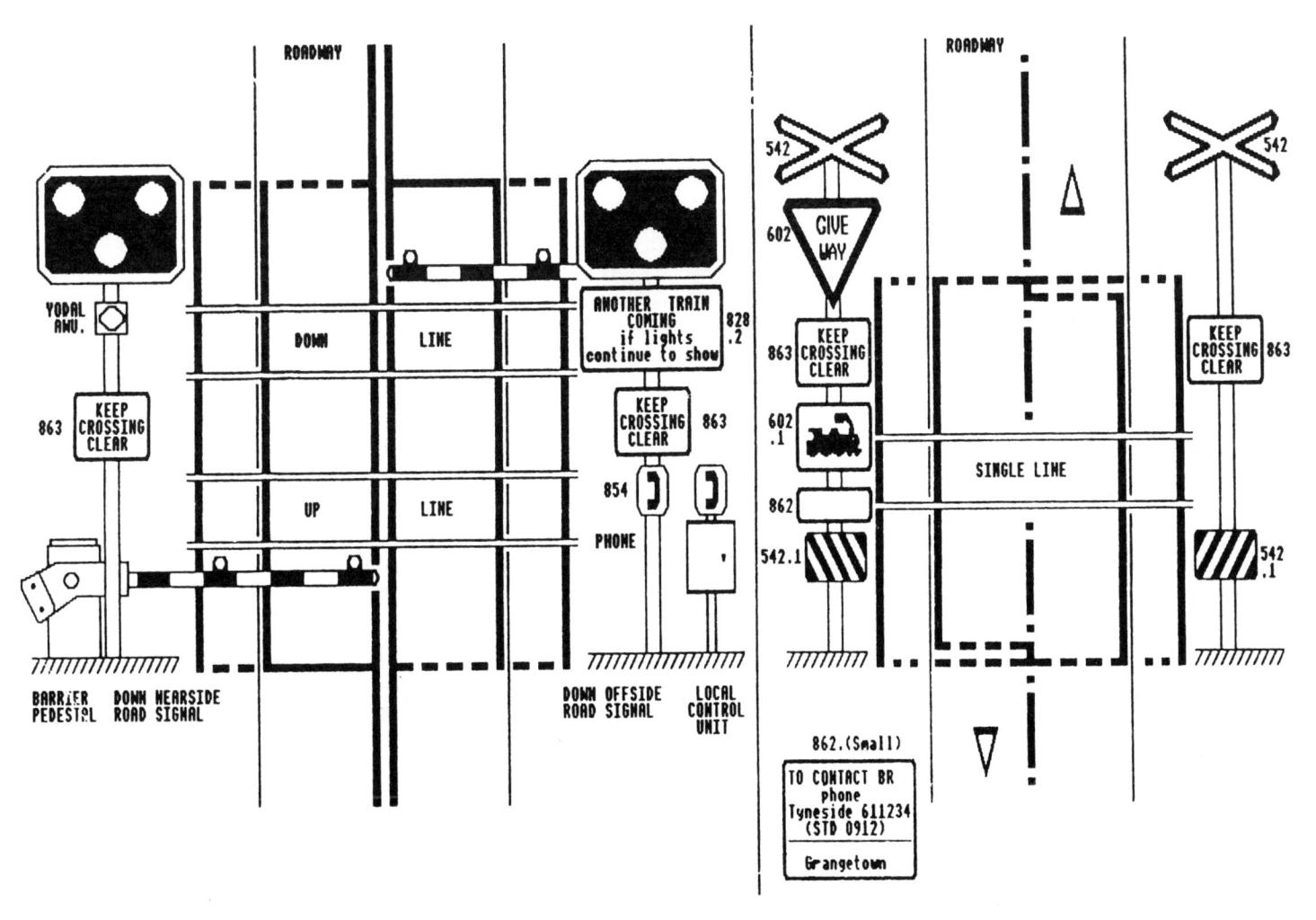

Fig. 6.1 Automatic half barrier crossing — layout

Fig. 6.2 Open crossing with no controls — layout

136 LEVEL CROSSINGS

Fig. 6.3 Automatic open crossing locally monitored — layout

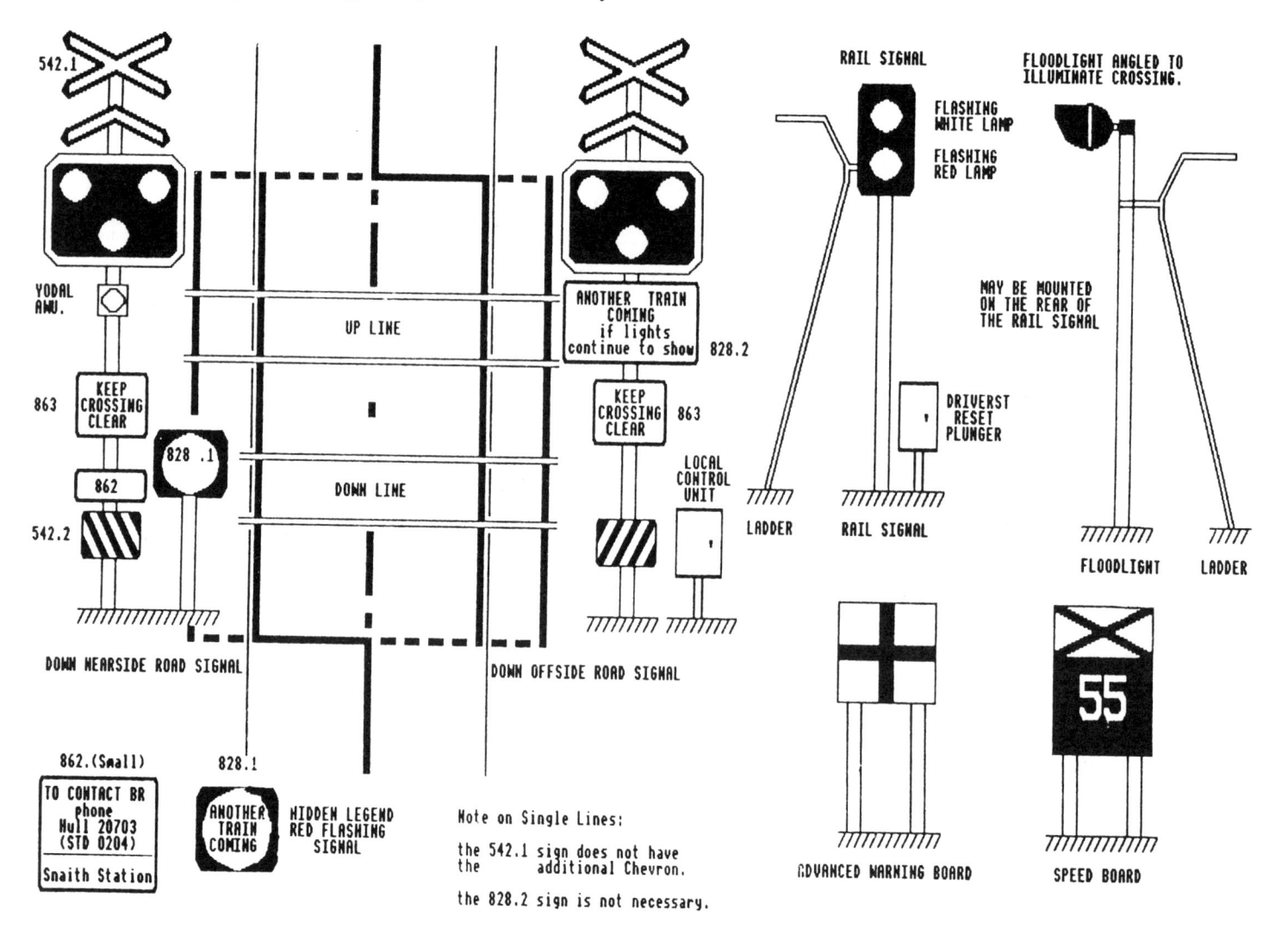

This was essentially an AHB but without barriers, where the protection was by the road users' obedience of audible and visual road signals. Many AOCRs were installed, the first being at Naas on the line between Gloucester and Newport. However, in July 1986 there was a serious accident at Lockington, Humberside at an AOCR crossing. This resulted in almost all of this type of crossing being converted to AHB operation, and it is now unlikely that any other AOCRs will be installed. The principle of the AOCL (see **Fig. 6.3**) has been further amended, so that the flashing white rail signal is now supplemented with a flashing red rail signal. This is a marker to remind the driver of his responsibility to monitor the crossing. The red aspect flashes to differentiate it from an absolute stop signal. AOCLs are now widely installed on lines where simplified signalling infrastructure exists. A further type of crossing is under development, to be known as an automatic half barrier locally monitored by the train driver (ABCL), and is for use where the road conditions make an AOCL unsuitable.

The principles of manned barriers (MCB) (see **Fig. 6.4**) have remained largely unchanged, except for the addition of the yellow road signal, and the removal of the skirts where local conditions allow. The main development is the extensive use of closed circuit television to supervise such crossings from a remote point (MCB/CCTV), thus allowing the operation of several crossings from one signalbox. It is usual for these crossings to be raised automatically by the passage of trains, and in some instances to be lowered automatically by the approach of a train. In all cases the signalman remains responsible for ensuring that the crossing is unobstructed before allowing the passage of rail traffic. Full manual operation can be selected at each crossing if so desired. Another development of the manned barrier is the trainman-operated type (TOB), which is used where rural railways cross major roads and an automatic crossing is not suitable. The train driver is responsible for lowering the barriers by the operation of a plunger or other device which can be reached from the cab. A flashing rail signal is provided, similar to an AOCL, to indicate that the road signals are operating and that the barriers are fully lowered. The driver is then responsible for ensuring that the crossing is clear before proceeding. The barriers are raised automatically after the passage of the train. A special signal is located in advance of the crossing to indicate to the driver that the barriers have actually raised. If they have not, then the train must be stopped and the barriers raised manually.

There are other types of modern crossing. Red/green miniature warning lights (MWL) (see **Fig. 6.28**) are used at footpath crossings or minor level crossings, which can either be fitted with gates or user-operated lifting barriers (rural barriers). The lights are automatically operated by the passage of trains. This type of protection is also used at some barrow crossings. There are also on call barrier crossings (OCB), where the user requests the supervising signalman to raise the barriers. These then lower automatically after a determined period. Such barriers are not fitted with road signals, but their lowering is preceded by an audible warning. There are only a few remaining examples of this type of crossing and it is unlikely that any more will be installed.

On very lightly used lines, uncontrolled open crossings (OC) are permitted (see **Fig. 6.2**). These are protected by road signs requiring the road user to give way to trains, which are limited to a speed of approach not exceeding 16 km/h. Such crossings are not allowed where the daily number of road vehicles exceeds 200. The road user should be able to see the approach of trains clearly. Unless the train driver has an unobstructed view of the road and approaching traffic, the train must be stopped before proceeding over the crossing.

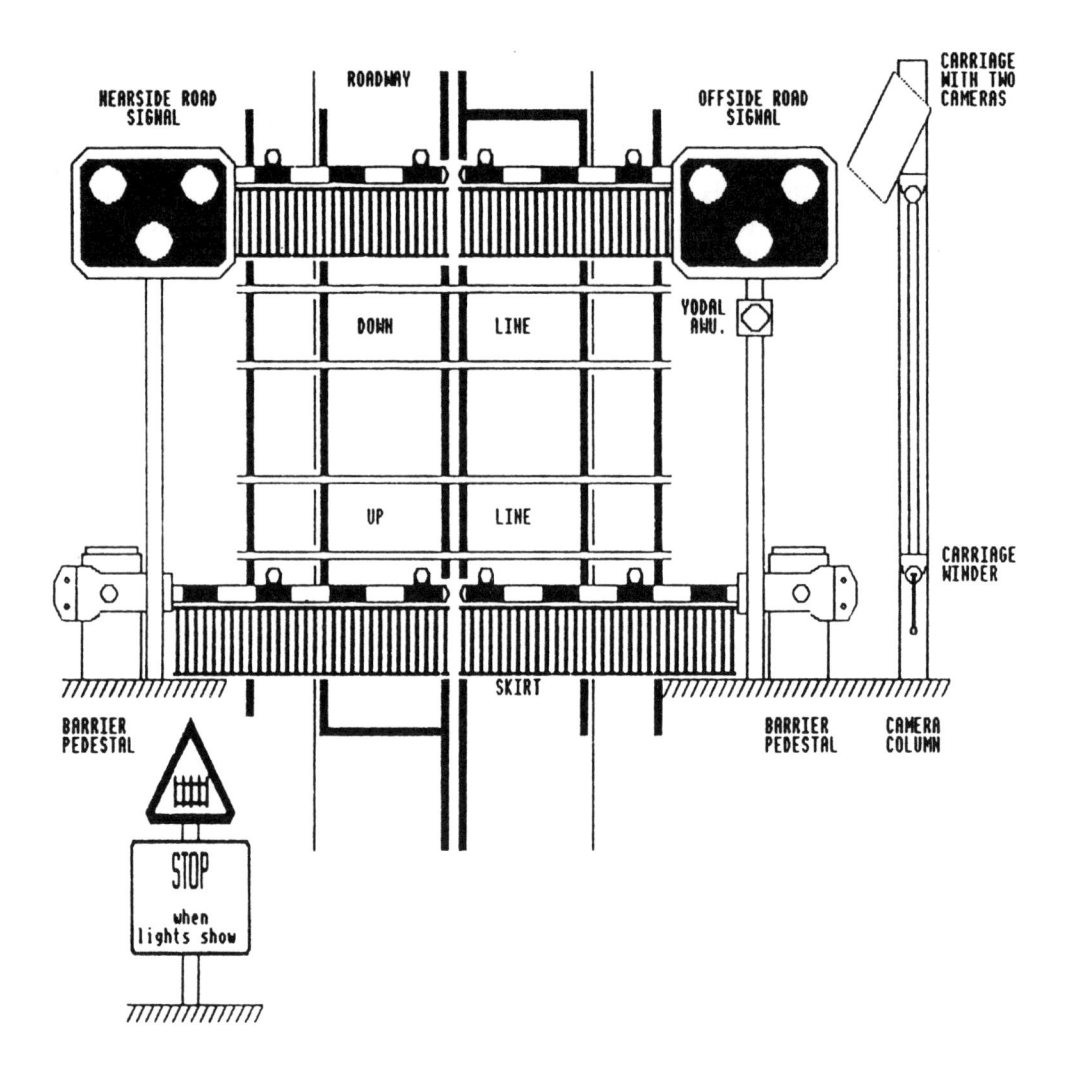

Fig. 6.4 CCTV monitored remote barrier crossing — layout

Automatic Half Barriers

GENERAL DESCRIPTION

Automatically operated crossings must give a warning for at least 27 s before the arrival of a train. Of all the trains using the line, 50% must arrive at the crossing within 50 s and 95% within 75 s. Thus on a double line, the operation of the crossing is initiated at 39 s from the crossing. This allows a delay of 10 s for the second train situation, 3 s for the yellow road signals, 7 s for the red road signals, 5 s for the barriers to lower and 12 s for the fastest train to arrive at the crossing. A further 2 s is added for relay operation time. Furthermore, if the roadway is wide, then 1 s is added for every 3 m of width exceeding 15 m. There is no specific signal to indicate to the rail driver that the crossing has correctly closed to the roadway.

One barrier is provided on each nearside road approach to the level crossing (see **Fig. 6.1**). These barriers will normally close off half the road, except in the case of very narrow roads, where less than the half width of the road may be closed. Each barrier (known as a boom) is conspicuously marked with alternate red and white stripes and is fitted with two red boom lamps showing a light in both directions. These are illuminated at all times unless the boom is fully raised. The booms must be capable of being freely lifted by hand. Each road approach is provided with two road signals, that on the nearside being known as the primary signal and that on the offside as the secondary signal. Both signals carry a reflective road sign '*keep crossing clear*' and the secondary signal is fitted with an additional road sign '*if the lights continue to flash another train is coming*'. Each signal consists of two red lamps and a single yellow lamp, arranged on a black backboard surrounded by a conspicuous white stripe. Each primary signal is fitted with an audible signal for warning pedestrian users of the operation of the crossing.

When the level crossing sequence is initiated by the approach of a train, the yellow lamp shows steadily for a period of 3 s, followed by the red lamps flashing alternately at a rate of 80 flashes/min. After a further 4–8 s the booms commence to descend. This takes 6–8 s and the booms then remain down for 12 s before the arrival of a train travelling at line speed. The red flashing signals continue to operate all the time that the barriers are down. The audible warning signal (usually a yodalarm) begins to sound when the yellow light shows, and continues to sound throughout the crossing operation. The loudness of the audible signals can be adjusted by means of a time clock to be softened during the night hours (2330 to 0730). After the train has passed, and the barriers are able to rise, the red road signals will extinguish once the booms have risen to 42 degrees, but if the booms fail to complete their rise in 7 s, the red road signals are re-initiated until the rising of the barriers is complete. The arrangement is such that should a further train approach the crossing, then the road must either not re-open, or if re-opening has commenced, the road must remain open for 10 s before the crossing sequence recommences. In the event of the crossing not re-opening before the passage of a second train, the audible signal doubles its frequency until the second train has passed and the barriers are able to rise. There are no limitations on the type of road or the traffic, except that the crossing must be situated in such a place that the traffic will flow freely, and the profile of the roadway must be such that there is no risk of a long wheelbase vehicle grounding on the crossing surface.

For rail traffic the maximum line speed must not exceed 160 km/h and there must be no more than two running lines. The operation of the crossing is arranged such that if both the red lamps fail in one road signal, then the booms will operate on an extended sequence, lowering immediately when the yellow sequence has been completed and not rising again until the lamps in that signal have been restored. Furthermore, the booms will not rise after the passage of a train unless both booms have fully lowered. The crossing is provided with an emergency telephone to the supervising signalbox, which is arranged so that it will overcall any other telephone conversation that the signalman may be engaged upon. The fact that a call has been made from the crossing will be stored at all times that the signalbox is unmanned. The normal condition of the crossing is repeated to the supervising signalbox in the form of an indication showing *barriers raised*; if the barriers are operating, this is replaced by the indication *barriers working*, whilst should the barriers remain lowered for more than a specified period, then the indicator will show *barriers failed* and the signalman will be alerted by an audible alarm. The crossing operation is supported by a high capacity battery; should the charging of the battery cease, this is alarmed to the signalbox.

CROSSING INITIATION

A track circuit is provided commencing at a distance which will give the prescribed 39 s, between a train initiating or striking in at line speed and arriving at the crossing. Operation of this track circuit is reinforced by a treadle at the running-on end to guard against the maloperation of the track circuit by lightweight vehicles. It has been found desirable to provide a short track circuit actually over the crossing to guard against the crossing being wrongly initiated by a track-laying road vehicle occupying the strike-in track circuit. Although this track circuit does not form part of the crossing initiation, its occupation will prevent the crossing re-opening after the passage of a train and it must be included in any signal controls. The track circuit may be omitted in urban areas where agricultural vehicles are unlikely. A further track circuit is provided at the running-out side of the crossing.

It is desirable for bidirectional controls to be applied, as this obviates the need for an attendant during single line working, but the feature may be omitted if the signalling arrangement is likely to require trains to stand on the running-out track circuit. The re-opening of the crossing is initiated by means of a treadle at the running-out side, separate strike-out treadles being provided at each side of the crossing for bidirectional running. Controls are applied to guard against the re-initiation of the crossing by a momentary wrong-side clearance of the running-out track circuit. In the event of an intermittent right-side failure of a strike-in track circuit, the crossing will be initiated; however once the failure clears, the crossing will reset after a time period without other intervention. Should the running-out track circuit be occupied for more than 2 min by a slow train, the crossing will be re-initiated, re-opening only when that track circuit has cleared.

AHB — DESCRIPTION OF OPERATION

Consider an up train approaching with the crossing in automatic operation (see **Figs. 6.5–6.12**).

Up Train Strikes In

The train occupies track circuit A, releasing A.TPSR which operates A/B.TCJR, thus operating A.SR which determines the direction of initiation.

Initiation of Sequence

A.TPSR in turn releases CON.YR, thus de-energising 1.TJR and 2.TJR. These circuits are identical and duplicated because of the characteristics of the relay. The contacts are not fail-safe, and it is therefore necessary to check the operation of both these relays for each operation. This is done by the JCR, JCSR and the TJPSR.

Minimum Opening Time

After a wait of 10 s, 1.TJR and 2.TJR will release, thus releasing the TJPSR and CON.SR, which will de-energise the HJR, CON.JPR and JCR.

Signalbox Indication

The CON.SR releases the BARR.KR, reversing the polarity of the indication line and changing the signalbox indicator from *raised* to *working*.

LEVEL CROSSINGS 141

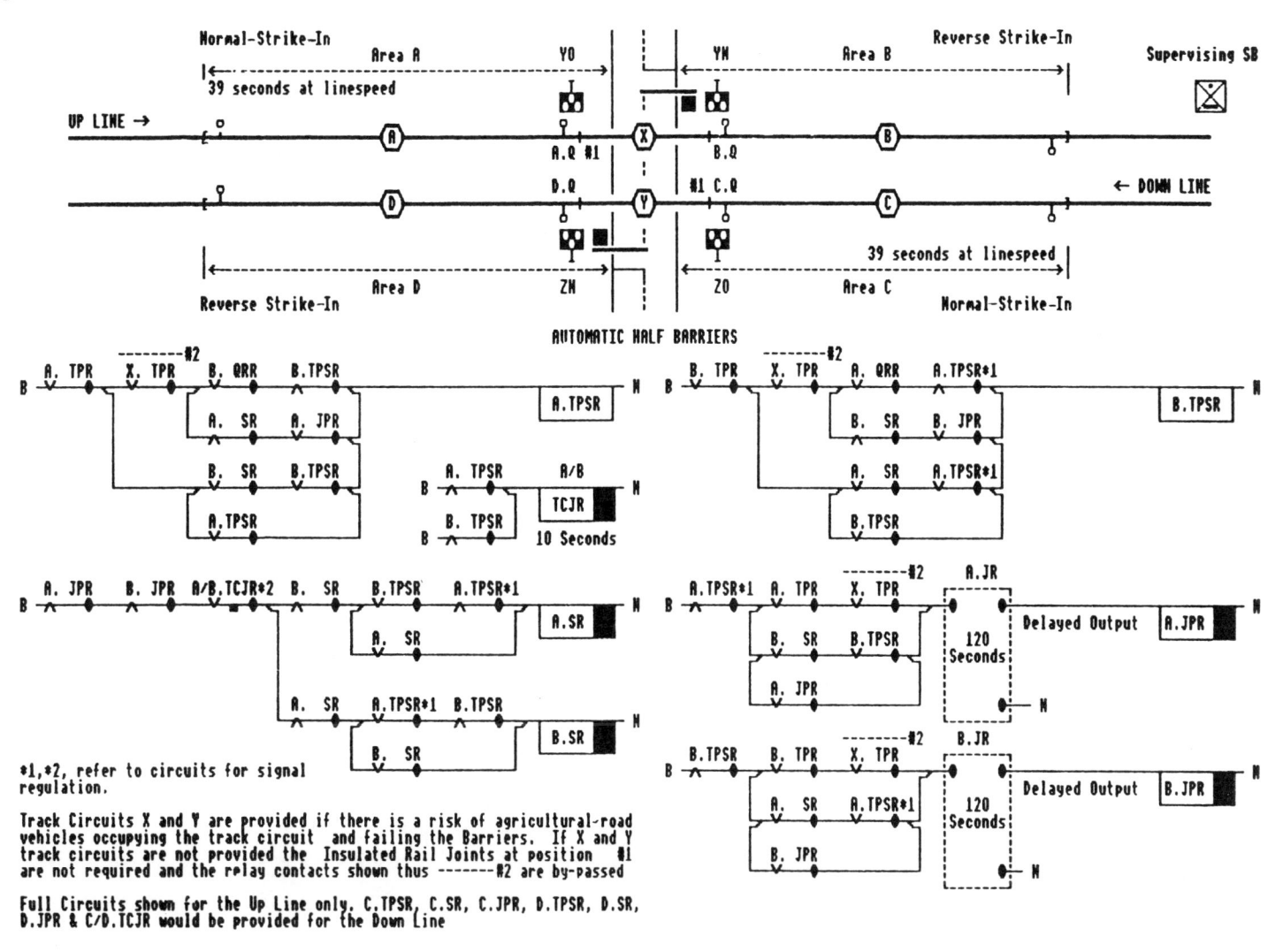

Fig. 6.5 Automatic half barrier crossing — bidirectional strike-in circuits

*1, *2, refer to circuits for signal regulation.

Track Circuits X and Y are provided if there is a risk of agricultural-road vehicles occupying the track circuit and failing the Barriers. If X and Y track circuits are not provided the Insulated Rail Joints at position #1 are not required and the relay contacts shown thus ------#2 are by-passed

Full Circuits shown for the Up Line only. C.TPSR, C.SR, C.JPR, D.TPSR, D.SR, D.JPR & C/D.TCJR would be provided for the Down Line

142   LEVEL CROSSINGS

Fig. 6.6   Automatic half barrier crossing — double line control circuits

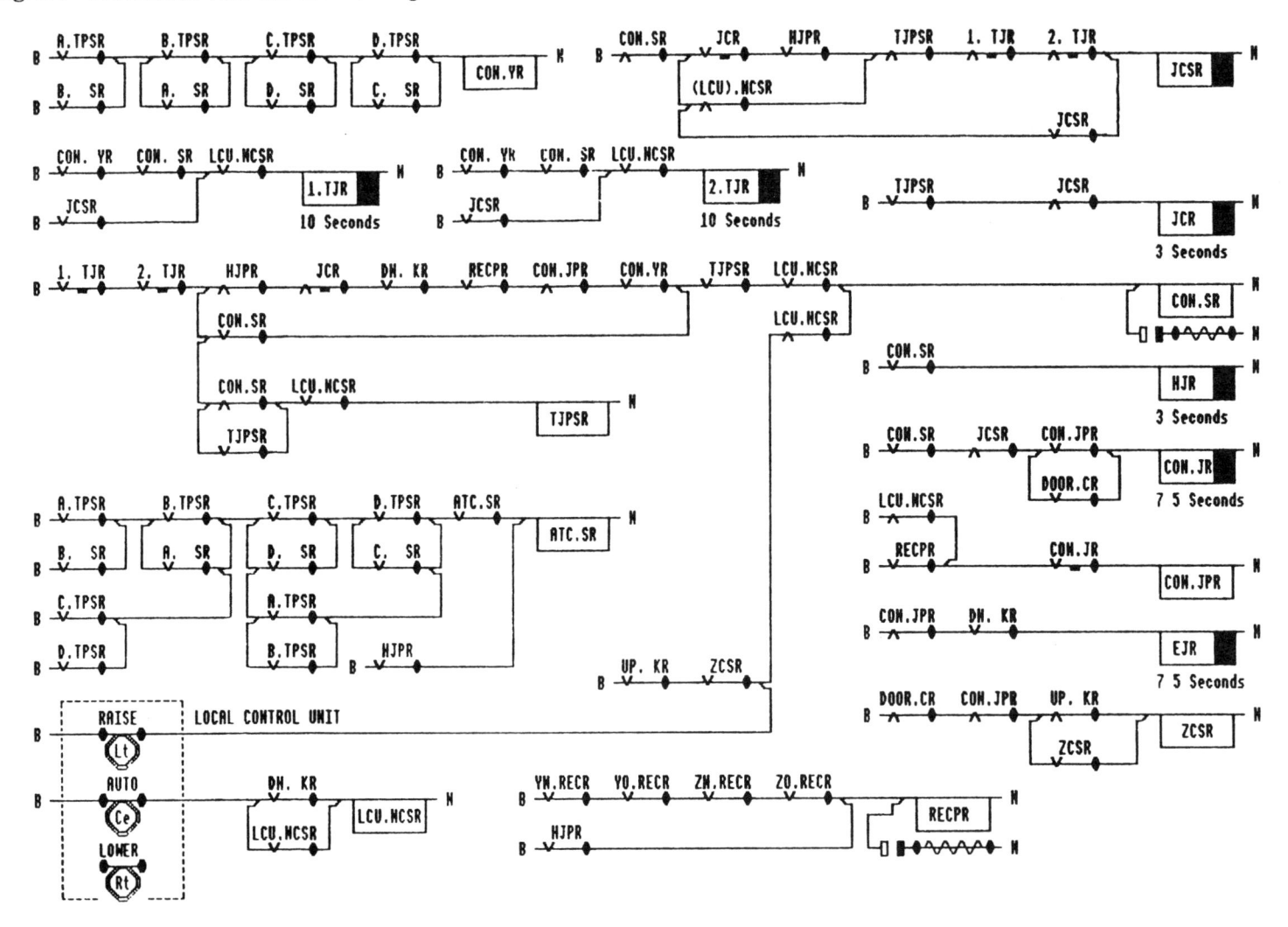

Yellow Road Signals
While the HJPR is operated, the release of the CON.SR will cause the yellow road signals to illuminate. The CON.SR will also start the audible warning.

Resetting for Another Train
With the HJPR and JCR still operated, and the TJPSR, CON.SR, and 1 and 2 TJRs released the JCSR will operate. The purpose of the JCR is to ensure the correct operation of the JCSR immediately after the release of the TJPSR. The JCSR resets 1 and 2 TJRs, which then reset the TJPSR. This ensures that if a further train strikes in, the 10 s minimum crossing open time will still apply.

Red Road Signals
After 3 s the HJR will release, followed by the HJPR, which extinguishes the yellow road signals and starts the static flasher, causing the red road signals to flash. Release of the HJPR will also de-energise the RECPR, but providing that one red lamp in each road signal is illuminated, the RECPR will be maintained by the RECR relays.

Lower Booms
After a further 4.5 s, the CON.JR will release (7.5 s) followed by the CON.JPR, which de-energises the hydraulic valves enabling the booms to commence lowering. When one boom falls below 81 degrees, the release of the UP.KR illuminates the boom lights. When both booms reach 4 degrees the DN.KR operates, setting the EJR.

Failure of a Red Road Signal
If upon the release of the HJPR, any one of the four road signals fails to illuminate, the RECPR will release, causing the CON.JPR to release without waiting for the CON.JR thus causing the booms to lower immediately after the end of the yellow road signal sequence.

Up Train Arrives
The train passes the crossing occupying X and then B track circuits. B.TPSR releases; however B.SR is prevented from operating by A.SR being operated.

Up Train Strikes Out
As the train passes over the crossing, it strikes treadle B, thus operating B.QRR. As the train clears track circuit A, and then X, with B.TPSR released, A.TPSR operates. A/B.TCJR is maintained by B.TPSR.

Normalisation of Crossing
Operation of the A.TPSR, with the A.SR operated causes the CON.YR to operate. Providing that the booms are proved down (DN.KR) and at least one red lamp is showing in each road signal (RECPR), the CON.YR operates the CON.SR. This resets the HJR and causes the release of the JCSR, which allows the operation of the CON.JR and CON.JPR.

Failure of a Red Road Signal
If any one of the four red road signals fails to illuminate, the RECPR will be released. It will then not be possible to reset the CON.SR, and thus raise the booms until the failure has been rectified.

Boom Obstructed or Fails to Lower
If either boom has failed to lower fully, the DN.KR will prevent the CON.SR resetting, thus failing the barriers.

Booms Start to Rise
The CON.JPR energises (closes) the hydraulic valves. The CON.JPR also operates the contactor relays Y.MR and Z.MR which cause the hydraulic pumps to operate, commencing the raising of the booms. As the booms reach 4 degrees above the horizontal the DN.KR releases, de-energising the EJR. The MR relays are fitted with a thermal cut out, such that if a boom fails to rise due to an obstruction, the feed to the hydraulic pump will be disconnected after a short period.

144 LEVEL CROSSINGS

Fig. 6.7 Barrier and open crossings — road signal lighting and proving circuits, audible warning circuits

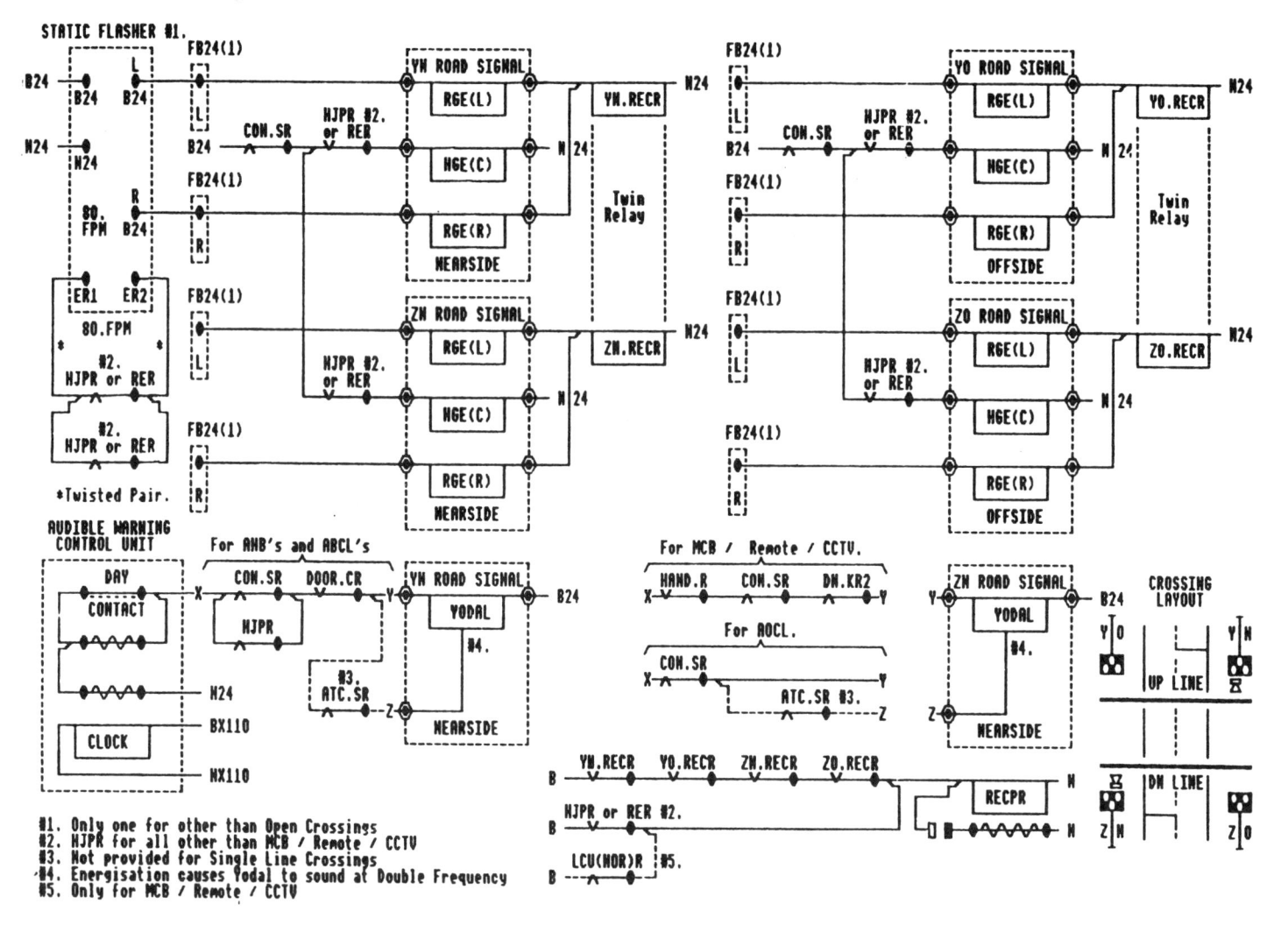

Fig. 6.8 Automatic half barrier crossing — boom operating and indication circuits

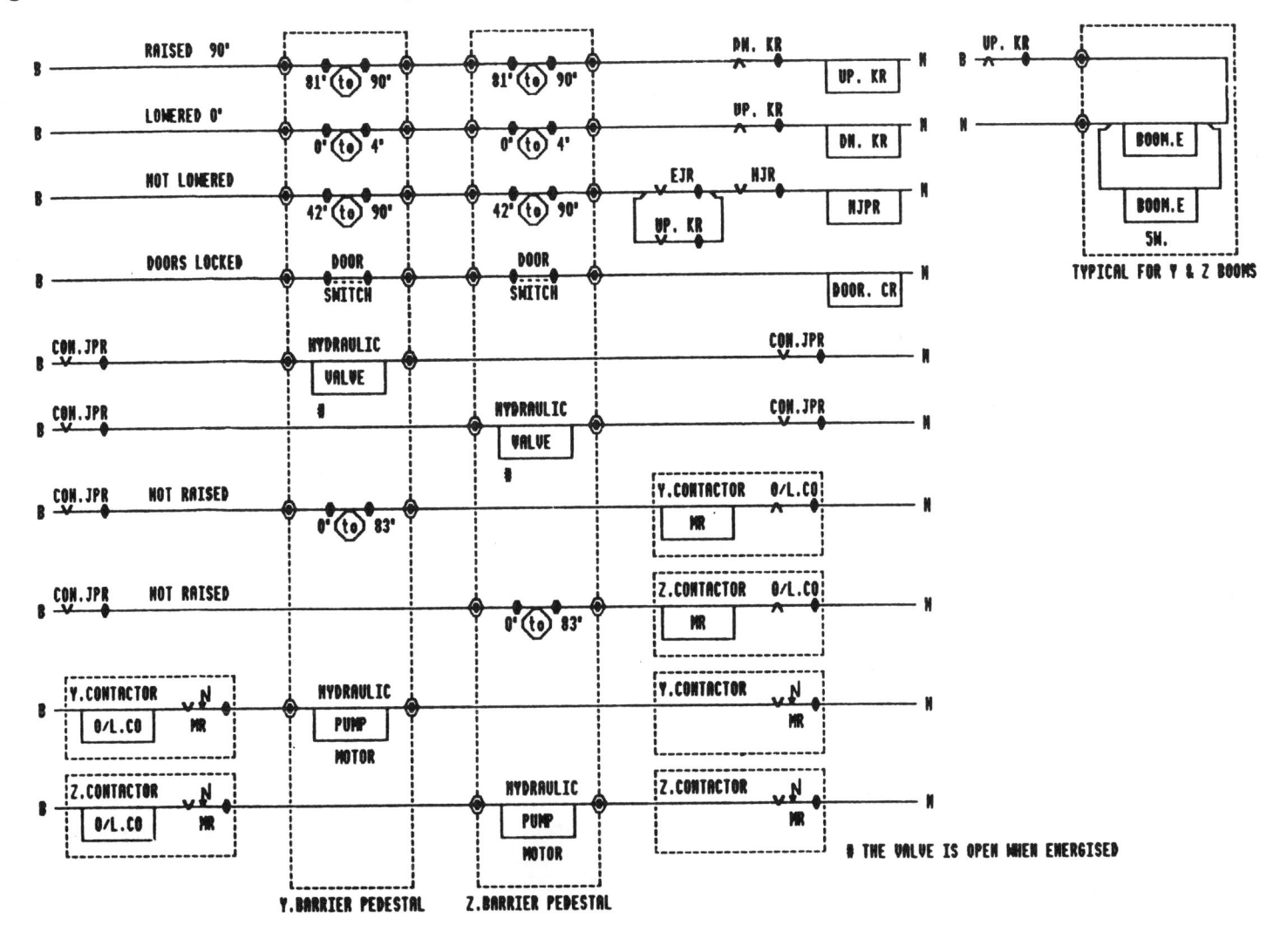

Extinguish Red Road Signals
When both the booms reach 42 degrees above horizontal, the HJPR operates, switching off the static flasher thus extinguishing the red road signals.

Booms Slow to Rise
If the booms have not fully raised within 7.5 s of having started, the release of the EJR will interrupt the HJPR circuit, thus restarting the red road signals until such time that the circuit is re-established by the booms being fully raised (UP.KR).

Booms Complete Rising
When both the booms reach 81 degrees above horizontal, the UP.KR operates, which maintains the HJPR after the EJR times out. The UP.KR also extinguishes the boom lights. When each boom reaches 83 degrees above horizontal, the respective contactor is released and the pump for that boom ceases to operate. The booms are maintained raised by the hydraulic valve energised (closed). Should a boom droop and remake the 83 degree contact, the respective MR will be energised so returning the boom to the fully raised position.

Signalbox Indication
The UP.KR operates the BARR.KR which again reverses the indication line, causing the indicator to show barriers *raised*.

Normalisation
As the train clears the run-off track circuit B with A.TPSR and A.SR operated, B.TPSR operates de-energising A/B.TCJR. After 10 s, A/B.TCJR releases followed by A.SR. The circuitry is normalised for the next move.

Train Long Time Clearing
As the train strikes out releasing B.TPSR and resetting A.TPSR with A.SR still operated, B.JR is energised which commences timing. If B track circuit remains occupied for more than 2 min B.JPR is operated which releases A.SR. This interrupts the CON.YR circuit re-initiating the barrier sequence. When B track circuit eventually clears, B.TPSR will not be reset in the normal way because A.SR has been released, but will reset by B.JPR operated. This will reset the CON.YR circuit so initiating the raising of the barriers. The operation of B.TPSR releases B.JPR.

Run-off Track Circuit Clears Momentarily
If track circuit B falsely clears momentarily, because of poor rail surface conditions for instance, B.TPSR will be reset in the normal way (A.SR, A.TPSR); however, A.SR will not release because A/B.TCJR remains operated for 10 s after de-energisation. Thus providing that track circuit B re-occupies within this time, it will not re-initiate the crossing. (This is important for lightly used lines or lines where light-weight stock is in use.)

Intermittent Track Circuit Failure
If, for example, track circuit A fails intermittently, then it will release A.TPSR and cause the initiation of the barriers by the release of the CON.YR. A.SR will be operated. However once the track circuit clears, A.JR will start to time out operating A.JPR after 2 min, thus releasing A.SR, and operating A.TPSR. This will reset CON.YR and cause the normalisation of the barriers.

Another Train Coming while Barriers Still Down
Consider an up train approaching the crossing, when a down train strikes in. As the up train approaches, the stick path of the ATC.SR will be interrupted by A.TPSR, but the feed will be maintained by C.TPSR/D.TPSR. As the down train approaches, the stick path will be further interrupted by C.TPSR, but the feed will still be maintained by B.TPSR/D.TPSR. As the up train passes the crossing and occupies the run-off track circuit, the ATC.SR stick path will be interrupted by B.TPSR. Or if the down train arrives first, the interruption will be by D.TPSR, releasing the ATC.SR. This will cause the yodalarm to sound with a double tone indicating to road users that another train is approaching the crossing. The ATC.SR will be reset by the HJPR.

LEVEL CROSSINGS

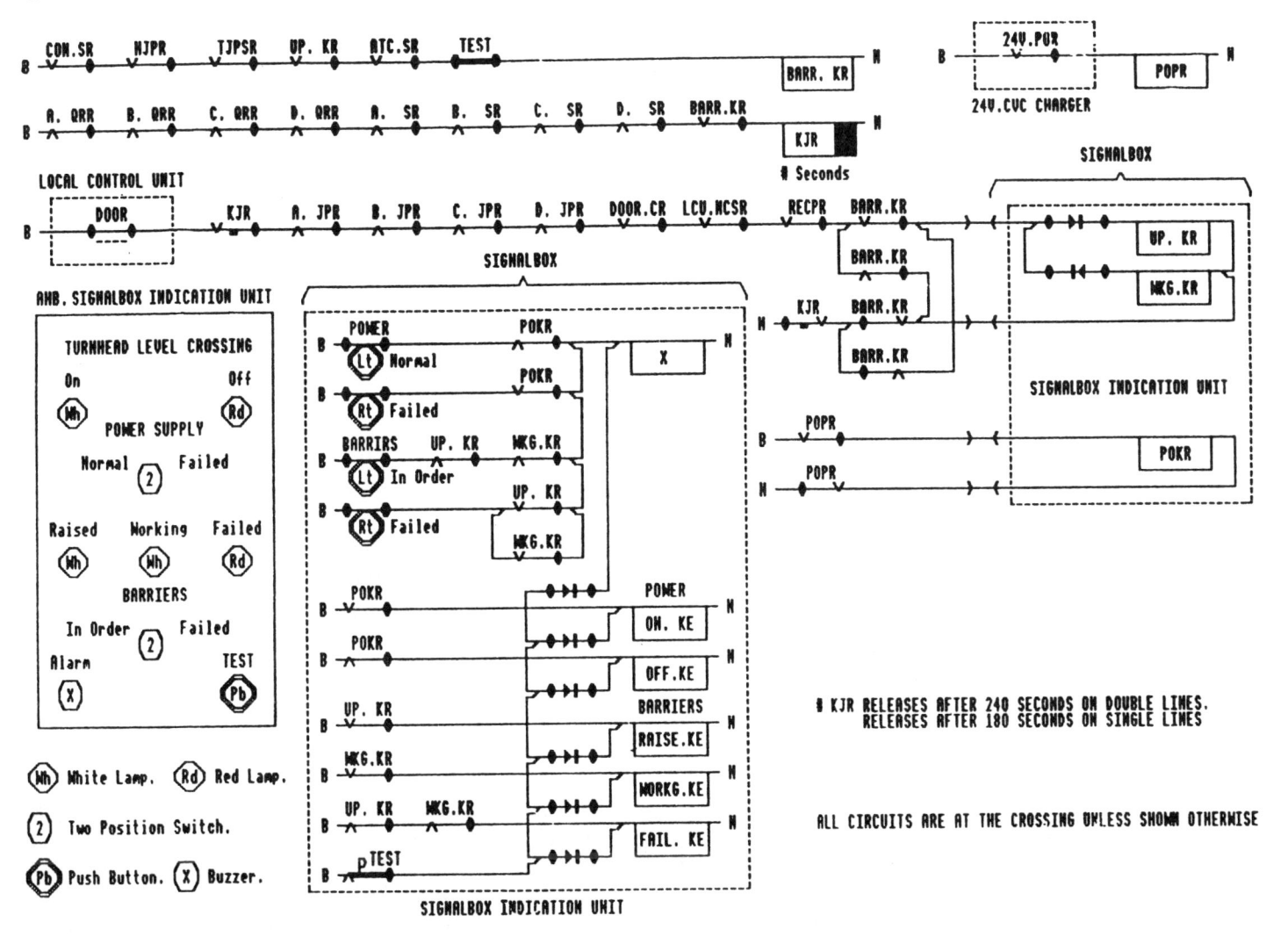

Fig. 6.9 Automatic half barrier crossing — indication circuits to signalbox

148 LEVEL CROSSINGS

Fig. 6.10 Automatic half barrier crossing — single line basic control circuits

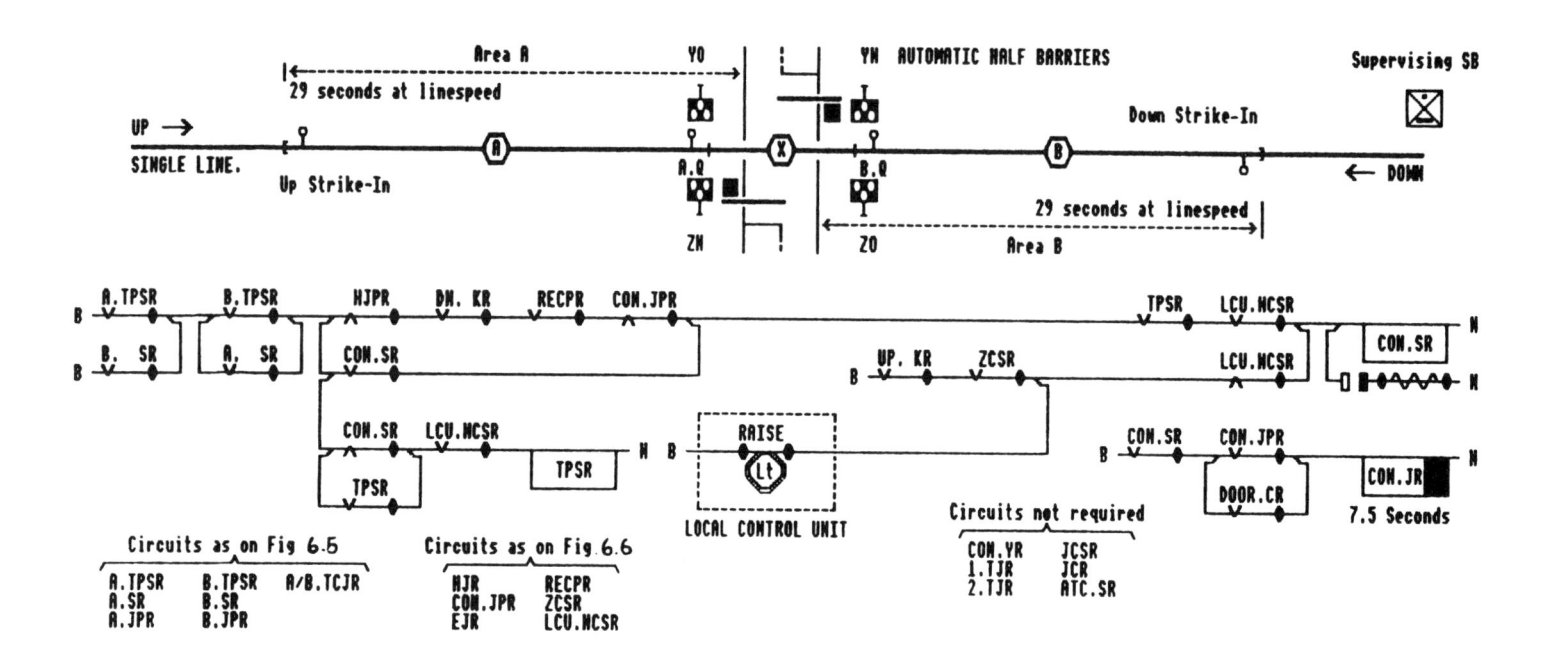

Another Train Coming as Barriers Start to Rise

Consider an up train having just cleared the crossing, when a down train strikes in. Having reached the stage that the CON.SR has operated, the barriers will rise in the normal way. However the releasing of C.TPSR will release the CON.YR which will de-energise 1.TJR and 2.TJR. After 10 s, these will release, once again initiating the lowering sequence by releasing the CON.SR. This ensures that the crossing will remain open for road traffic for at least 10 s.

Operation with Local Control Unit

Opening the local control unit door interrupts the indication line to the signalbox, thus immediately showing a *failed* indication. The control switch is placed at the *lower* position releasing the LCU.NCSR, followed by the CON.SR so commencing the lowering of the barriers. To raise the barriers, the switch is placed at *raise*, directly operating the CON.SR so raising the barriers. The crossing may only be returned to automatic control by the energisation of the LCU.NCSR, which requires the switch placed at *auto* when the booms are down (DN.KR). Closing the LCU door restores the indication to *working*.

Hand Operation

On local control, opening a barrier pedestal door to obtain access to the hand pump releases the DOOR.CR, operating the ZCSR. When both booms have been pumped to the raised position, the UP.KR operates followed by the CON.SR, so causing the road signals to extinguish. If a boom is lowered by hand, the UP.KR interrupts the CON.SR so lighting the road signals.

Barrier Indication

If the system is not normalised within a prescribed time, the KJR is released interrupting the indication line, thus causing the indicator to show *failed*. This is alarmed until acknowledged by the signalman.

CROSSING WITH SIGNAL STAGING
(Figs. 6.11 and 6.12)

If there is a signal within the strike-in area and that signal is at danger for traffic purposes, the approach of a train is prevented from initiating the crossing. If the train is within the strike-in area when the signal becomes able to clear, the crossing must then be initiated and have reached a stage of operation (determined by the distance between the signal and the crossing) before the signal may clear. This is to ensure that the train does not arrive at the crossing before the barriers have been lowered for a safe period. If there is no train within the strike-in area, the signal is able to clear at any time without having an effect on the crossing.

If the signal is at danger, an up train occupying track circuit AA will release A1.TZSR. This will not be followed by the release of A1.TPSR because that relay is maintained by 101.ALSR operated (101 at danger and free of approach locking). Thus the barriers will not be called. When the signal is able to clear, 101.UCR operates releasing A1.TPSR, followed by the CON.YR thus initiating the crossing. The CON.SR energises the XJR, which is followed by the XJPR after the prescribed time. 101.UCSR then operates releasing 101.ALSR causing the signal to clear. The train passes the signal releasing A2.TPSR, which resets A1.TZSR. (A1.TZSR is provided with a resetting timer to cover an intermittent failure situation.) As the train passes the crossing, 101.ALSR is reset by the sequential occupation of each track circuit. 101.ALSR resets A2.TPSR, allowing the barriers to rise. If the signal is placed to danger after the crossing has been initiated, the barriers will remain lowered, 101.ALSR will be operated after timing out, resetting A2.TPSR, followed by the CON.YR allowing the barriers to raise. If the train runs by 101 signal at danger, the barriers are called by AB track circuit. If the signal is cleared with no train in the strike-in area, 101.UCSR is operated immediately by the contact of A1.TZSR.

150 LEVEL CROSSINGS

Fig. 6.11 Automatic half barrier crossing with signal regulation

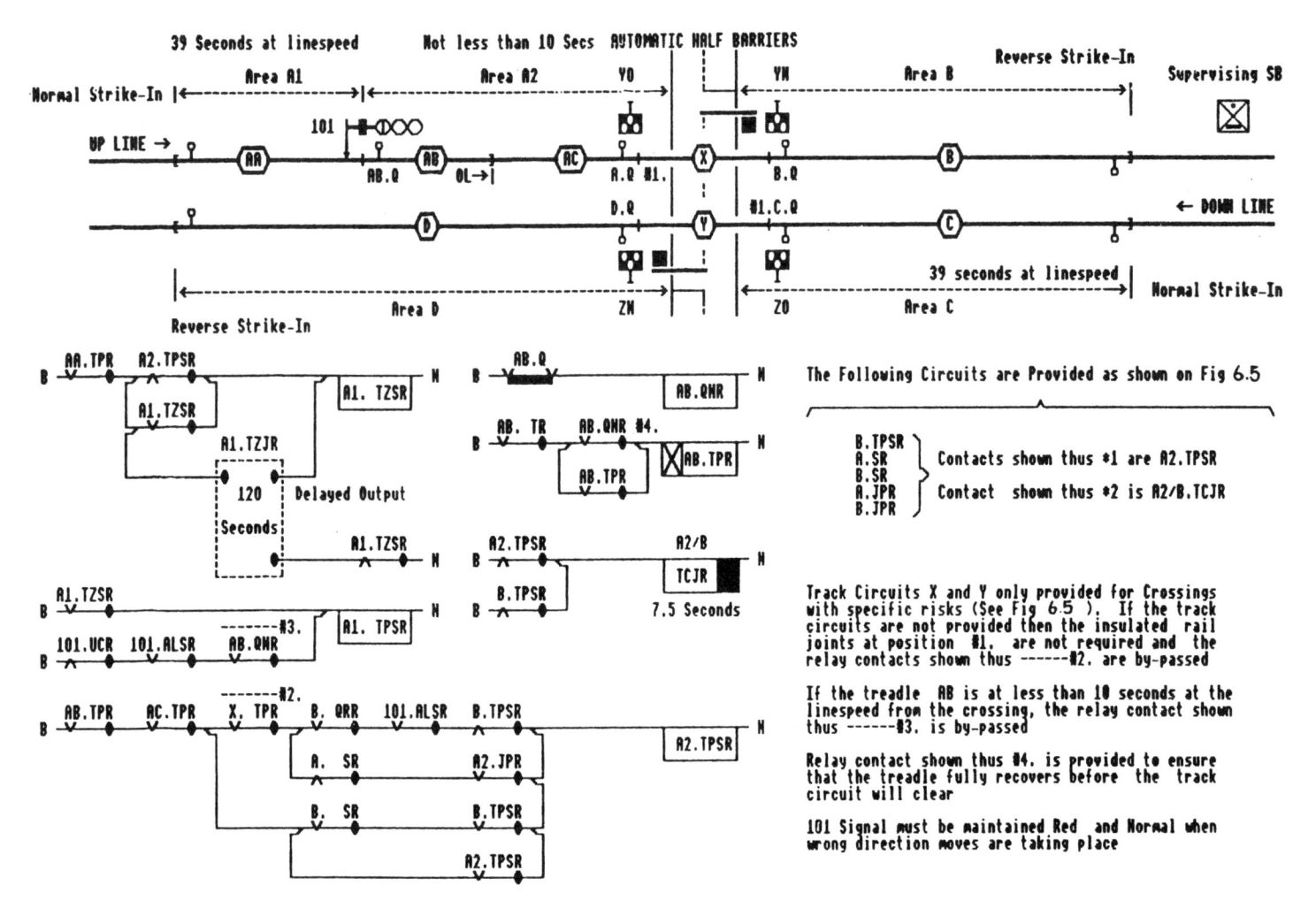

LEVEL CROSSINGS 151

Fig. 6.12 Automatic half barrier crossing with signal regulation, continued

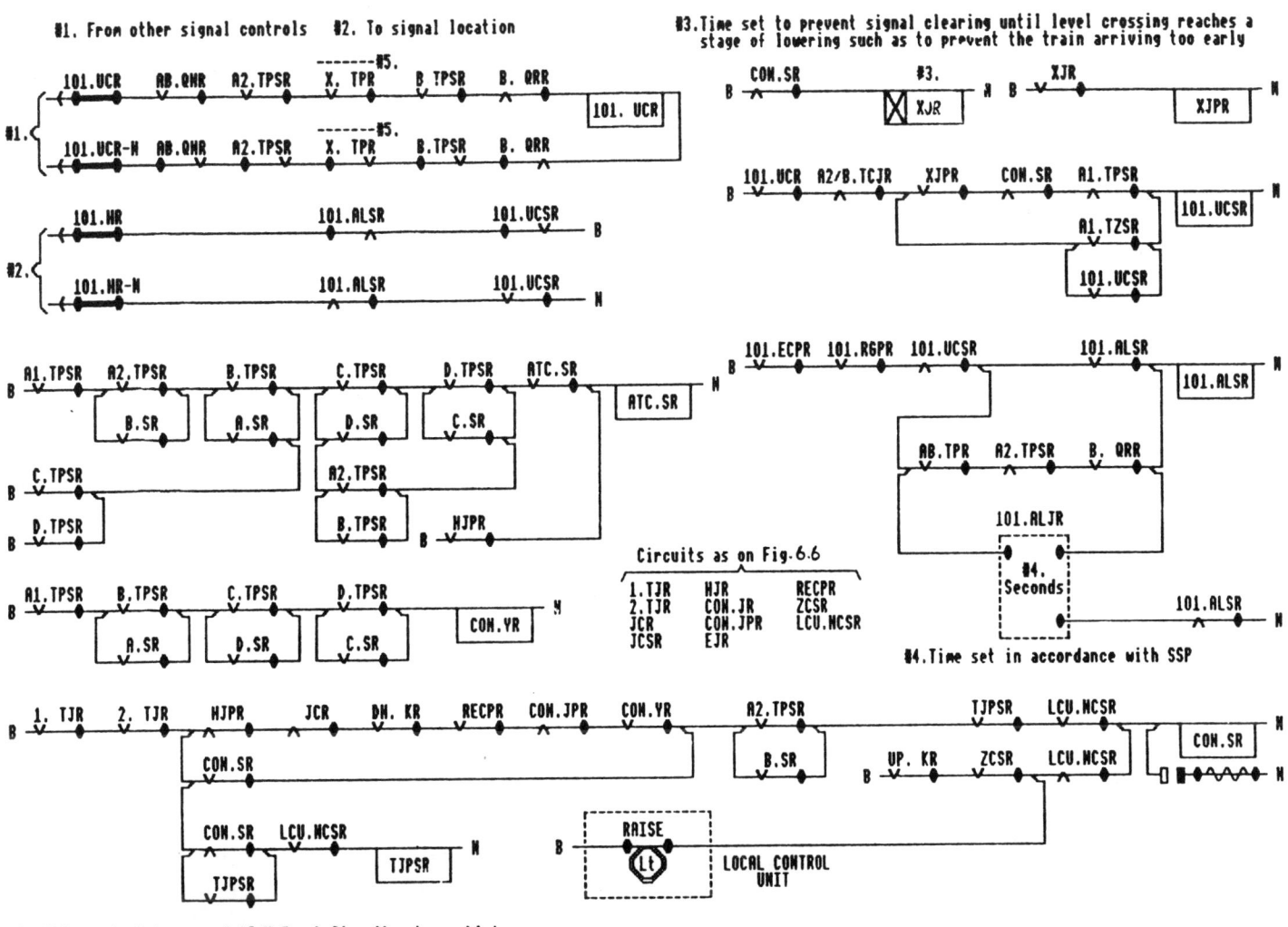

Automatic Open Crossing Locally Monitored

GENERAL DESCRIPTION

Automatically operated open level crossings must give a warning to road users for at least 27 s before the arrival of the fastest train. Of all trains using the line, 50% must arrive at the crossing within 50 s and 95% within 75 s. Therefore on single track lines, the initiation of the crossing is set to 29 s from the crossing, which allows 2 s for the operation of relays, 3 s for the yellow road signals, and then 24 s of red road signals before the fastest train arrives at the crossing. On double lines, because of the possibility of a second train coming in the opposite direction, the initiation of the crossing is set to 39 s, with the effect of initiation being delayed for 10 s to allow for a minimum road opening time if another (opposite direction) train strikes in just as the crossing opens to road traffic. Furthermore, if the roadway is wider than 15 m then 1 s is added for every 3 m in excess of this width.

On approaching the crossing, the driver will first encounter an advanced warning board, consisting of a black St. George's cross on a reflective white background. This is located at the required braking distance at line speed to a further board displaying the crossing speed. This distance is subject to a minimum of 100 m. The crossing speed board displays the speed at which a train may cross the roadway (different speeds are permitted for passenger and freight trains). This board consists of a black St. Andrew's cross on a reflective white background, mounted above the permissible speed (or speeds) for the crossing, which are shown in white figures on a black background. The crossing speed board must be placed in such a position that the crossing and its associated rail signal can be clearly seen by the driver of an approaching train. It must be at such a distance from the rail signal that it is possible for the driver to stop in the event of the rail signal failing to show a proceed aspect, subject to a maximum distance of 600 m. Finally the rail signal, which is usually placed at 5 m from the edge of the crossing, normally displays a red flashing aspect (this being to denote that the red aspect does not constitute an absolute stop signal). When the crossing sequence has been initiated, and the red road signals are being correctly displayed, the rail signal shows a flashing white aspect, indicating to the driver that the crossing may be traversed at the allowable crossing speed. Once initiation has taken place, the crossing is floodlit from lamps which are usually mounted at the rear of the rail signal. For rail traffic, the maximum line speed must not exceed 88 km/h. There must be no more than two running lines.

Each road approach is provided with two road signals. That on the nearside is known as the primary signal and carries a reflective road signal *'keep crossing clear'*, and a sign indicating the telephone number to be used to contact the railway operating control centre. The road signal on the offside is known as the secondary signal and carries an identical road sign and, if the crossing is double track, an additional sign *'if the lights continue to flash another train is coming'*. Both road signals are surmounted by a white reflective St. Andrew's cross on a red background, this cross having an additional lower chevron if the line is double track. Both signals are also provided with a reflective red striped plate on a white background, which indicates that there are no barriers. On double track railways, an another train coming signal is provided adjacent to the primary road signal. This flashes to indicate to road users that a further train is approaching if the crossing has not re-opened after the first train has passed. The legend is secret behind a red filter. Each road signal consists of two red lamps and a single yellow lamp, arranged on a black backboard surrounded by a conspicuous white stripe, whilst the primary signals are fitted with audible warning devices for warning pedestrian users of the operation of the crossing. There are no limitations on the type of road or traffic, except that the crossing must be situated in a place where the traffic flows freely, and the road profile must be such that there is no risk of a long wheelbase vehicle grounding on the crossing.

There are, however, limitations on the frequency of usage of the crossing. If the crossing line speed is 88 km/h then the daily traffic moment (number of trains multiplied by the number of road vehicles within a specified period) must not exceed 4,000; if the crossing line speed is only 40 km/h then the traffic moment may be 15,000.

When the level crossing sequence is initiated by the approach of a train, the yellow lamp shows steadily for a period of 3 s, followed by the red lamps flashing alternately at a rate of 80 flashes/min. This continues for 24 s before the arrival of a train travelling at line speed. The audible warning signal (usually a yodalarm) commences when the yellow light shows, and continues to sound until the crossing re-opens. It is usual for the volume of the audible alarms to be adjusted by means of a time clock, so that the sound can be softened during the night hours. Once the train has cleared the crossing, the red road signals will cease to operate. The arrangement is such that should a further train approach the crossing, then the road must either not re-open, or if re-opening has commenced, the crossing must remain open for 10 s before the sequence recommences. In the event of the crossing being maintained closed by the approach of a second train, then the another train coming road signal will operate and the yodalarms will change tone until the crossing re-opens after this second train has passed.

In the event of the failure of all the red lamps in one road signal, or the failure of the main power supply to the installation, the rail signal will not display a proceed aspect when the crossing sequence is initiated. Under these circumstances, the driver may pass the rail signal and cross the roadway, but only at extreme caution after ascertaining that it is safe to do so. If there is a very slow train or a failure which causes the crossing to be initiated with no train approaching, the rail signal will revert to danger after a period of 3 min with the road signals still operating. After a further 30 s, the crossing will normalise and the road signals will cease to show. A reset plunger is provided in a locked cupboard at the foot of the rail signal, accessible by a driver's key. This plunger may be used by a train driver to re-initiate the complete operation of the crossing sequence for a further period of 3 min. In the event of the lamp failing in the flashing white rail signal, the red signal is re-initiated.

The ground equipment used for initiation of the crossing may be identical to that already described for the AHB crossing. Some installations of AOCL crossings are initiated and cleared entirely by the operation of track circuits, without the addition of treadles. This is because in the event of the maloperation of a track circuit failing to initiate the crossing correctly, the rail signal will fail to show a proceed aspect and the driver will stop before arrival at the crossing. In view of the continuing problems with unreliability of track circuit shunt with most classes of diesel multiple unit train, it has proved essential to provide a treadle to ensure that the striking-in track circuit has operated. In many cases, a treadle is provided at the strike-out point. Under these circumstances, it is not possible for a combination of wrong- and right-side track circuit failures to produce an erroneous early re-opening of the crossing. It has also proved necessary to prevent the momentary wrong-side failure of the running-off track circuit re-initiating the crossing after the passage of a train, as this encourages road users to disobey the road signals.

AOCL — DESCRIPTION OF OPERATION
Consider an up train with the crossing in automatic operation (see **Figs. 6.13–6.15**).

Up Train Strikes In
Train occupies track circuit A, releasing A.TPSR thus operating A.SR which determines the direction of initiation.

Initiation of Sequence
A.TPSR releases the CON.YR, thus de-energising 1.TJR and 2.TJR.

154 LEVEL CROSSINGS

Fig. 6.13 Automatic open crossing locally monitored — bidirectional strike-in circuits

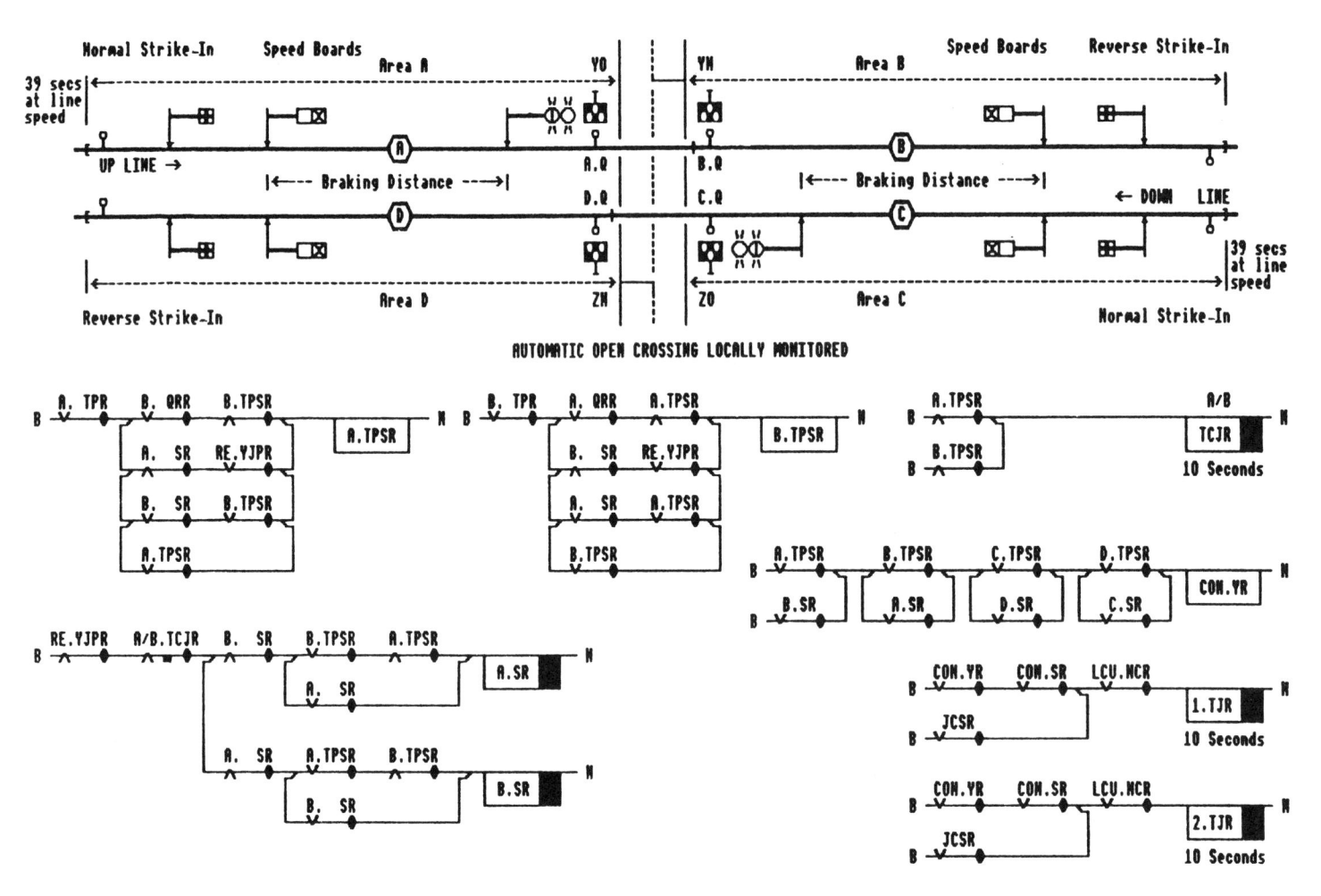

Fig. 6.14 Automatic open crossing locally monitored — control circuits

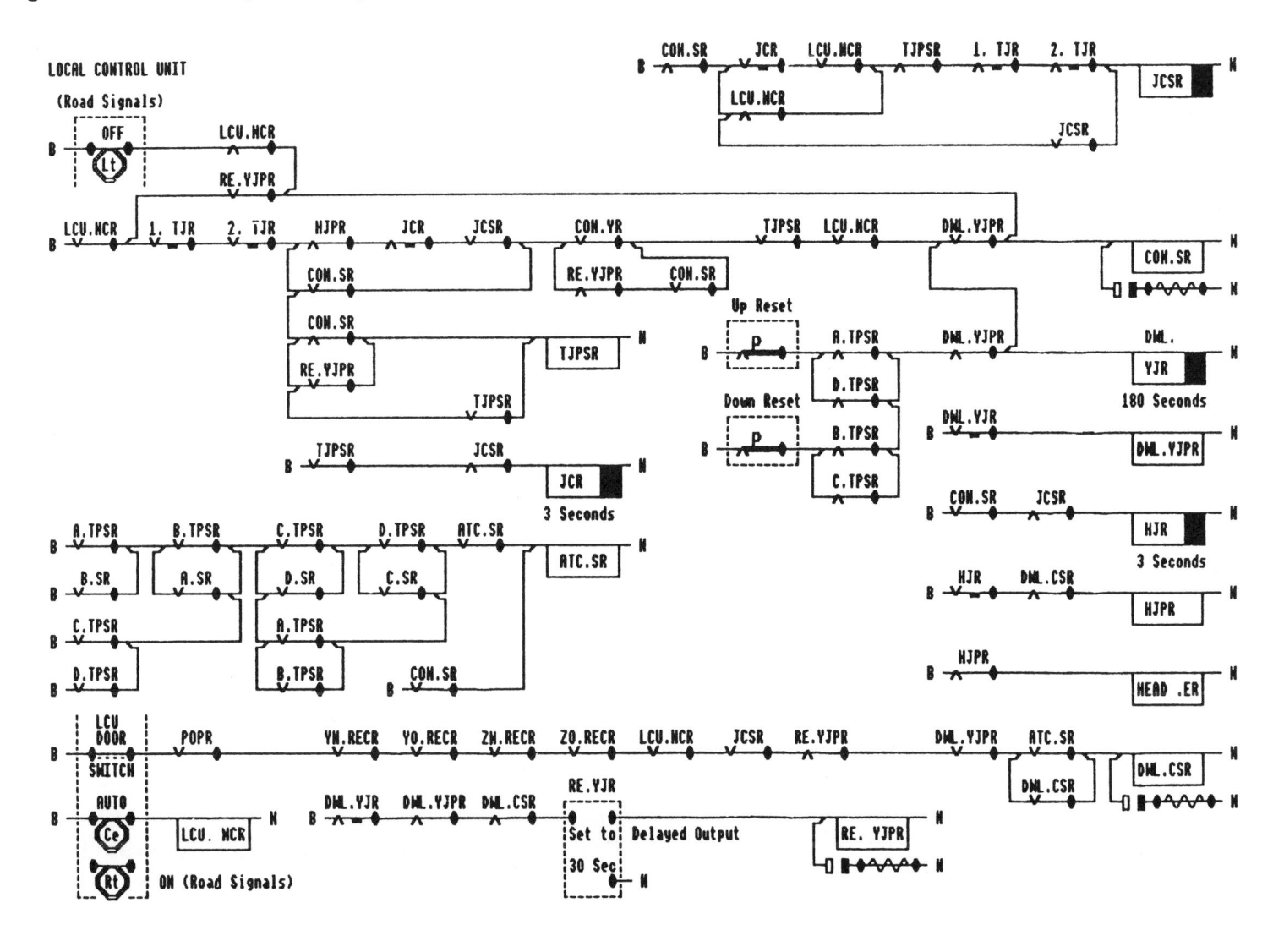

Minimum Opening Time
After 10 s, 1.TJR and 2.TJR will release thus de-energising the DWL.YJR and releasing the TJPSR and CON.SR which will de-energise the HJR and JCR.

Yellow Road Signals
While the HJR remains operated, the release of the CON.SR will cause the yellow road signals to illuminate and start the audible warning.

Resetting for Another Train
With the JCR still operated and the CON.SR and 1 and 2 TJRs released, the JCSR will reset. The purpose of the JCR is to ensure the correct operation of the JCSR immediately after the release of the TJPSR. The JCSR releases the JCR and resets both TJRs which is followed by the TJPSR. This is a self-checking procedure and ensures that should another train strike in, the necessary 10 s minimum open time will still apply.

Floodlights
After 3 s, the HJR will release followed by the HJPR causing the operation of the HEAD.ER, which illuminates the crossing floodlights enabling the train driver to see and monitor the road surface of the crossing.

Red Road Signals
Release of the HJPR also extinguishes the yellow road signals and starts the static flasher, causing the red road signals to flash. Providing that all lamps are intact in each red road signal unit, the respective RECR will operate.

Driver's Rail Signal
Providing that the RECR is operated for all red road signals, that the power is on to the battery charger (POPR), and that the ATC.SR has operated after the last train, the DWL.CSR operates. Because the A.SR is operated and A.TPSR is released, denoting an up train, the UP.DWL.R will operate, causing the up flashing white rail signal to show. Providing that this is alight (UP.DWL.ECR), the flashing red rail signal will be extinguished. This indicates to the train driver that the crossing is operating and that the correct road signals are being displayed.

Failure of White Rail Signal
Should the white flashing rail signal fail at any time that it should be showing, the rail signal will revert to the red flashing mode.

Failure of Red Road Signal
Should any one red road signal unit fail to show during the operation of the crossing, the respective RECR will release followed by the DWL.CSR and the respective DWL.R, which will cause the rail signal to revert to the red flashing mode.

Up Train Arrives at the Crossing and Strikes Out
The striking out sequence is similar to that for an AHB.

Normalisation of the Crossing
Operation of the A.TPSR with the A.SR still operated, will cause the CON.YR to reset. This causes the re-energising of the DWL.YJR (providing that the train has reached the crossing within 3 min of striking in, the DWL.YJR and hence the DWL.YJPR will still be operated) thus resetting the CON.SR.

Extinguish Red Road Signals
The CON.SR releases the JCSR which operates the HJR and hence the HJPR. This releases the HEAD.ER, extinguishing the floodlights. The HJPR also switches off the static flasher, thus extinguishing the red road signals.

Driver's Rail Signal to Red Flashing State
The JCSR releases the DWL.CSR followed by the UP.DWL.R, thus causing the rail signal to revert to the red flashing state.

LEVEL CROSSINGS 157

Fig. 6.15 Automatic open crossing locally monitored — signal circuits

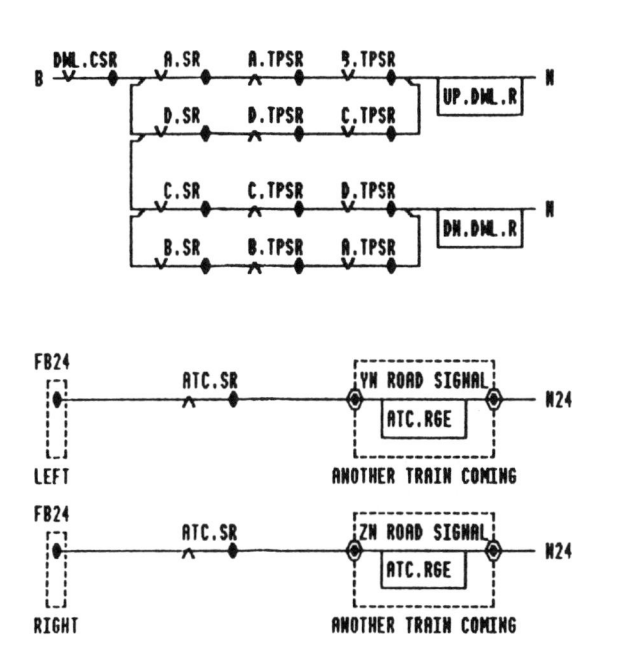

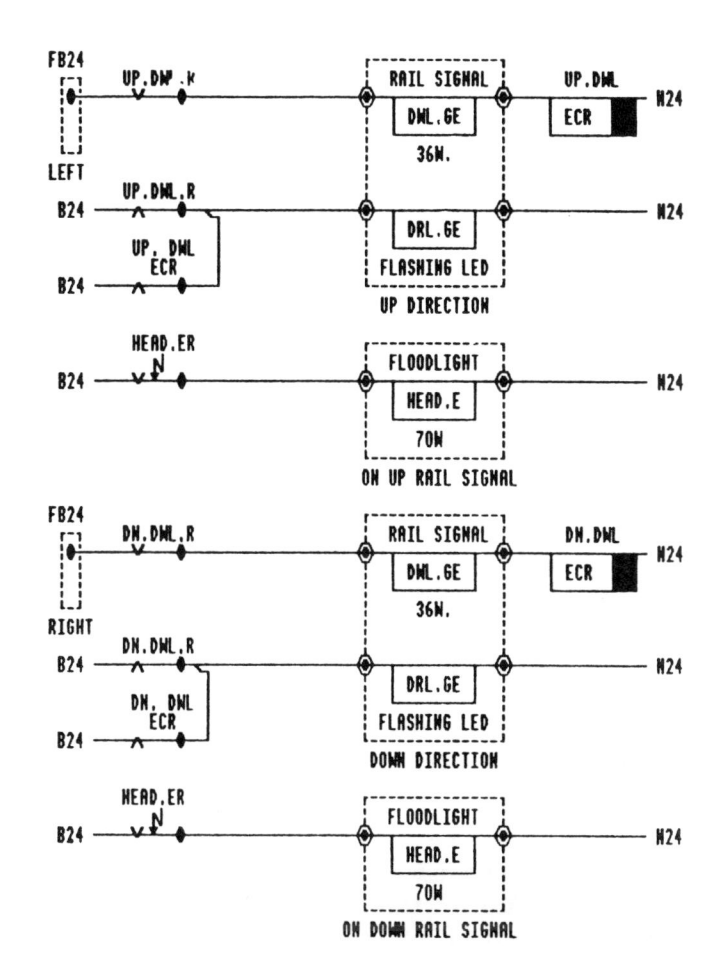

LEVEL CROSSINGS

Train Fails to Arrive at the Crossing
If after striking in, the train fails to arrive at the crossing within 3 min, the DWL.YJR will release, followed by the DWL.YJPR. This interrupts the DWL.CSR causing the rail signal to return to the red flashing state.

Resetting after a Failure to Arrive
The above conditions energise the RE.YJR which commences to time out, followed after 30 s by the RE.YJPR. This directly sets the CON.SR, initiating the normalisation of the crossing, extinguishing the red flashing road signals and thus re-opening the road.

Slow Train Arrives at Crossing
When the slow train does arrive at the crossing, it will come to a stand at the red flashing rail signal. The driver can then re-initiate the crossing by the operation of a plunger, which is contained in a locked box accessible by means of a driver's standard key. This plunger is fixed to the post of the rail signal. The plunger re-energises the DWL.YJR which interrupts the RE.YJR and RE.YJPR, thus restarting the crossing sequence, causing the red flashing road signals to show, as indicated by the white flashing rail signal.

Another Train Coming while the Crossing is Still Operating
Consider an up train approaching the crossing as a down train strikes in. The ATC.SR operates as described for AHBs. This alters the tone of the audible warning signal and also illuminates the another train coming road signal. The CON.SR will be prevented from resetting by the CON.YR.

Another Train Coming as Crossing Clears after Last Train
Consider an up train having just cleared the crossing when a down train strikes in. Having reached the stage that the CON.SR has operated, the crossing will normalise. However the crossing will re-initiate as already described for the AHB, after a minimum road open time of 10 s. (Delay of TJRs.)

Crossing Set to Local Control
The local control cupboard is accessible by a standard operator's key. The unlocking of the door interrupts the DWL.CSR which maintains the rail signal at flashing red. The three-position switch is normally in the centre *auto* position. (There is a mechanical interlock to ensure that the door cannot be closed unless this switch is at *auto*.) Turning this key to the left hand *off* position interrupts the LCU.NCR thus disabling the normal controls and maintaining the CON.SR operated. To cause the crossing sequence to initiate, the key is turned to the right hand *on* position, which releases the CON.SR, starting the normal sequence, with the exception that the rail signal will remain at flashing red. It is then the crossing keeper's responsibility to hand signal the train over the crossing.

Intermittent Track Circuit Failure
If, for example, a track circuit fails intermittently, the crossing will be initiated. After 3.5 min the crossing will time out and the road lights will extinguish. The RE.YJPR will reset A.TPSR, which initially called the crossing, and release A.SR.

Automatic Half Barrier Crossing Locally Monitored

GENERAL DESCRIPTION
This type of crossing is a direct development of the AOCL. As a result of the Stott report on level crossings following the 1986 Lockington accident, certain restrictions were put on the use of AOCL crossings, depending on the traffic moment. The automatic half barrier locally monitored (ABCL) does not have such traffic restrictions, but it must satisfy the road conditions which are placed on AHB crossings in that road traffic must not be likely to stand on the crossing and the profile of the crossing must be such that a long vehicle will not ground on the road surface. The crossing is similar to an AOCL except that it also has half barriers. It does not have the

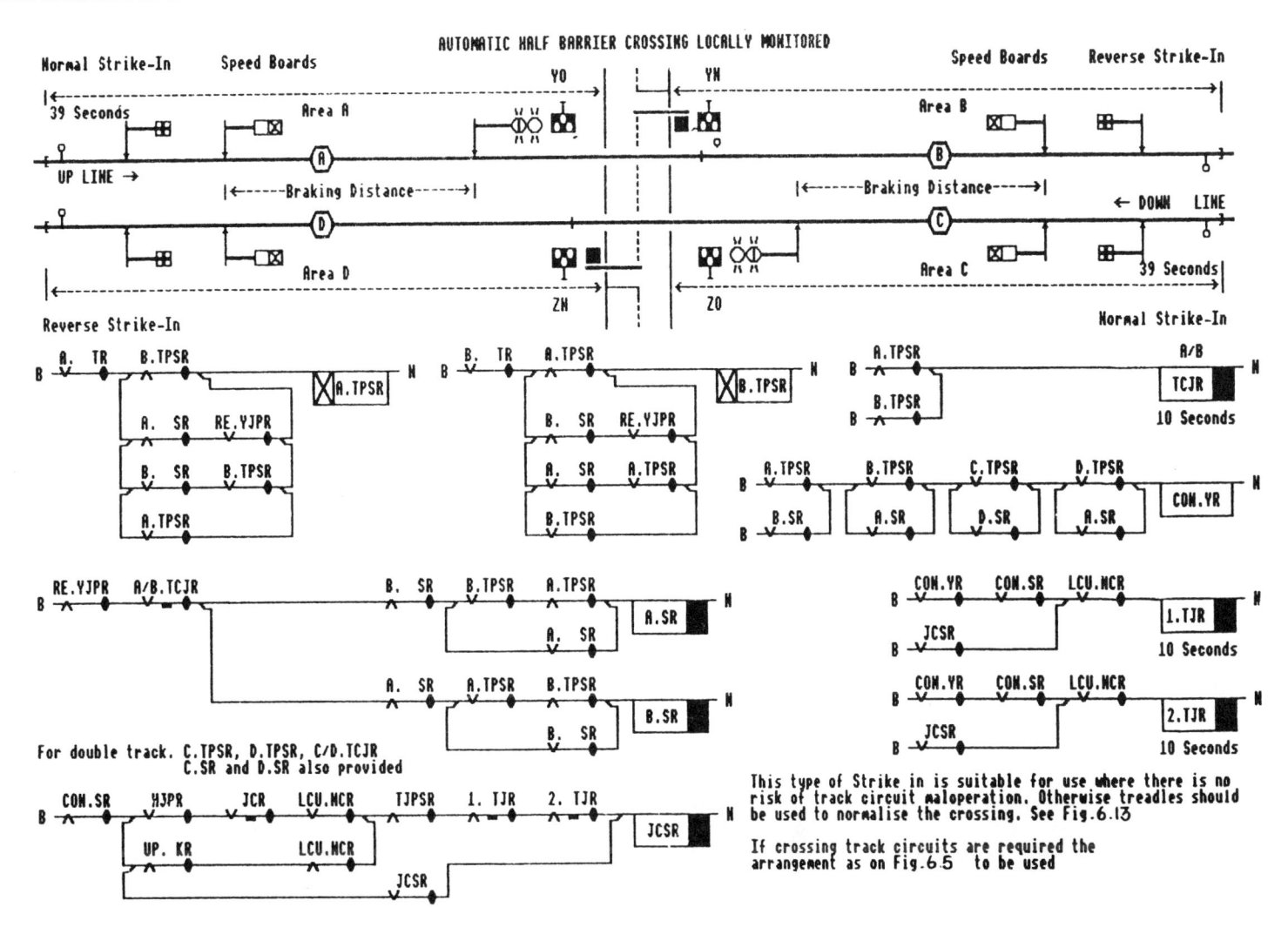

Fig. 6.16 Automatic half barrier crossing locally monitored — bidirectional strike-in circuits

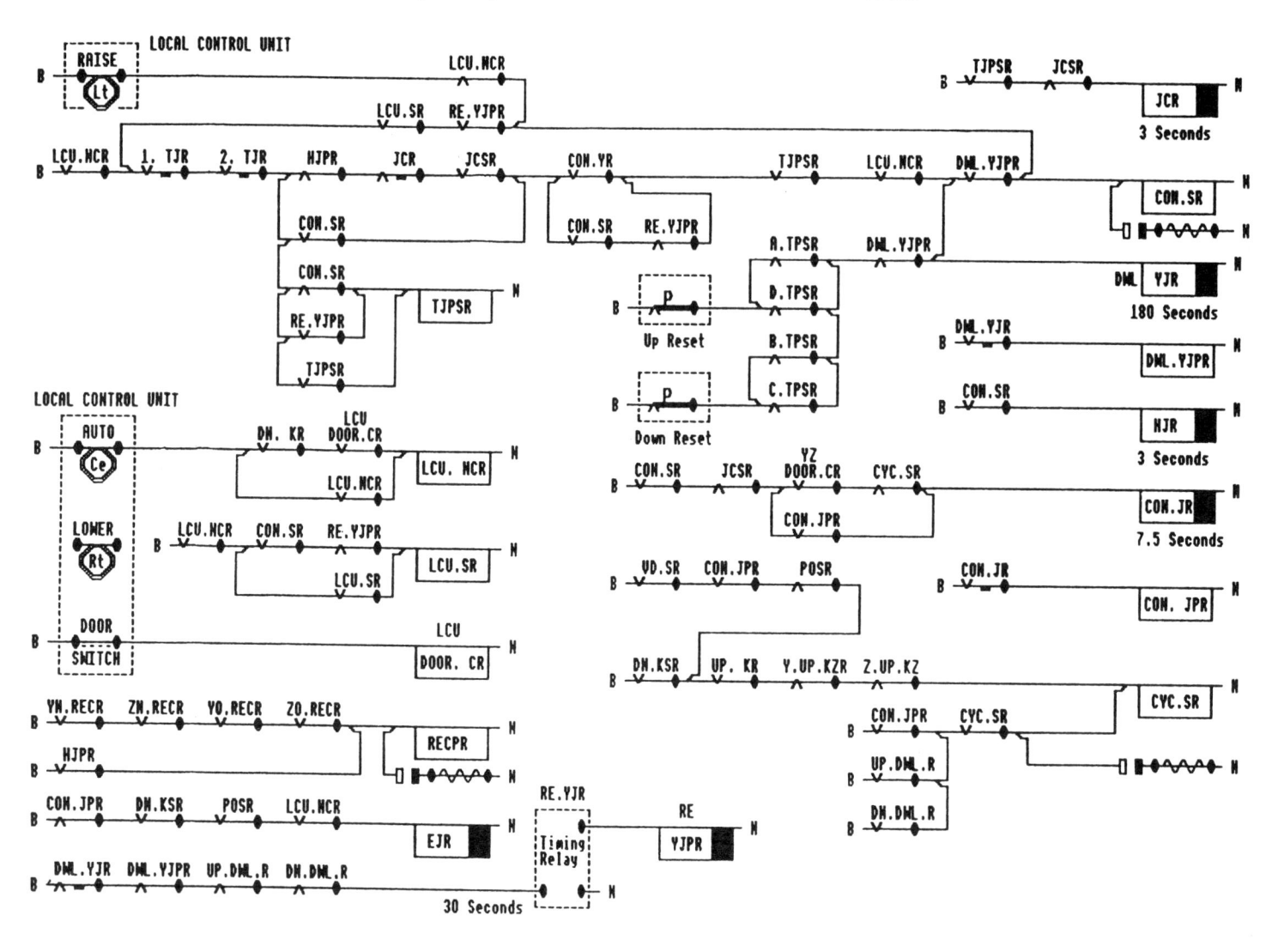

Fig. 6.17 Automatic half barrier crossing locally monitored — double line control circuits

another train coming road signal, because in this event the barriers are prevented from re-opening the road.

From the point of view of the train driver, the crossing is identical to an AOCL, having the same signs on the rail approach and being protected by special red/white flashing rail signals. The rail speed, like an AOCL, is limited to 88 km/h. As far as a road driver is concerned the crossing is identical to an AHB, except that there are no emergency telephones. There is an additional sign on the nearside road signal post, giving a telephone number on which the railway operations centre can be contacted.

The principal difference between an ABCL and an AHB is that in the event of any system failure, the barriers are maintained in the raised position. Unlike an AHB, the barriers do not fall upon the loss of an electrical feed to the hydraulic valves. This is necessary because the state of the crossing is not indicated to any supervising signalbox.

In the case of a system failure, the driver may operate a plunger in a locked box at the base of the rail signal to initiate the lowering of the barriers. If this also fails, the driver is authorised to pass over the crossing at extreme caution with the barriers in the raised position, providing that the train crew ascertain that the road traffic is stopped.

ABCL — DESCRIPTION OF OPERATION

Consider an up train with the crossing set to automatic operation (see **Figs. 6.16–6.19**).

Up Train Strike In
The mode of strike in is similar to that already described for an AOCL. In this example, circuits have been shown for a crossing where the striking out is entirely dependent upon the operation of track circuits.

Initiation of Sequence
A.TPSR releases the CON.YR, thus de-energising 1.TJR and 2.TJR.

Minimum Opening Time
After 10 s, 1.TJR and 2.TJR will release, thus de-energising DWL.YJR and releasing the TJPSR and CON.SR which will de-energise the HJR and JCR.

Yellow Road Signals
While the HJR remains operated, the release of the CON.SR will cause the yellow road signals to illuminate and start the audible warning.

Floodlights
After 3 s the HJR will release followed by the HJPR thus causing the operation of the HEAD.ER which illuminates the crossing floodlights enabling the train driver to monitor the road surface of the level crossing.

Red Road Signals
Release of the HJPR also extinguishes the yellow road signals and starts the static flasher, causing the red road signals to flash. Providing that all lamps are intact in each red road signal unit, the respective RECR will operate, thus operating the RECPR. Failure of any road signal to show will stop the barriers from lowering by disabling the valve circuits.

Booms Commence to Lower
After 7.5 s the CON.JR will release, followed by the CON.JPR. Providing that the red road signals are flashing (RECPR) and the battery supply is healthy (POSR), this causes the valves to be energised in both barrier pedestals. Unlike automatic half barriers these valves are energised to open. The respective VALVE.CRs also operate. As the booms commence to lower, the UP.KR releases, thus de-energising the CYC.SR. The valves are maintained open by the release of the UP.KR. Once the booms have lowered to 83 degrees above horizontal, the UP.KZRs operate. The release of the UP.KR illuminates the boom lights.

Driver's Rail Signal

Once the barriers have commenced to lower, as detected by the valves being open (VALVE.CRs) and the booms not being fully raised (UP.KZRs), also provided that the power supply is intact (POPR) and the red road signals are operating (RECPR), the DWL.CSR operates. This in turn selects either the UP or DOWN.DWL.R, dependent upon the direction of the strike in. The DWL.R energises the respective driver's white rail signal. If this signal is alight, the ECR operates, which extinguishes the driver's red flashing rail signal. Operation of the DWL.R maintains the CYC.SR.

Booms Fully Lowered

Once the booms are fully down the DN.KR operates, which in turn operates the DN.KSR setting the EJR ready to time the subsequent raising of the booms.

Failure of Red Road Signal

Should any one red road signal fail to show during the operation of the crossing, the RECPR will release followed by the DWL.CSR and DWL.R, causing the rail signal to revert to red flashing mode.

Train Arrives at the Crossing and Strikes Out

The striking-out sequence is similar to that of an AOCL. As the train occupies track circuit B, B.TPSR is released, which maintains A/B.TCJR and operates A.TPSR as that track circuit clears. B.SR is prevented from operating by A.SR. As the train passes the crossing, the respective DWL.R is disengaged causing the CYC.SR to release and the rail signal to revert to flashing red.

Normalisation of the Crossing

The resetting of the CON.YR causes the re-energisation of the DWL.YJPR (providing that the train has arrived at the crossing within 3 min of striking in, the DWL.YJR and hence DWL.YJPR will not have released) resetting the CON.SR, followed by the HJR, the CON.JR and CON.JPR. Operation of the CON.SR resets the HEAD.ER thus extinguishing the crossing floodlights.

Booms Commence to Raise

The operation of the CON.JPR de-energises (closes) the valves and operates both the MRs. This causes the operation of the hydraulic pumps thus raising the booms. The CON.JPR de-energises the EJR.

Extinguish Red Road Signals

When both the booms have reached 42 degrees above horizontal, the HJPR operates. This switches off the static flasher thus extinguishing the red road signals. The yellow road signals will not restart because the CON.SR is operated.

Booms Slow to Raise

If the booms have not fully raised within 7 s of having started, the release of the EJR will interrupt the HJPR circuit, thus restarting the red road signals until such time that the circuit is re-established by the booms being fully raised (UP.KR).

Booms Fully Raised

When both booms reach 81 degrees above the horizontal, the UP.KR operates, which in turn operates the CYC.SR, thus releasing the DN.KSR. When each boom reaches over 83 degrees above the horizontal, the respective UP.KZR relay de-energises, releasing the respective contactor so that the boom ceases to raise. Operation of the UP.KR extinguishes the boom lights.

Train Fails to Arrive at the Crossing

If after striking in, the train fails to arrive at the crossing within 3 min, the DWL.YJR will release, followed by the DWL.YJPR. This interrupts the DWL.CSR thus causing the rail signal to return to the normal red flashing state. This also starts the RE.YJR timing sequence.

LEVEL CROSSINGS 163

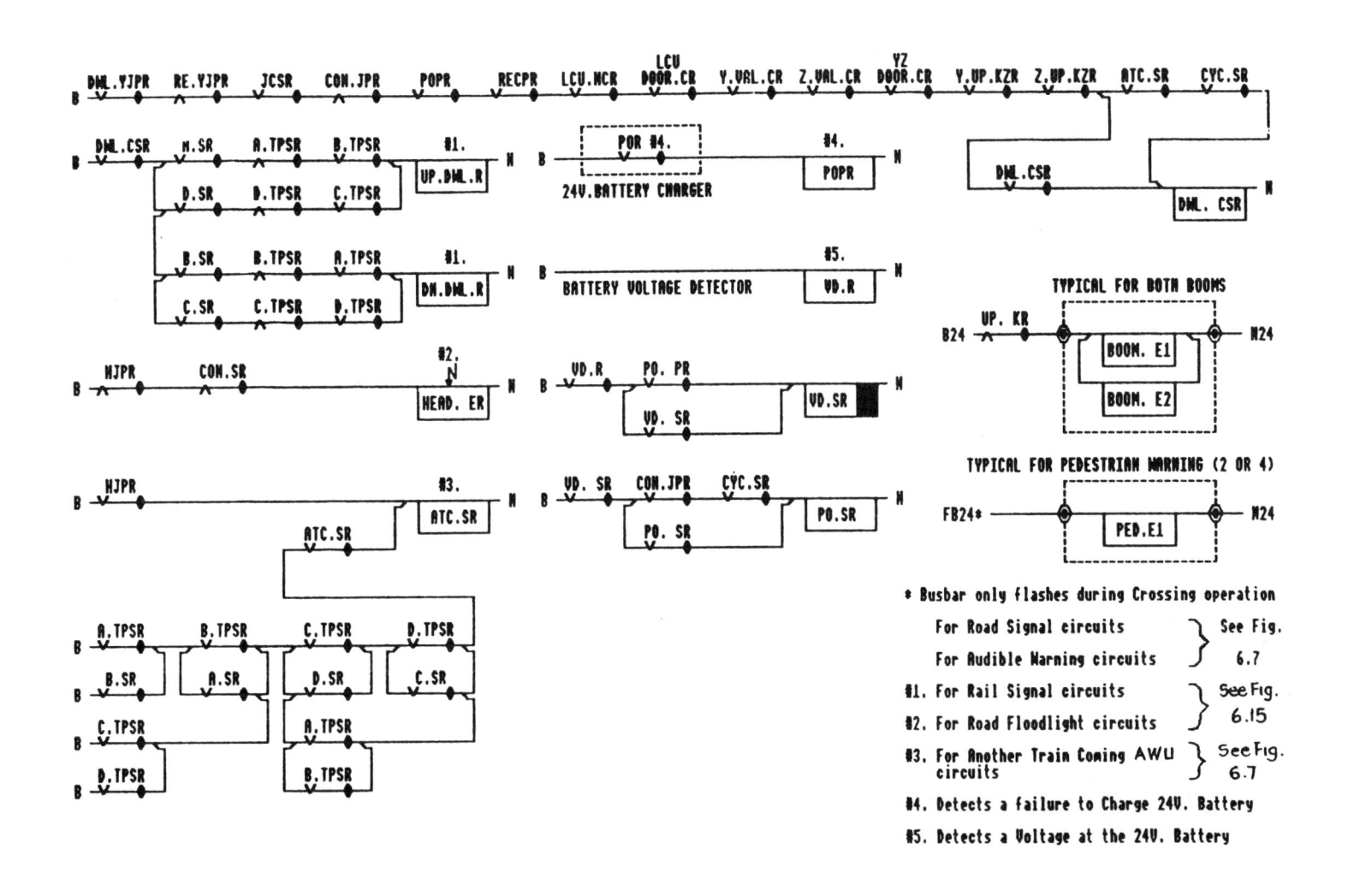

Fig. 6.18 Automatic half barrier crossing locally monitored — rail signal circuits

164 LEVEL CROSSINGS

Fig. 6.19 Automatic half barrier crossing locally monitored

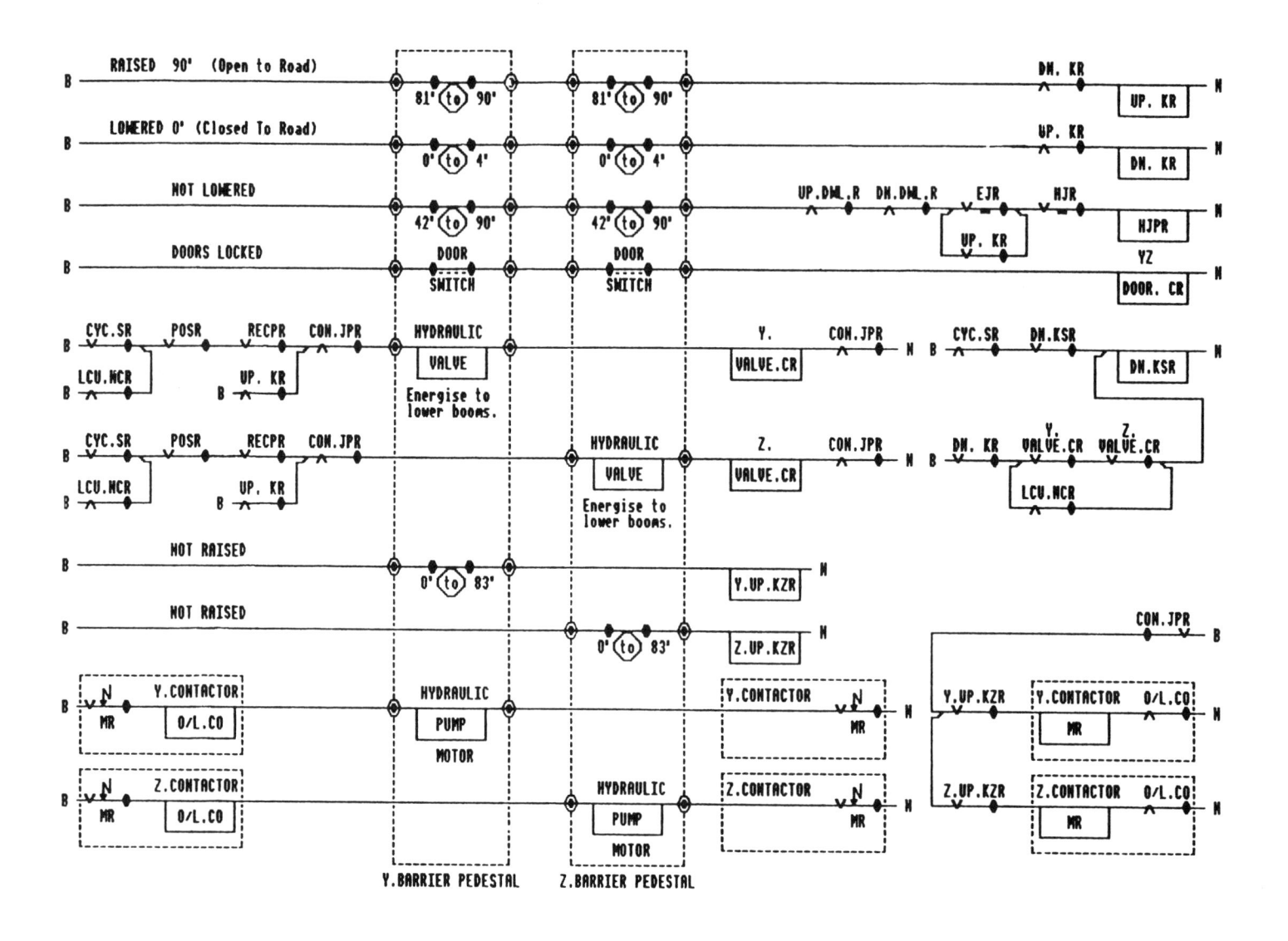

Resetting After a Failure to Arrive

After the driver's red rail signals have been showing in both directions for 30 s, the RE.YJR completes timing and operates the RE.YJPR thus operating the CON.SR and commencing normalisation.

Slow Train Arrives at the Crossing

Once a slow train arrives at the crossing, it will come to a stand at the flashing red rail signal. To re-initiate the crossing the driver must operate a plunger situated in a locked box at the foot of the signal. Providing that the track circuit is occupied (an anti-vandalism feature) operation of the plunger energises the DWL.YJR, followed by the DWL.YJPR. This interrupts the RE.YJR and RE.YJPR thus releasing the CON.SR and restarting the barrier sequence. Furthermore, the operation of the DWL.YJPR interrupts the operation of the plunger, thus preventing the plunger from holding the CON.SR. This makes the reset plunger one shot.

Another Train Coming while the Crossing is Still Operating

If another train strikes in on the opposite approach to the crossing, this will cause the operation of the DWL.R for the opposite direction, thus causing the driver's rail signal for that direction to show a flashing white aspect. In this event as the first train clears the crossing, the DWL.R for that direction will be disengaged by the occupation of the run-off track circuit, causing the rail signal for that direction to revert to flashing red. Also as the first train clears the crossing, the ATC.SR will release causing the audible warning signals to change tone.

Another Train Coming just after the Crossing Clears

If another train strikes in just after the crossing normalisation sequence has commenced, the re-initiation of the crossing will be delayed for 10 s, by the release of 1.TJR/2.TJR to allow for the minimum road opening time.

Boom Obstructed and Fails to Lower Fully

In the event of a boom being obstructed and failing to lower fully, the sequence will continue as described, but the DN.KR will fail to operate, thus the DN.KSR will also not operate. After the passage of the train the booms will raise as normal but with the DN.KSR not having operated, the CYC.SR will therefore fail to operate, so registering that the full cycle of operation has not taken place and denoting a system failure. (This is not, however, indicated to anywhere.) Once a further train strikes in with the CYC.SR not operated, the operation of the DWL.CSR will be inhibited, thus maintaining the rail signal at flashing red.

Boom Droops

In the event of a boom drooping to 83 degrees above the horizontal during the time that the barriers are up due, for example, to a leaking hydraulic pack, the respective UP.KZR operates, which is followed by the MR contactor, thus pumping the barrier up again.

Local Control

The local control unit is accessible by a standard operator's key. The unlocking of the door releases the LCU.DOOR.CR, which interrupts the DWL.CSR and so prevents the rail signal from showing a proceed aspect. Once the three-position control switch is moved from the *auto* position the LCU.NCR is released, allowing the *raise* position to control the CON.SR, enabling direct working of the barriers irrespective of track circuit occupation. When it is required to resume automatic operation, the barriers must be down and the local control unit door locked. This will cause the operation of the LCU.DOOR.CR, followed by the LCU.NCR; the CON.SR will then operate to raise the barriers, dependent upon track circuit control. Then the LCU.SR will operate. This relay is to ensure that the crossing normalisation timing out circuit (RE.YJPR) is not operated at the time that automatic control is selected, which would bypass the track circuit controls.

Hand Operation

If it is necessary to hand operate the barriers, the action of opening a barrier pedestal door releases the YZ.DOOR.CR, thus interrupting the DWL.CSR and preventing the rail signal from showing a proceed aspect.

Power Failure

If the battery ceases to charge, the system will still operate except that the rail signal (DWL.CSR) is inhibited by the POPR. Once the battery ceases to hold the VD.R, the VD.SR will release, followed by the POSR. This prevents the barriers from being lowered until such time that the power is restored and the battery voltage recovers.

Manned Barrier Crossings

GENERAL DESCRIPTION

For each road approach, two barriers are normally provided, one mounted on the nearside of the road and the other on the offside. When closed to the roadway, the barriers meet at the centre of the road to close the highway fully. Each boom is conspicuously marked with alternate red and white stripes and carries two boom lamps (three on longer booms), showing a red light in both directions. Each barrier is normally fitted with a skirt, which when the barriers are lowered, blocks the space between the boom and the surface of the road. When the crossing is closed to the roadway, the booms should form a total barrier to animals and small children. Exceptionally, if the road is narrow, only one boom may be provided, which can be mounted on either side of the road, although it is preferable that it be situated on the left. It must also fully close the roadway. (Two booms must be provided, however, if the barriers are automatically lowered.) Also exceptionally, the skirt may be omitted if the local traffic does not involve animals on the hoof or there is not a significant pedestrian usage. Each road approach is provided with two road signals, that on the nearside of the road being known as the primary signal, and that on the offside as the secondary signal. Each signal consists of two red lamps and a single yellow lamp, arranged on a black backboard surrounded by a conspicuous white stripe. Each primary road signal is fitted with an audible warning device for warning pedestrian users of the operation of the crossing. When the level crossing sequence is started, initially the yellow lamps show steadily for a period of 3 s, followed by the red lamps flashing alternately at the rate of 80 flashes/min. After a further 4–8 s, the nearside booms commence to lower with all boom lamps being illuminated once the leading barriers have commenced to fall; full lowering taking approximately 7 s.

When both nearside booms are fully down, the offside booms commence to lower, also taking the same time. Any road traffic still on the crossing at the time that the booms commence to lower can thus escape before the offside booms fully enclose the railway. Once the booms are lowered, they are hydraulically locked in that position and may not be lifted by hand. The audible warning device will commence to sound as the sequence is initiated and will continue until all barriers are fully lowered. The sound of the audible warnings may be softened during the night hours. Once it is safe to re-open the roadway, all barriers rise simultaneously and the red road signals cease to flash once all barriers have reached 42 degrees above the horizontal. If, after a period of 20 s, all the barriers have not fully risen, the road signals will be re-initiated and the crossing will be classed as failed. There are no limitations on the type of road, or the traffic using the road. The only limitation on the use of the crossing is that once the barriers have risen, they may not be re-initiated until the road has been open for at least 10 s. However this control is not enforced except where automatic lowering is used; re-initiation of the crossing is at the discretion of the controlling operator.

Manually operated barrier crossings are always provided with protecting signals to prevent the passage of a train until the barriers are fully lowered and the crossing has been observed to be clear of road traffic, pedestrians or animals.

The protecting running signals are desirably placed at a minimum of 185 m from the crossing, but may be as little as 51 m, except where station platforms are involved, in which case a distance of 23 m is allowable.

Track circuits are provided between the protecting signals and the crossing to prevent the barriers from being raised until the train has passed clear. In certain cases, if a train runs by the protecting signal at danger, these intervening track circuits initiate the road signals, but do not lower the barriers.

TYPES OF MANNED BARRIER LEVEL CROSSING

MCB/Local
These are locally operated from a position adjacent to the crossing, where the operator is placed so that he can clearly see the whole of the enclosed road surface of the crossing (signalbox or gatebox). At such crossings, a train running by the protecting signals does not initiate the road signals.

MCB/Remote
These are remotely operated from a position away from the crossing, but where the operator is in such a position that the whole of the enclosed road surface of the crossing can be clearly seen by line of sight. Such instances are exceptional, the distance between the signalbox or gatebox and the level crossing not usually exceeding 277 m with no blind spots or obstructions which may obscure the sight of a person or animal. At such crossings, a train running by the protecting signals automatically initiates the road signals.

MCB/CCTV
These are remotely operated from a position away from the crossing, where the operator has to view the whole of the enclosed road surface of the crossing by means of a closed circuit television system. The television equipment usually consists of a single mast with duplicated cameras, the crossing being viewed on a single monitor with a second monitor available as a back-up. Either camera can serve either monitor.

On some long skew crossings, it has been found necessary to use more than one CCTV system, having two masts viewing different parts of the crossing with the pictures being simultaneously displayed on separate monitors in the signalbox. At such crossings, a train running by the protecting signals automatically initiates the road signals.

TOB
These are locally operated from a position adjacent to the crossing, but by the train crew of the movement which requires to use the crossing. In such cases, the train crew must be capable of viewing the whole of the enclosed road surface before proceeding. The operation of the crossing is either from a control pedestal mounted at ground level, or by a device which can be reached by the driver from the cab. The protecting signals are similar to those used at an AOCL/ABCL, normally displaying a flashing red aspect replaced by a flashing white aspect when the road signals are operating and the barriers are down.

MODES OF OPERATION
Barriers can be lowered by the actions of the operator, in the case of a CCTV crossing by selecting a picture, then pressing the *lower* button on the control console. Once the barriers are fully down, the operator has to ensure that the enclosed road surface is unobstructed, either by direct observation or the use of CCTV. He will then signify that no such obstruction exists by the operation of the *crossing clear* button. Once this is done, it is then possible to clear the protecting signal.

At busy signalboxes, where the number of crossings controlled is large, or where the other duties of the operator are onerous, the barriers may be lowered automatically by the approach of the train. In these cases the strike-in point is set at such a distance from the crossing that it is possible to clear the protecting signal in time to avoid the outermost distant signal from displaying a restrictive aspect. Providing that the route is set towards the crossing, that all intervening signals are showing a proceed aspect, and that the protecting signal would otherwise be capable of showing a proceed aspect, the approach of the train initiates the road signals and then the lowering of the barriers, simultaneously causing the CCTV picture to be displayed. When the barriers are fully lowered, an alarm is sounded in the signalbox to remind the operator to check that the crossing surface is unobstructed. If this is so, the operator presses the *crossing clear* button and is able to clear the protecting signal.

Once the protecting signal has been cleared the barriers may only be raised once the signal is at danger and free of approach locking, and all intervening track circuits are clear. Raising is by operating the *raise* button, providing *manual raise* has been selected. It will only be possible to raise the barriers if the *locked* indication is not showing.

If *automatic raise* has been selected, the barriers will rise automatically by the action of the train placing the protecting signal to danger and then occupying in sequence the approach track circuit to the crossing and the running-off track circuit. Once the approach track circuit and all other intervening track circuits from the protecting signal are clear, the barriers will rise. However, this will be prevented unless the protecting signal in the other direction is also at danger and free of approach locking.

If it is necessary to place the protecting signal to danger before the train has arrived, after the approach locking to the signal has normalised, it is also necessary to select *manual raise* in order to open the roadway.

In an emergency it is possible to stop either the lowering or the raising of the booms at any time by the operation of the *stop* button. To restart the booms it is necessary to operate the *lower* or *raise* with *manual raise* selected.

PROTECTING SIGNALS

Non-interlocked crossing-only signals are operated from the signalbox by means of a relay circuit following the switch or push buttons. This circuit has no other interlocking controls. All approach locking is achieved locally to the level crossing.

Non-interlocked, otherwise automatic signals are operated as for crossing-only signals, but additional controls in the aspect level are included.

Interlocked signals are fully interlocked at the main signalbox or remote relay room, the condition of the signals at danger and free of approach locking being repeated by relay circuits to the level crossing.

Shunting signals are always fully interlocked. Shunting over the level crossing requires the automatic raise facility to be inhibited.

Semaphore signals may be used in any of the above modes, with automatic raising, providing that they are motor operated and capable of being placed to danger by the passage of a train. If semaphore signals are mechanically operated, then the crossing may not be fitted with automatic raise facilities.

OPERATION OF THE CLOSED CIRCUIT TELEVISION SYSTEM

The display of the picture of the crossing is obtained by pressing the *picture* button. On manually lowered installations, the *lower* button for the barriers will not be effective unless a picture has been selected. When automatic raising is selected, the picture ceases to be displayed after the *crossing clear* button has been operated. If *manual raise* is selected, then the picture is displayed at all times whilst the barriers are down. If the picture is selected for any other purpose, the display is extinguished after a period of approximately 50 s. If the crossing is automatically lowered, the picture is displayed at the time that the lowering sequence is initiated and is extinguished by the operation of the *crossing clear* button.

Fig. 6.20 CCTV monitored remote barrier crossing — typical layout — signal and train approaching circuits

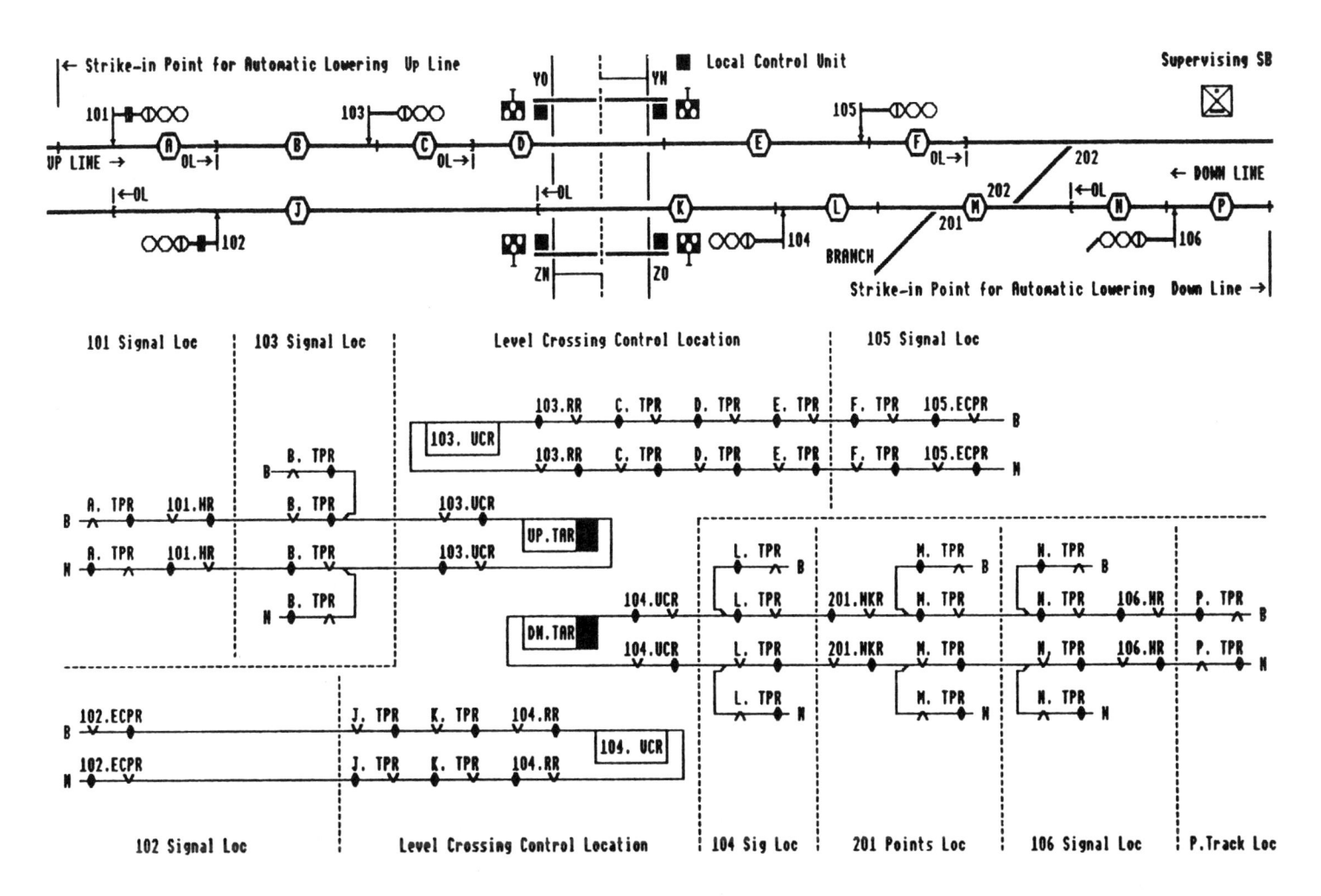

MANNED BARRIERS — DESCRIPTION OF OPERATION
(Figs. 6.20–6.27)

Selection of CCTV Picture
The signalman operates the *picture* plunger on the control console, which interrupts the picture line circuit, thus releasing the PICTURE.NPR. This releases the PICTURE.SR, causing the display of a picture on the CCTV monitor. When the signalman releases the picture button, the PICTURE.NPR will re-operate.

To Close the Roadway
The signalman operates the *lower* plunger, which releases the MAN.LOWER.NPR. The release of the PICTURE.SR has removed the hold from the LOWER.SR (ensuring that the picture is selected before the barriers are lowered) enabling the MAN.LOWER.NPR to release the LOWER.SR.

Initiation of the Barrier Sequence
The LOWER.SR releases the CON.SR thus de-energising the HJR. The CON.SR interrupts the UP.CPR line circuit, so extinguishing the *up* indication at the signalbox.

Yellow Road Signals
While the HJR remains operated, the release of the CON.SR will cause the yellow road signals to illuminate and start the audible warning.

Red Road Signals
The HJR will release 3 s after de-energisation, followed by the HJPR, which thus releases the RER, and de-energises the CON.JR and the RECPR. The RER extinguishes the yellow road signals and starts the static flasher, causing the red road signals to flash. Providing that all lamps are intact in each red road signal unit, the respective RECR will operate, thus maintaining the RECPR. Release of the RER, providing that the RECPR remains operated, completes the REKR line circuit to the signalbox, so indicating that the red road signals are operating.

Emergency Stop Before Lowering Commences
The signalman operates the *stop* plunger, which energises the STOP.NPR. This operates the RAISE.SR, and the LOWER.SR, which is maintained by the MAN.LOWER.NPR. Both these relays reset the XNSR, thus resetting all control relays. To restart it is necessary to operate the *lower* plunger.

Lower Nearside Booms
The CON.JR releases 5 s after de-energisation, followed by the CON.JPR and the LOWER.R. The LOWER.R releases the valves on the nearside booms, thus allowing the booms to commence to lower. The LOWER.R also releases the RAISE.R, disabling the pump circuits to prevent the barriers from attempting to drive back up. As soon as one of the booms falls below 81 degrees above horizontal, the UP.KR and UP.KPR are released, thus illuminating the boom lamps on all booms.

Lower Offside Booms
As soon as both nearside booms have fallen below 4 degrees above horizontal, the DN.KR1 is operated. This disables the nearside boom valve circuits, and releases the offside boom valves. The offside booms commence to lower. Once both offside booms have fallen below 4 degrees above horizontal the DN.KR2 operates. This disables the offside valve circuits.

Emergency Stop During Lowering
The signalman operates the *stop* plunger, which energises the STOP.NPR. This operates the LOWER.SR, followed by the LOWER.R which closes the valves, thus arresting the lowering of the booms. The LOWER.SR will be maintained by the MAN.LOWER.NPR. In order to restart the lowering, it is necessary to re-operate the *lower* plunger.

Fig. 6.21 CCTV monitored remote barrier crossing — signal control and interlocking circuits

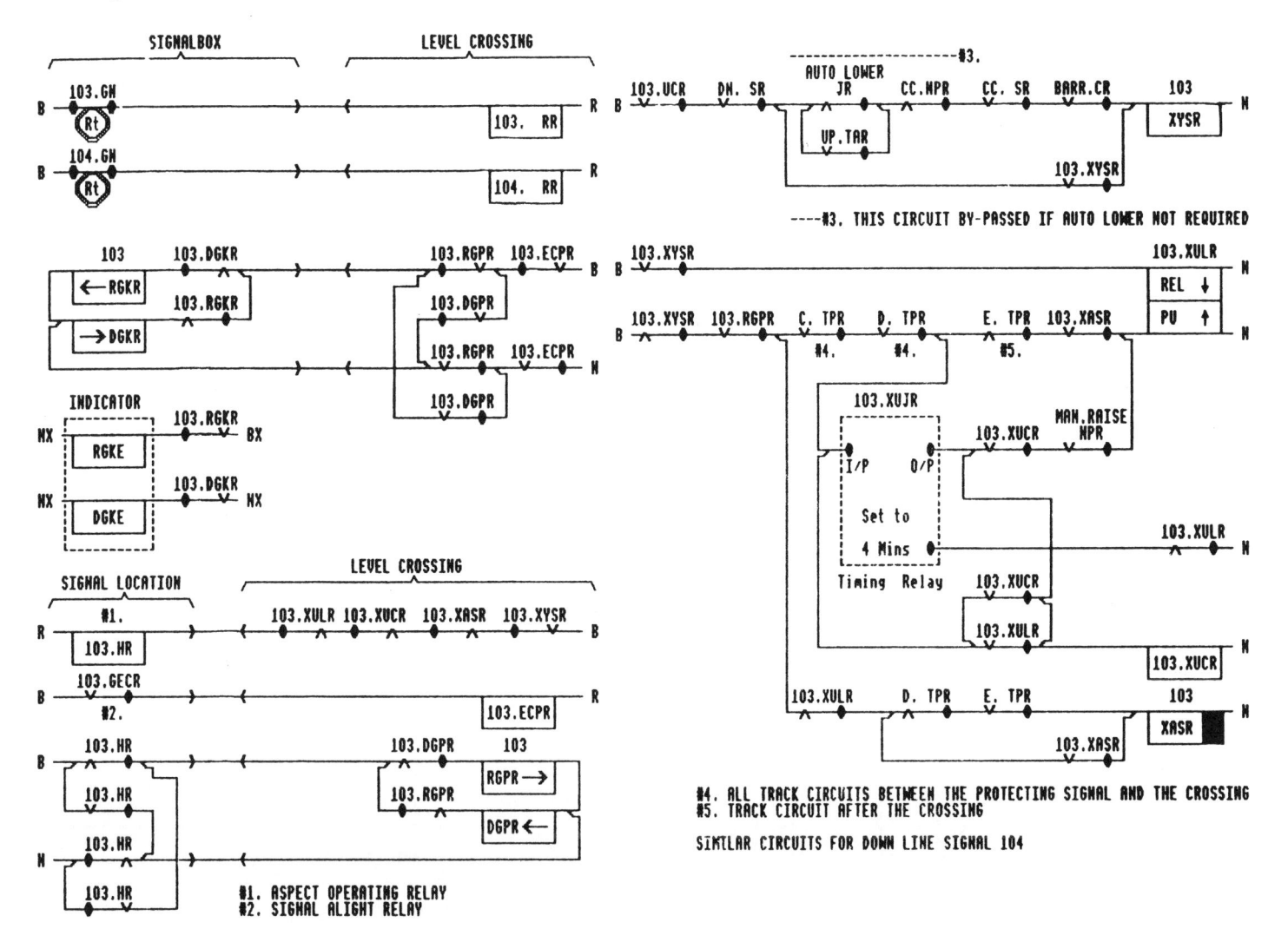

172   LEVEL CROSSINGS

Fig. 6.22 Manned barrier crossing — typical control console — indication and alarm circuits

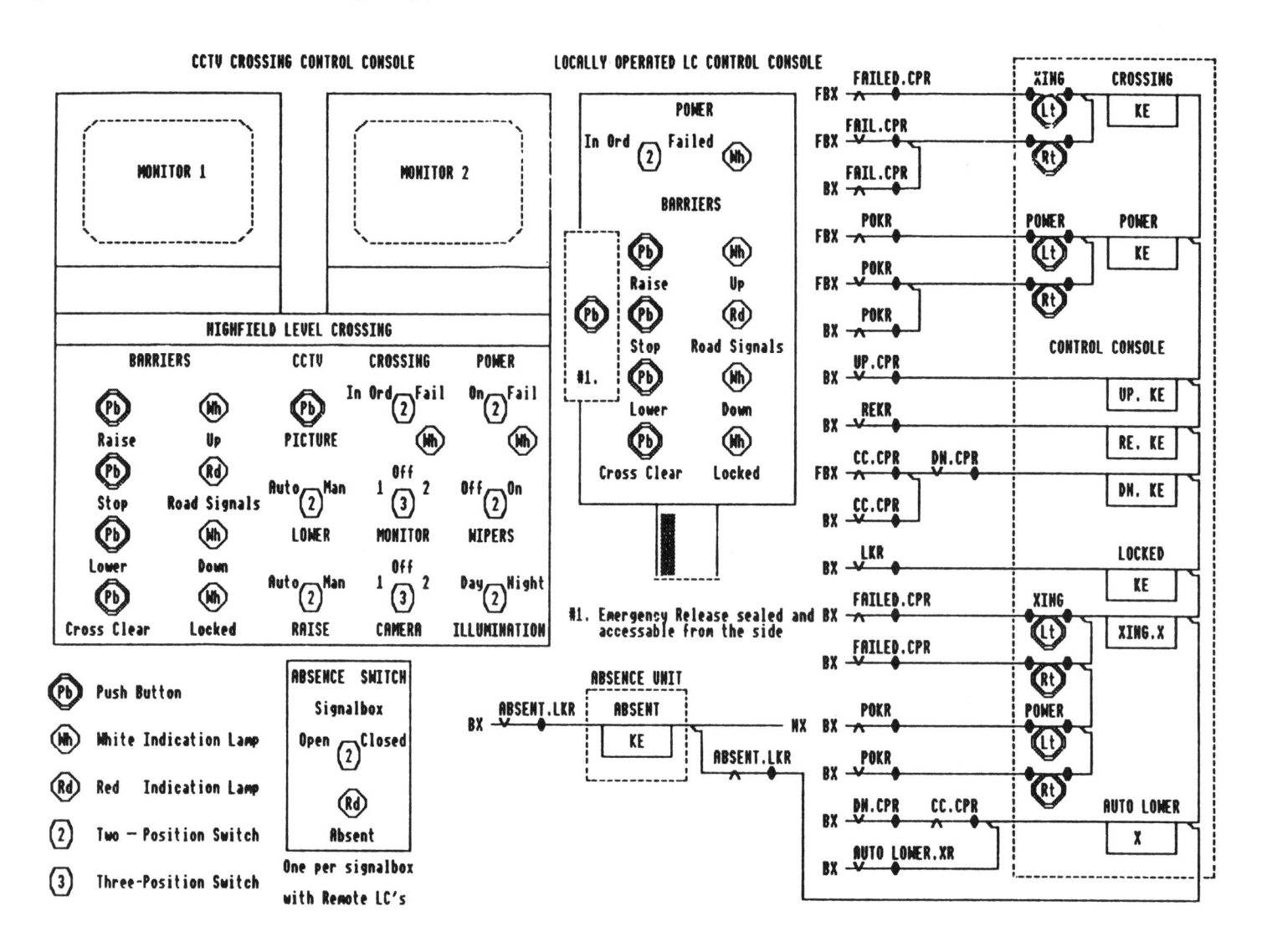

Fig. 6.23 CCTV monitored remote barrier crossing — control and indication circuits

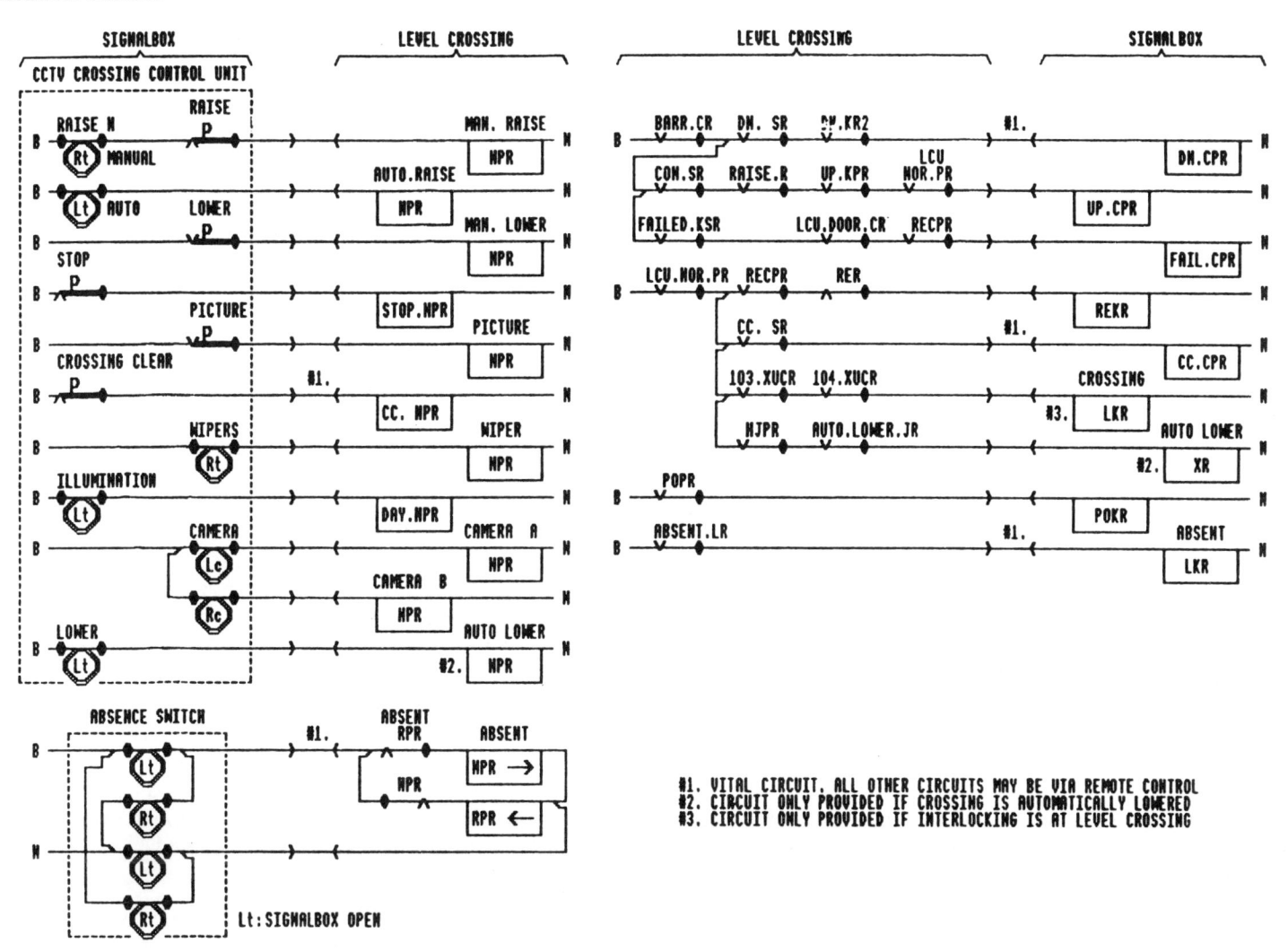

Completion of Lowering
Operation of the DN.KR2 operates the DN.SR. This relay checks the operation of all the control relays and ensures that the crossing clear sequence has not been pre-selected. The DN.SR completes the DN.CPR line circuit to the signalbox, indicating that the barriers are fully lowered by flashing the *down* indication.

Crossing Clear
The signalman must now look at the monitor to ensure that the crossing is unobstructed. If so, the *crossing clear* plunger is operated, which energises the CC.NPR line circuit and operates the CC.SR. The crossing clear indication line circuit to the signalbox is energised by the CC.SR, operating the CC.CPR, which causes the barriers *down* indication to steady.

To Open the Roadway
The *raise* selection switch is placed to the *manual* position, which interrupts the automatic raise line circuit. The signalman operates the *raise* button, which with the *raise* switch in the *manual* position energises the manual raise line circuit, operating the MAN.RAISE.NPR. This operates the LOWER.SR which is then maintained by the MAN.LOWER.NPR. The MAN.RAISE.NPR, with the signal interlocking normal and the red road signals operating (RECPR), operates the RAISE.SR. Providing that the interlocking is normal and all track circuits are clear, this operates the XNSR. The XNSR releases the DN.SR operating the CON.SR, which resets in sequence the HJR, HJPR, CON.JR and CON.JPR. The DN.SR interrupts the DN.CPR circuit, causing the *down* indication at the signalbox to extinguish. The signalman may then release the *raise* button, and thus the MAN.RAISE.NPR. The RAISE.SR is maintained by the DN.SR released.

Raising of the Booms
Operation of the CON.JPR releases the DN.KR1, followed by the DN.KR2. The CON.JPR, providing that all the red road signals are operating (RECPR) also causes the operation of the LOWER.R. This energises (closes) all valves in readiness for the raising. The operation of the LOWER.R completes the circuit for the RAISE.R. This operates the barrier MRs causing the hydraulic pumps to start, so commencing the raising of the barriers.

Extinguishing of the Red Road Signals
Once all barriers have risen beyond 42 degrees above the horizontal, the RER circuit is completed. Operation of the RER halts the static flasher causing the red road signals to cease. The individual RECRs release, but the RECPR is maintained by the RER.

Booms Complete Rising
Once all booms have reached 81 degrees above the horizontal, the circuit for the UP.KR is completed and is followed by the UP.KPR. This causes all the boom lights to extinguish and energises the UP.CPR circuit, so causing the indication at the signalbox to show *up*. As each individual boom reaches 83 degrees above the horizontal, the individual MR circuit is interrupted, thus causing each motor to stop. Each barrier is maintained in the raised position by the valve being closed. Should any boom droop, reclosing the 83 degree contact, the respective MR will operate so re-operating the boom to the fully raised position. Operation of the UP.KPR releases the RAISE.SR, which in turn releases the XNSR, interrupting the feed to the CON.SR which is maintained by its own slugging path, until the release of the XNSR re-energises the CON.SR by means of the UP.KR operated.

Extinguishing of the CCTV Display
The UP.KR, UP.KPR and RAISE.R reset the PICTURE.SR so extinguishing the picture.

LEVEL CROSSINGS 175

Fig. 6.24 CCTV monitored remote barrier crossing — miscellaneous circuits

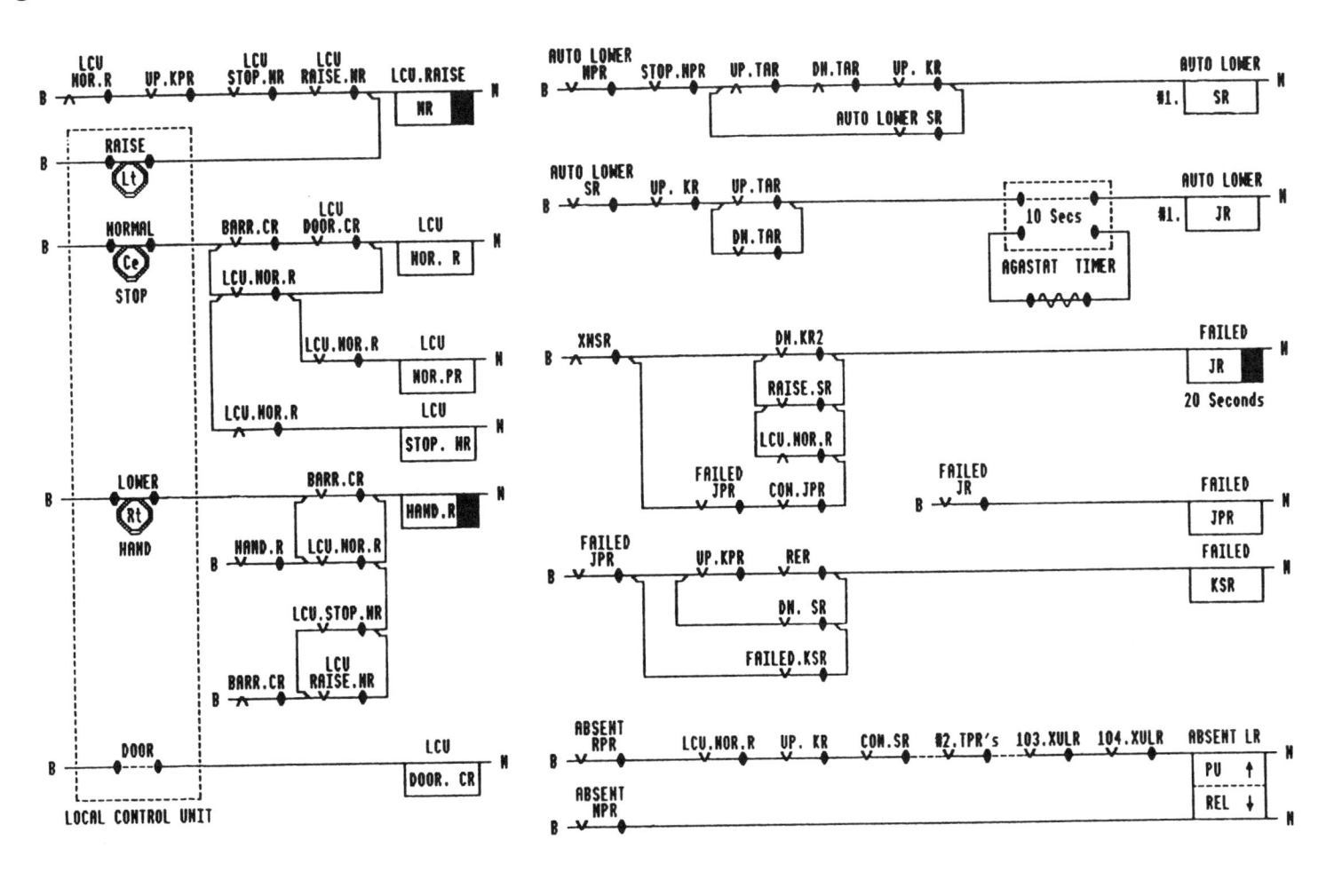

#1. RELAYS ONLY PROVIDED IF AUTO LOWER REQUIRED
#2. ALL TRACK CIRCUITS BETWEEN PROTECTING SIGNALS AND THE CROSSING

176 LEVEL CROSSINGS

Fig. 6.25 CCTV monitored remote barrier crossing

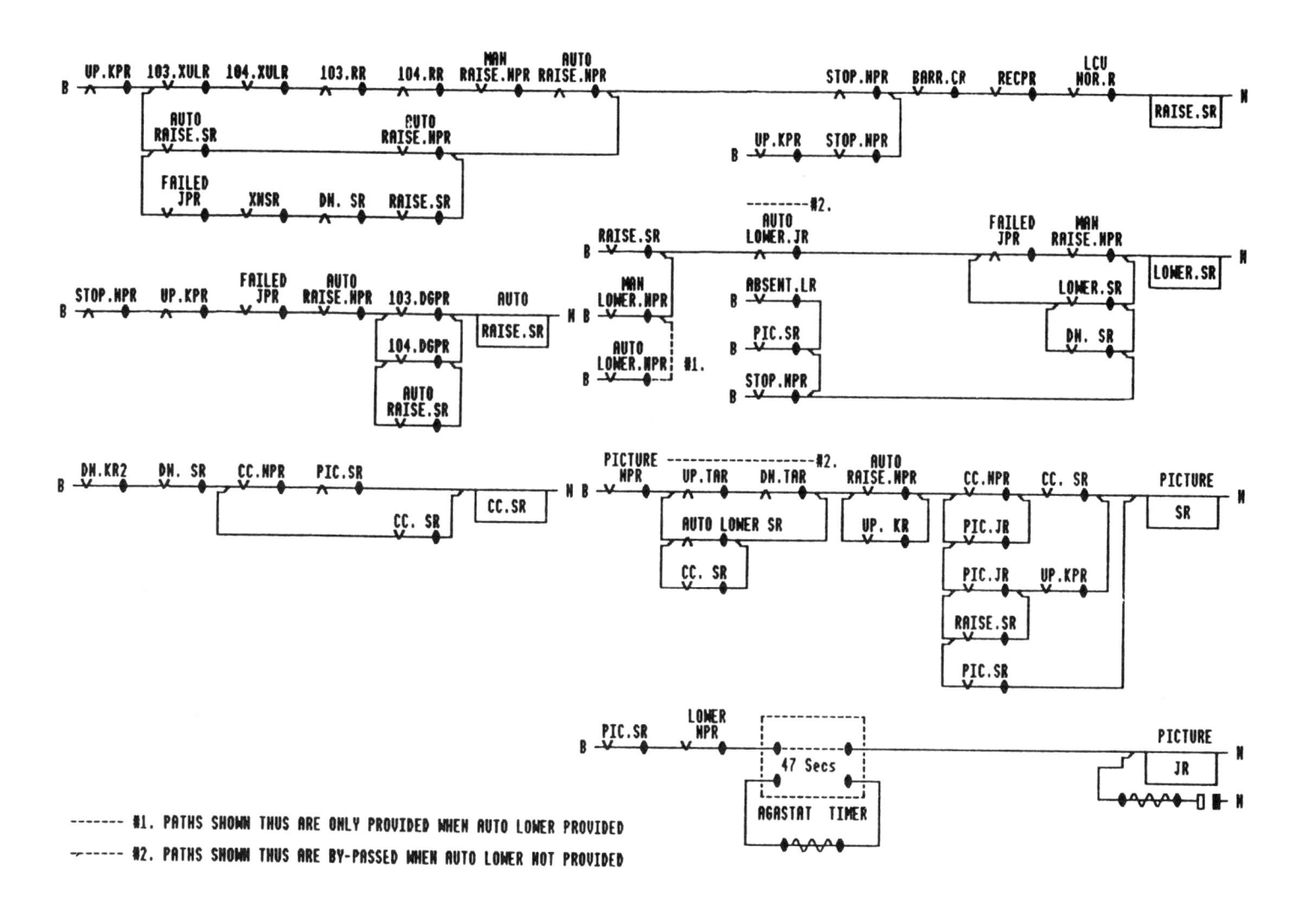

LEVEL CROSSINGS

Operation of the Protecting Signal
To clear a protecting signal, the signalbox switch must be operated so energising the 103.RR circuit. This is followed by 103.UCR which checks all the normal signal controls (signal ahead, track circuits, etc). The barriers have been lowered and the *crossing clear* plunger operated (CC.SR) and released (CC.NPR) thus operating 103.XYSR, which unlatches 103.XULR. The signal clearing circuit 103.HR is now completed, causing the signal to show a proceed aspect, followed by the release of 103.RGPR and the operation of 103.DGPR. This is indicated to the signalbox by means of the 103.RGKR/DGKR circuit.

Automatic Raising of the Barriers
With the *raise* switch on the control console set to *automatic*, the AUTO.RAISE.NPR line is energised. With the barriers not raised (UP.KPR released) and a protecting signal showing a proceed aspect (103.DGPR), the AUTO.RAISE.SR is operated, followed by the RAISE.SR.

Extinguishing of the CCTV Display
Providing that the raise selection switch on the console is in the *automatic* position, the automatic raise line circuit is energised thus maintaining the AUTO. RAISE. NPR. This together with the operation of the CC.NPR and CC.SR, causes the operation of the PIC.SR, thus extinguishing the picture.

Train Passes Through
The train passes 103 signal and occupies C.TC, thus releasing 103.UCR followed by 103.XYSR, causing the signal to return to danger and re-operating 103.RGPR. On occupation of D.TC, 103.XASR operates. On occupation of E.TC with C.TC and D.TC having cleared, 103.XASR causes the relatching of 103.XULR, followed by 103.XUCR and the release of 103.XASR. This completes the circuit for the XNSR providing that the opposite direction 104 signal controls are normal. Thus the XNSR operates and commences the raising sequence.

The Train does not Proceed
The signalman replaces the control switch of 103 signal, releasing 103.RR, followed by 103.UCR and 103.XYSR. Thus the signal reverts to danger, re-operating 103.RGPR. With the intervening track circuits C.TC and D.TC clear, 103.XUJR is energised and proceeds to time out, being completed by the operation of 103.XUCR. This energises the LKR line circuit, indicating to the signalman that the barriers are free. It is then necessary to select *manual raise* and operate the *raise* plunger; the MAN.RAISE.NPR then relatches 103.XULR completing the XNSR circuit and commencing the raising sequence.

Automatic Lowering
The *lower* switch must be set to *automatic*, thus energising the AUTO.LOWER.NPR line. With no train having struck in and the barriers raised, the AUTO.LOWER.SR operates. Assuming a down train, and providing that the route is set, the signals reading towards the crossing are clear (106.HR, 201.NKR) and the signal protecting the crossing is pre-set (104.UCR), a train occupying P.TC will operate the initiation relay DN.TAR, energising the AUTO.LOWER.JR. The DN.TAR also releases the PICTURE.SR thus displaying a picture at the signalbox. After 10 s, the AUTO.LOWER.JR operates. The purpose of this relay is to ensure that if the barriers have just risen (UP.KR) the automatic lowering is not re-initiated until the road has been open for 10 s. The AUTO.LOWER.JR releases the LOWER.SR thus commencing the lowering sequence, and also energises the AUTO.LOWER.XR line. This sounds an alarm in the signalbox to remind the signalman to watch the barriers lowering. The alarm continues until such time as the line is interrupted by the release of the HJPR. But once the barriers are down, the DN.CPR recommences the alarm as a reminder

178 LEVEL CROSSINGS

Fig. 6.26 CCTV monitored remote barrier crossing — barrier control circuits

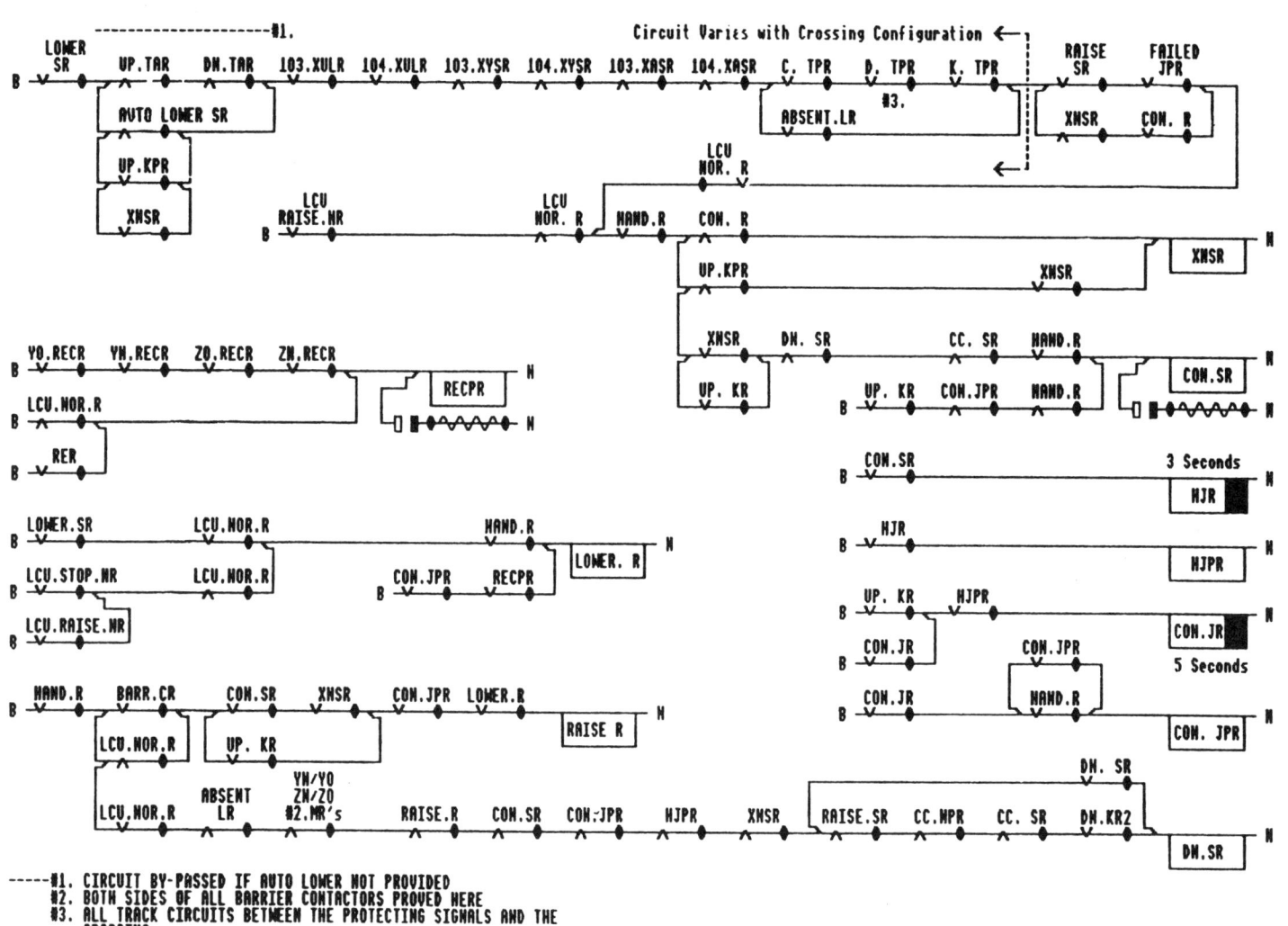

Fig. 6.27 Remote barrier crossing — boom operating and indication circuits

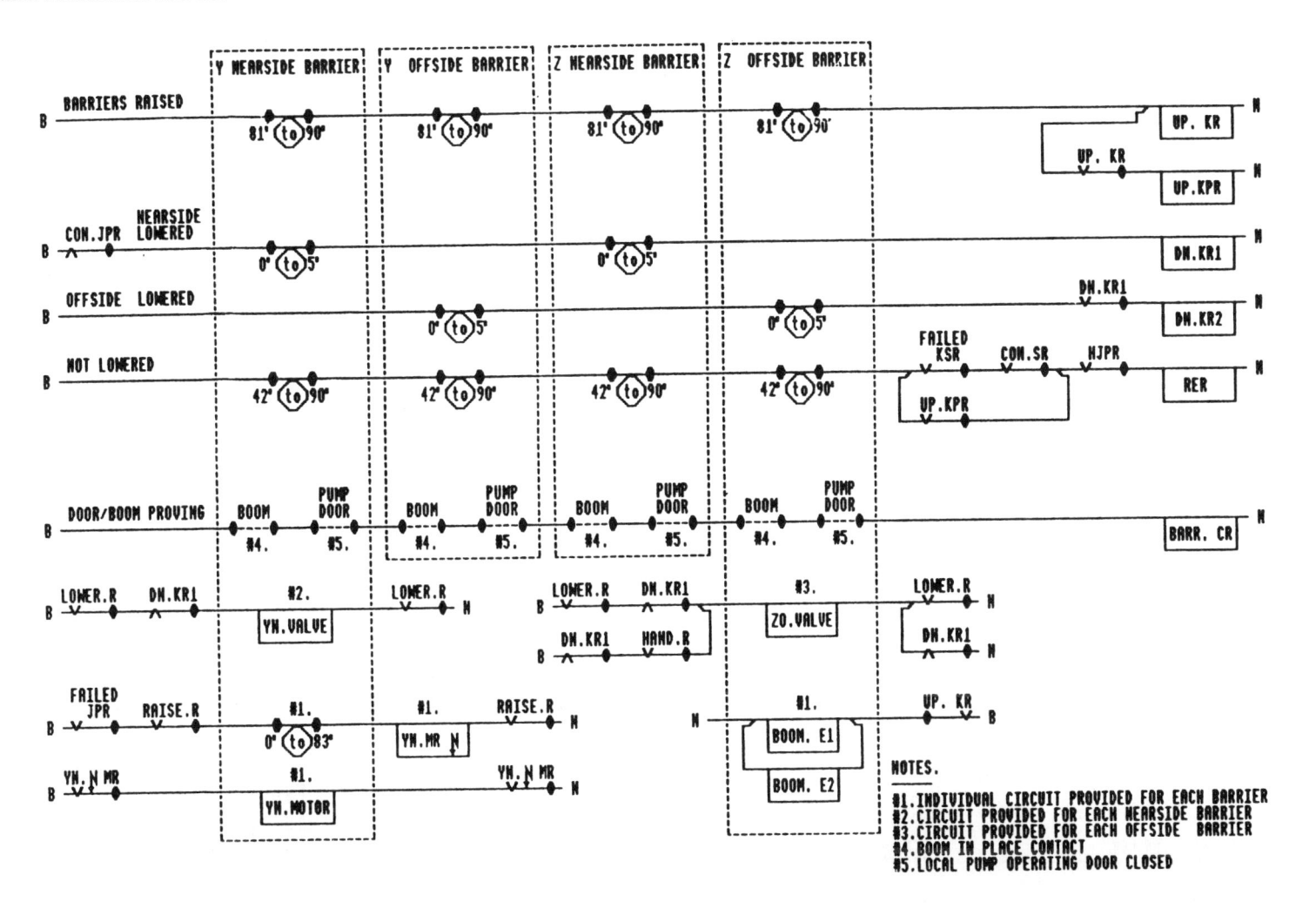

180 LEVEL CROSSINGS

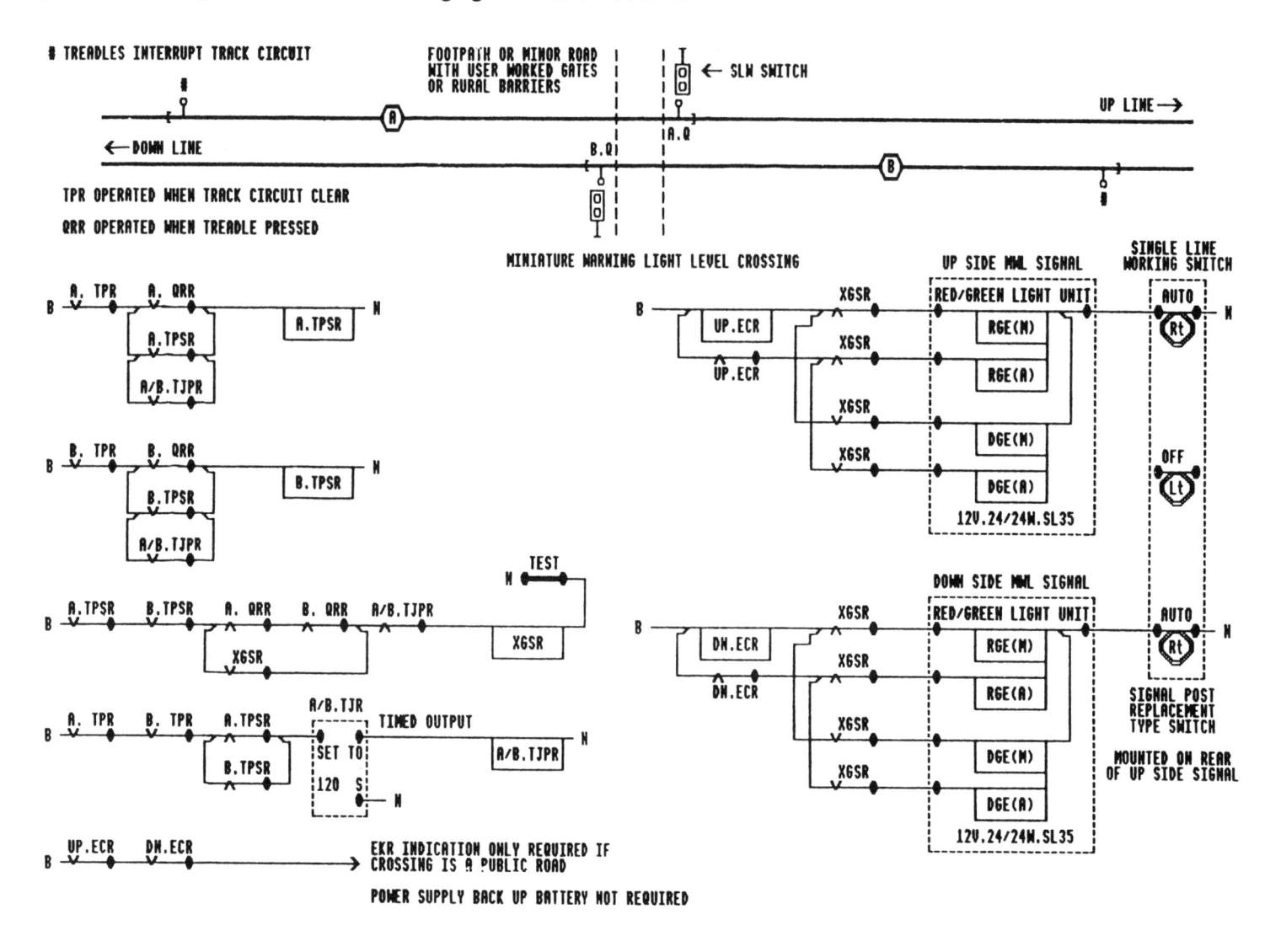

Fig. 6.28 Crossing with miniature warning lights — control circuits

to the signalman to check that the crossing is clear. This is cancelled by the operation of the CC.CPR.

Protecting Signals with Automatic Lower Selected

If both signals are pre-set, only the signal applicable to the direction in which the train is approaching will clear; this is done by the inclusion of the TAR operated contact in the XYSR circuit. Should the signal for the opposite direction clear when not required for a train, then the automatic raising of the crossing would be unnecessarily prevented.

OPERATION OF CROSSING WITH LOCAL CONTROL UNIT (LCU)

Operation on Power

Opening of the LCU door releases the LCU.DOOR.CR interrupting the FAILED.CPR line, showing the *failed* indication at the signalbox. Initially the barriers should be maintained in the raised position by selecting *raise* on the three-position switch. This operates the LCU.RAISE.NR which maintains the CON.SR thus bypassing all interlocking controls. Selecting *raise* releases the LCU.NOR.R and LCU.NOR.PR. To lower the barriers *lower* is selected; the LCU.RAISE.NR will be maintained through the *stop* position by the LCU.STOP.NR, but will release once the switch reaches *lower*. Thus the release of the LOWER.R will commence the lowering sequence. At this stage the HAND.R is maintained by the BARR.CR. To raise the barriers *raise* is selected, operating the LCU.RAISE.NR followed by the XNSR and LOWER.R, commencing the raising sequence. The lowering of the barriers may be stopped in an emergency by selecting *stop*, which operates the LCU.STOP.NR, followed by the LOWER.R which closes the valves and arrests the barriers. To restart, the switch is turned to *lower*. To stop during the raising sequence, *stop* is selected which releases the LCU.RAISE.NR, followed by the XNSR and RAISE.R. To continue, *raise* is selected again.

Operation by Hand

It is advisable to lower the barriers before selecting hand operation, in which case the road signals will be operating. With *hand* selected on the LCU, opening of any barrier pedestal door releases the BARR.CR, thus releasing the HAND.R. To raise the barriers, all doors must be opened and the pump handles extended. Under these circumstances, the hydraulic valves are mechanically closed. Once all barriers have been pumped up, the UP.KR operates, followed by the CON.SR, which operates all the control relays including the LOWER.R. Thus the valves will be electrically closed maintaining the booms raised when the pump handles are normalised. It is possible to return to local control on power with the pedestal doors still open by selecting *raise*, in which case the HAND.R is operated by the LCU.RAISE.NR, thus re-energising the XNSR and LOWER.R.

Return to Control from the Signalbox

If the barriers have been on hand operation and the pedestal doors are still open, selection of *normal* will operate the LCU.STOP.NR, followed by the HAND.R. Operation of the LCU.NOR.R requires that all pedestal doors are closed (BARR.CR) and the LCU door is closed and locked (LCU.DOOR.CR). If the barriers have been on local control, once the LCU door has been locked, the LCU.NOR.R will operate, releasing the LCU.STOP.NR and the LCU.RAISE.NR, restoring control to the signalbox and energising the FAILED.CPR line, showing a *normal* indication.

Barriers Failed Indication

As the barriers are lowered, the FAILED.JR is de-energised by the release of the CON.JPR. If the barriers fail to reach the fully lowered position, and the FAILED.JR is not re-energised by the DN.KR2, within 20 s, the FAILED.JR releases, followed by the FAILED.KSR, thus indicating *failed* at the signalbox.

When the barriers are raised, the FAILED.JR is de-energised by the XNSR. If the barriers then do not fully rise within 20 s, allowing the FAILED.JR to become re-energised by the RAISE.SR, with the XNSR released, the FAILED.JR releases, followed by the FAILED.KSR, again indicating *failed* at the signalbox. The road signals are re-initiated but the barriers are not lowered.

Signalbox Unmanned
When the line is closed the signalman operates the *absence switch*, energising the ABSENT.RPR circuit, which providing that the barriers are raised and the protecting signal interlocking and track circuits are normal, latches the ABSENT.LR. This bypasses the over-run track circuit controls in the road signal initiation circuit (CON.SR) and holds the LOWER.SR. Thus if a track circuit failure occurs while the signalbox is unmanned, it will not cause the road signals to operate, or if a line circuit failure occurs releasing the MAN.LOWER.NPR, the barriers will not lower inadvertently.

Failure of the Red Road Signals
If, before the raising sequence, all red lamps fail in one road signal, the RAISE.SR is inhibited, thus preventing the barriers from starting to rise.

CHAPTER SEVEN

Equipment

BR 930 Series Relays

The first volume of the Textbook introduced the BR 930 series of miniature plug-in relays and tabulates a number of the variations available.

The purpose and method of achieving certain of these variations is described briefly below.

GENERAL CONSTRUCTION

The basic and most commonly used form is the DC Neutral Line Relay BR 930.

The relay base fits into its appropriately coded plugboard, to which all associated wiring is terminated. Up to four contact stacks may be fitted, each of which may have up to four pairs of fixed and moving independent contacts. The variations commonly available are defined in the table previously mentioned.

The iron circuit consists of a heel piece (or yoke), a core with its coil and an armature which is pivoted from the heel piece, as shown in **Fig. 7.1(a)**.

The travel of the fixed contacts is limited by adjustment cards fixed to the heel piece, while the moving contacts are positioned by operating arms driven from the armature. The fixed contacts are pre-set against their stops on the adjustment card so that the correct pressure is obtained as soon as the moving contact starts to lift the fixed one. Contact wear during the life of the relay therefore has little effect on contact pressure.

The relay will function irrespective of the direction in which the DC is applied to the coil.

The BR 930 series of relays must always be fail-safe and the following essential mechanical and magnetic design features ensure this:

Fig. 7.1(a) DC neutral line relay BR 930

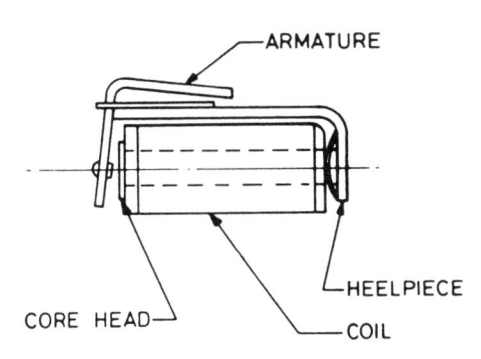

- Armature bearings are designed to prevent sticking due to friction, for example in knife-edge form.
- Armature return springs, if incorporated, are of the compression type, so that they will still operate if broken.
- In the event of spring failure, the armature will return by gravity.
- Magnetic circuits are of low retentivity iron. Iron-to-iron contact between armature and coil is prevented by the use of a non-magnetic residual pin on the armature to maintain a minimum air gap.
- Welding of contacts is prevented by using silver-impregnated graphite fixed and silver moving contacts.
- Materials are chosen to minimise corrosion or deposits which might cause moving parts to stick.
- Adequate contact separation is provided to ensure integrity of open contacts, with appropriate electrical clearance to isolate independent circuits within the relay.
- Silver or silver plated components are not used in combination with insulating materials susceptible to silver migration.

AC IMMUNE RELAY (BR 931)

With the advent of AC railway electrification, the possibility arises of faults in DC circuits causing unwanted AC to appear by induction in the DC supply to a relay. This could cause false operation with the danger of a wrong-side failure. Relays in AC areas, and particularly those operated through cables adjacent to electrified lines, therefore have to be immunised against AC operation.

This is achieved by the insertion of a copper sleeve or slug over the core adjacent to its pole face and a magnetic shunt between the coil and slug, as shown in **Fig. 7.1(b)**. When voltage is first applied to the coil, eddy currents induced in the slug create a magnetic flux in opposition to that in the coil, with consequent delay in the build-up of the resultant flux. This in turn increases the response time of the relay, so that the armature will not respond to alternating pulses of coil flux at 50 Hz.

BIASED AC IMMUNE LINE RELAY (BR 932)

Whereas the standard BR 930 line relay will respond irrespective of the polarity of DC applied to the coil, there are instances in which operation of the relay must be dependent upon the direction of the supply current, ie it will only pick up when DC is applied in one direction, but not when the polarity is reversed. Such relays are also required to be immune up to 1,000 volts 50 Hz.

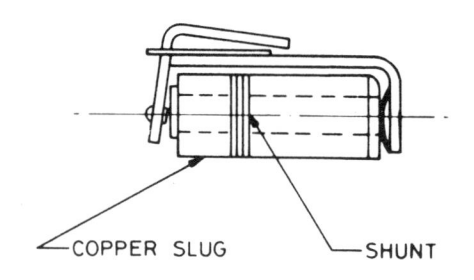

There are three ways of producing the required characteristic, each of which introduces a permanent magnet into the magnetic circuit:

- A series magnet is attached to the core pole face.
- A series magnet is attached to the armature.
- A shunt magnet is used to bridge the core and heel piece. In this method, a magnetic shunt, to which a permanent magnet is fitted, is inserted between heel piece and core, behind its pole face, as shown in **Fig. 7.1(c)**. When the coil is energised in the operate direction, the main flux cannot pass through the shunt, which is saturated by the opposing magnet flux, and it therefore passes through the armature to operate the relay. When energised in the reverse direction, the main flux reinforces the magnet flux and is not available to operate the relay.

Fig. 7.1(c)   AC immune line relay BR 931

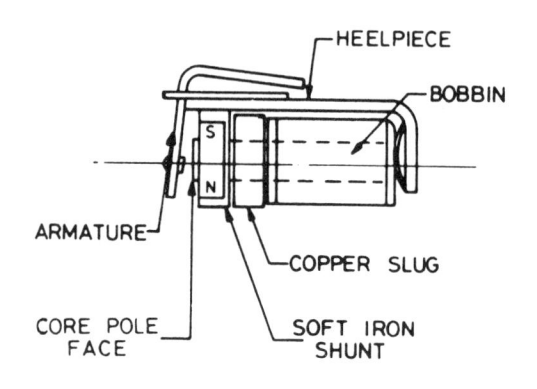

Fig. 7.1(b)   Biased AC immune line relay BR 932

AC immunity is achieved by the addition of a copper slug, and additional magnetic shunt if necessary, as previously described.

SLOW PICK-UP NEUTRAL RELAY (BR 933)

The slow pick-up relay is used primarily as a track relay repeater, and is designed to bridge the gap which can occur due to a temporary loss of shunt of the track circuit, or slow drop-away of the track relay. It is mainly used in trackside location cases and has therefore to be AC immune.

The timing requirement is that the pick-up shall be as long as possible with the drop-away as short as possible.

Fig. 7.1(d) Slow pick-up neutral relay BR 933

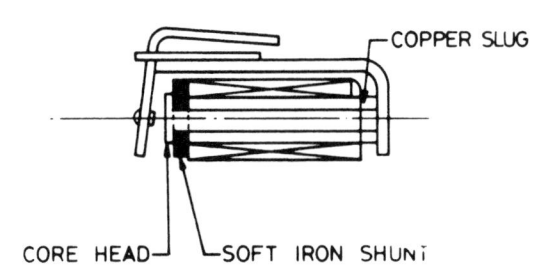

In one method, this is achieved by introducing a magnetic shunt path in parallel with the air gap. A copper slug is fitted on the core over which the coil is wound, with a magnetic shunt adjacent to the pole face, as shown in **Fig. 7.1(d)**. The flux generated by the coil is driven through the slug to build up in the shunt. When the shunt path becomes saturated, the flux builds up across the air gap to operate the relay. On release, the air gap flux collapses quickly, releasing the armature, whilst the remaining flux dies away in the coil and slug section of the magnetic circuit.

SLOW RELEASE NEUTRAL RELAY (BR 934)

The need arises for a relay having a slow release feature, particularly to cover the transit time of an AC immune relay which is inherently slow acting due to its copper sleeve.

One method of achieving slow release is by using a much larger copper sleeve on the core, the sleeve occupying the maximum space possible compatible with the minimum space required for the coil.

In a fully energised relay, the flux is at its highest when the air gap between armature and core is at a minimum. The height of the residual pin is therefore reduced to the minimum possible consistent with correct release of the relay when the applied voltage drops to a specified level.

MAGNETICALLY LATCHED RELAY (BR 935)

In a magnetically latched relay, the armature is held down in the conventional way by a return spring or springs. No residual pin is fitted, and when current of the correct polarity is applied, the relay will be energised to its fully operated (up to stop) position. It is held there by a permanent magnet on de-energisation of the coil. To release the relay, current of opposite polarity is applied to a second winding to oppose the flux from the permanent magnet, which is then bypassed by a magnetic shunt, thus allowing the armature to return to the de-energised position. The shunt avoids demagnetisation of the magnet and also prevents this exerting excessive pull except when the armature approaches the fully operated position. The magnetic circuit is as shown in **Fig. 7.2(a)**.

The relay is used particularly in interlocking circuits where contacts must not change state in the event of a power supply failure.

POLARISED MAGNETIC STICK RELAY (BR 936)

This is also used where the operation of the relay is dependent on the polarity of the current, and where the contacts must remain in their last operated position until the relay is energised in the opposite direction.

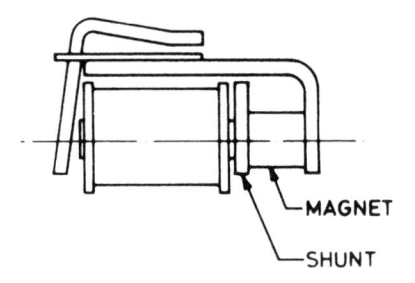

Fig. 7.2(a) Magnetically latched relay
BR 935

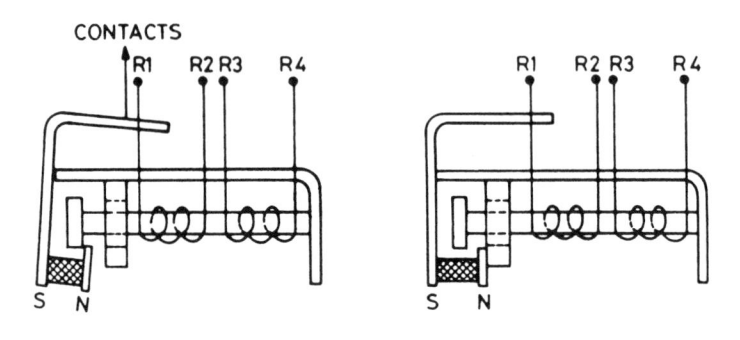

Fig. 7.2(b) Polarised magnetic stick relay
BR 936
[*released and attracted positions shown*]

It has a particular application in point control circuits, where it is vital that the points do not change state in the event of a power supply failure. The relay has normal and reverse contacts, covering the normal and reverse positions of the points, as distinct from the front and back contacts of other relays, which are made respectively in the energised and de-energised positions.

Two coils are provided so that the relay may be energised from electrically isolated circuits. In one method of obtaining stick conditions, a permanent magnet is fitted to the end of the armature such that in its two positions, the magnet forms two separate magnetic circuits, as shown in **Fig. 7.2(b)**.

In the normal position, the contacts bias the armature away from the core so that the magnet is in metallic contact with the shaped core head, the relay being then magnetically held in that position. When current of opposite polarity is applied to the second coil, the core flux repels the magnet and provides attraction to the core face. The armature is thus moved to the reverse position, further acceleration of the movement occurring during the transition as the magnet approaches the other pole face. When fully operated, the second magnetic circuit is completed, holding the contacts in the reverse position until the first coil is again operated.

DC TRACK RELAYS (BR 938, 939 AND 966F2)

While the function and electrical characteristics of track relays are described in the first volume of the Textbook, the method of construction is now obsolete.

Modern track relays are similar to those in the BR 930 series, but have the largest possible magnetic circuit and only two front contacts, which are used to operate a slow pick-up relay (BR 933) to provide the main circuit functions. Because of the need to operate some track circuits over as great a length as possible, maximum sensitivity is required with as high as possible percentage release to achieve good train shunt values.

LAMP PROVING RELAYS (BR 940 SERIES)

It is necessary to prove every main signal such that partial filament failure or total loss of illumination from a lamp which is intended to be lit results in a warning being given at the control centre and in the event of total failure, the signal in rear will be restored to, or remain at, red.

For junction indicators displaying a row of five lamps, at least three must be proved alight to preserve the identity of the indication.

A range of relays has been developed for use as current detectors for both AC and DC lamps, as listed in the first volume of the Textbook. In each case a slow release feature is incorporated to cover the transit time of the relay controlling the signal, for example the main signal lamp proving relay must not release with change of aspect, except when both lamp filaments have failed.

For AC applications a full-wave rectifier is built into the relay, its characteristics being such that it will withstand a severe overload caused by a fault occurring in the lamp circuit, the line fuse blowing before damage is caused to the relay. For flashing signals, the slow release feature is sufficient to prevent it releasing during the period of the flashing cycle when the lamp is momentarily extinguished.

BIASED CONTACTOR (BR 943)

This is an AC immune biased relay fitted with two pairs of heavy duty front contacts for switching electric point motors. Standard signalling back contacts are included for proving purposes.

In view of the heavy current required to drive the point machine and the relatively small contact opening possible with the BR 930 series relays, arcing can readily occur as the circuits are broken, causing consequential rapid damage to the contacts. To minimise this difficulty, the circuits are double-cut by connecting the two pairs of front contacts in series. At each contact position a blow-out magnet is fitted, polarised such that the arc is blown forward and out.

Separate contactors are required to operate the points normal and reverse.

TWIN RELAYS (BR 960 AND BR 961)

Twin relays are widely used where no more than eight pairs of contacts are required to be controlled by one relay. The twin relay therefore accommodates within the width of a single relay the equivalent of two single relays each carrying only half the usual number of contacts.

The DC Neutral Line Twin Relay (BR 960) is thus the counterpart of the single relay BR 930 and the AC immune version (BR 961) corresponds with its single counterpart BR 931.

The advantage of using twin relays lies in reduced cost compared with two single ones, and in the saving of space within the relay room, which in turn reduces the structural costs of the installation.

SSI INTERFACE RELAY (BR 966 F7)

With the introduction of SSI it has been necessary to incorporate a special AC neutral line relay in the BR 930 series as an interface between electronic circuitry and certain signalling functions. Its construction is such that it is basically a neutral relay incorporating a full-wave rectifier bridge for operation from AC. The relay also includes a copper slug which produces the desired delayed pick-up and drop-away times.

Main Signals

In the first volume of the Textbook, the general construction of main signals was described.

The construction details of the signal have not changed, except that where vandalism is prevalent, the clear glass outer lenses are replaced by polycarbonate ones.

It is also no longer a British Railways requirement that provision be made for junction indicators to be mounted directly on top of the signal head. The indicators are now installed on a separate mounting frame, which has the merit that they may be separately adjusted for sighting purposes. It also follows that a somewhat less robust design of signal head is acceptable, making fabricated construction possible as an alternative to a casting, resulting in a reduction in manufacturing cost.

Subsidiary Signals

The position light junction indicator, multi-lamp route indicator and position-light shunt and subsidiary signals were described in the first volume of the Textbook, together with a reference to the stencil indicator. The latter, having two lamps mounted on a central adjustable partition, enables a full width indication, or two independent half width ones, to be displayed. The indicator is capable of being built into multiple display blocks from one to seven units high, within limits imposed by the structure gauge at any particular location.

Limit of Shunt Signals

In locations where trains are allowed to proceed in the wrong direction along a section of track (eg in the down direction on an up line), it has been the practice to locate a limit of shunt notice board at the point beyond which such a movement may not proceed. As an alternative, an unworked mechanical disc signal, fixed permanently to display its stop aspect, has been used.

Such notice boards are now being superseded by a variation of the standard shunt/subsidiary position light signal. In this arrangement, two red lights in the horizontal plane are permanently displayed, whilst the third opening normally providing the off aspect for a shunt signal is blanked off.

This arrangement has the merit of following the standard practice of presenting a red aspect at a signal at which a train must stop.

Repeater Signals

Such signals are necessary where the view of the signal ahead may be restricted, such as on sharp curves, through bridges or in stations, and advance warning needs to be given if a stop aspect is being displayed.

The most common form of repeater is the banner signal, which consists of an illuminated circular background, against which a small black semaphore arm is displayed. In the horizontal position, the arm indicates that the signal it is repeating is at stop (red). When the main signal displays any other aspect, the repeater arm rotates to 45 degrees in the equivalent position to the left hand upper quadrant of a semaphore signal.

Where it is necessary to repeat conditions at a junction, a second banner signal has been used alongside the first, the one for the main route being higher than that for the divergence and thus being the equivalent of splitting semaphores. Under these circumstances, the banner for the main route would remain horizontal and the other would be cleared when a diverging route is signalled and the corresponding junction indicator is lit. Exceptionally, the second banner may repeat more than one divergence to one side of the main route, the actual route to be taken becoming apparent once the junction indicator is sighted.

Fig. 7.3 shows a typical arrangement of banner signals for a right hand diverging route.

Fig. 7.3 Typical arrangement of banner repeating signals with right hand route (West Hampstead BR–LMR)
[*photograph by courtesy of Westinghouse Signals Ltd*]

Position Light Speed Signals

The speed signal was originally introduced to control shunting movements over the hump in large marshalling yards. It is colloquially known as the 'Toton' signal, from the yard of that name near Nottingham where it was first used.

With the increasing popularity of bulk load freight trains, the need for shunting at such yards has largely disappeared. The signal has, however, found a relatively new application for the control of such trains in industrial sidings. In particular, it is used for the control of loading and unloading of 'merry-go-round' trains where, for example, the wagons of a complete train are loaded on the move at a colliery, and similarly unloaded at a power station.

The signal is illustrated in **Fig. 7.4** and consists of a circular array of eight lights surrounding a central pivot light. A row of three lights is lit in various angular positions. All lights are white, with the exception of the two outer ones forming a horizontal row, which may be red. A doublet optical system is used similar to that for the shunt/subsidiary signal.

A row of horizontal lights is used to indicate stop. When inclined at 45 degrees in the equivalent of the left hand upper quadrant, the train is allowed to proceed at slow speed. In the case of merry-go-round trains, this is the speed required for loading or unloading and is commonly about 1 km/h, for which special low speed controls are fitted to the locomotive. A vertical row of lights permits the train to proceed at a higher speed, usually after it has passed completely through the loading or unloading station. A fourth indication, with lights inclined at 45 degrees in the equivalent of the left hand lower quadrant may be used to reverse the train.

Because conditions may change as the train proceeds, it is necessary to place a series of such signals along the exit track, usually for the length of the train, and all will simultaneously show the same aspect. Thus, the train may be stopped, its speed altered, or it may be reversed as necessary by the controller or automatically through the associated control equipment.

For example, at an unloading station, if the bunkers are full, level sensors may be used to retract the automatic lineside units which unlock and open the wagon hopper doors and thus prevent further unloading. This action may be used to put the speed signals to stop until unloading may be safely resumed. The controller may also need to reverse the train if one or more wagons has passed through without releasing its load.

Fibre Optics

The increasing use of fibre optics for illumination purposes in many commercial and industrial applications has led to its introduction as an alternative to the traditional methods of illumination for railway signalling equipment.

The principle of the technique is that light from a single source is transmitted by a number of flexible glass fibre 'light guides', the outlets of which may be arranged to form a predetermined pattern. The usual arrangement is for the light guide outlets to be inserted into a metal plate having a matrix of holes, or the plate may be drilled to suit the configuration to be displayed. The plate may be covered by a suitable diffusing screen which can be of glass or polycarbonate sheet to resist vandalism. By superimposing the light derived from several separate sources, different patterns may be displayed or the colours changed.

For railway signalling purposes, the technique has a number of advantages. Firstly, where several indications are to be displayed, each is secret when not illuminated. Secondly, where separate units have hitherto been necessary for each indication displayed, a number of such displays may be generated in a single unit. Thirdly, for certain applications, an economy in light sources and therefore also in power consumption is achieved; and lastly, the source of light need not be within the unit itself.

Fig. 7.4 'Toton' position light speed signal

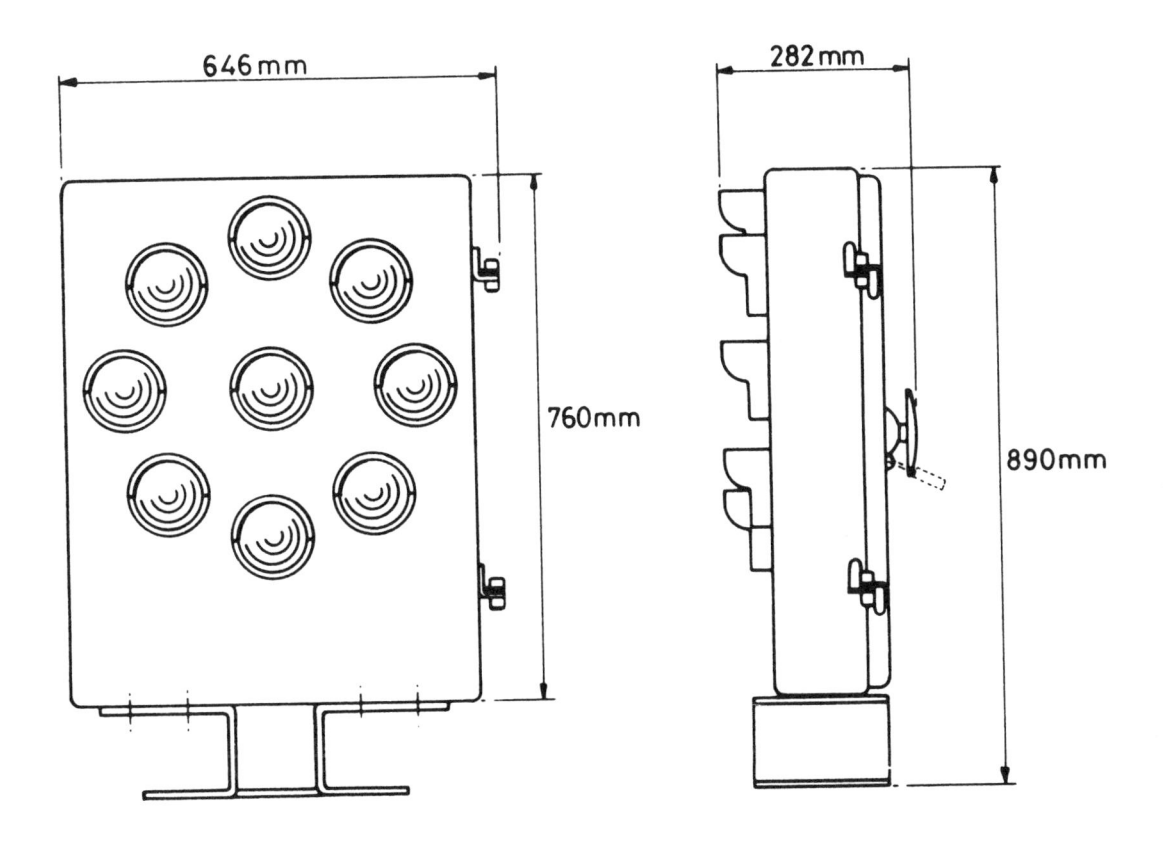

The most obvious application for fibre optics is in substitution for conventional methods of illumination in indicators such as the theatre and stencil types.

THEATRE-TYPE ROUTE INDICATOR

In this application, the array of lamps hitherto used is replaced by a network of small outlets into each of which an optical fibre light guide is coupled. Each indication is illuminated by its own single light source feeding an appropriate set of light guides to produce the required symbol.

In the conventional indicator, some lamps are often common to two or more characters and confusion of indications can occur should one or more lamps in a display fail to light. In the fibre optic equivalent, the outlets for each such symbol may be slightly displaced from one another without detriment to the indication displayed. There is thus no possibility of confusion of indications due to lamp failures; furthermore, each indication is totally secret when unlit. A further advantage lies in economy of power consumption and lamp requirements by using a single, instead of a multiple, light source for each indication displayed. This further simplifies lamp proving arrangements.

The requirements for indicators of this type are defined in BR Specification 1651 Part 1. Among other requirements, the characters specified approximate to 'Gill Sans' script and units must be able to display up to ten alpha-numeric characters.

The units commonly use tungsten halogen lamps rated at 50/55 watts at 12 volts, with appropriate housings or reflectors, each lamp having its own transformer, usually fed at 110 volts 50 Hz. The visibility range under good weather conditions is specified as 350 m in daylight and 210 m at night.

STENCIL-TYPE INDICATOR

In the conventional stencil indicator, one unit is required for each indication displayed, with the exception that where single characters only are to be shown, two such indications may be housed in a single unit. As previously mentioned, for multiple indications, the units have otherwise to be built into multiple display blocks within the structure gauge or other physical limitations.

In the fibre optic equivalent, the stencil is replaced by a network of small light outlets in a similar manner to that described for the theatre type.

The requirements for such indicators are defined in BR Specification 1651 Part 2. The characters specified again approximate to 'Gill Sans' script but are smaller than those for the theatre type. The units must be able to display up to four alpha-numeric indications each of up to three characters. Within the limits of available space to accommodate light sources and light guides, additional indications may be possible where less than three characters per indication are needed.

The method of illumination is similar to that previously described, but because the characters are smaller, there is a corresponding reduction in range of visibility to about 125 m in daylight and 100 m at night.

These arrangements cover the majority of applications, and the use of more than two such units associated with any one signal would thus be a rarity. Where two units are needed, they may be assembled one above the other as for the conventional stencil indicator.

The advantage of using fibre optics lies again in the secrecy of the display when not illuminated, with economy in the housing requirements for multiple indications. When associated with ground shunt signals, there is consequently less difficulty in accommodating the unit within structure gauge limitations.

REPEATER SIGNAL

The banner signal is another application for which a fibre optic equivalent is being evaluated.

In its current form, the moving semaphore arm is replaced by illuminating appropriate parts of the background, to reproduce the equivalent of the black arm in either of its two positions represented by unlit portions of the display. It has the advantage over its electromechanical counterpart that, having no moving parts, control circuits and maintenance are simplified.

In one prototype version, the signal uses the case and other components of the theatre-type indicator previously described, with similar illumination arrangements and visibility range. The lighting arrangements are duplicated, however, with the outlets from two sets of light guides interspersed. In the event of failure of one lamp the remaining one will therefore continue to provide a satisfactory indication at slightly reduced brightness. If necessary, one set may be switched off at night to reduce the brightness of the signal.

While improved visibility may be claimed, it is of course unusual for the required indication to be observed as an unlit area, with the risk of the brightness of the background itself reducing the visibility of the unlit band.

Because this application is at present in a relatively early stage of evaluation, some changes may be necessary before a fibre optic alternative is generally accepted.

FUTURE DEVELOPMENTS

Main Signal

Fibre optics have been developed for this application in the United States and elsewhere, and some experimental work has been done in the UK.

As at present conceived, red, yellow and green light sources are brought to a common focal point in the signal head from which a near-parallel beam is produced through a suitable clear lens arrangement. Separate lens systems for each aspect are unnecessary and the fibre optic signal becomes the visual equivalent of the searchlight signal. A second lens system is of course also necessary for the double yellow indication.

An advantage of such an arrangement lies in the reduction in size of the signal head required, although secondary outputs are needed to reproduce the equivalent of the auxiliary filament of the conventional signal in the event of main lamp failure. A difficulty arises with the accuracy of the optical system and acceptable colour reproduction, which must be controlled to the same close limits as for conventional colour light signals. These problems have to be resolved before this application of fibre optics becomes generally acceptable.

Remote Light Sources

At the present stage of development, the light sources for the equipment previously described are housed within the units themselves. However, with the introduction of fibre optic light guides, it follows that such arrangements are not essential and the light source may be removed from the unit itself and be located, for example, at the base of the signal instead of in an elevated position, or in an adjacent apparatus case. This has the merit of enabling maintenance, or lamp replacement, to be carried out more conveniently, with the added advantage that any associated electrical controls can also be more readily accommodated and serviced.

It is apparent that the signal unit itself may be of reduced size and weight if the illuminating source and associated control equipment is no longer locally sited. For elevated arrangements, such a reduction would have an obvious advantage during installation, and with reduced wind and direct loading, a corresponding economy in size of the supporting structure could follow.

Such an arrangement would be of particular advantage with theatre-type indicators, which are almost invariably associated with elevated main signals, but it would be less significant with the stencil type which generally relate to ground shunt signals. Remote location of the light sources

194 EQUIPMENT

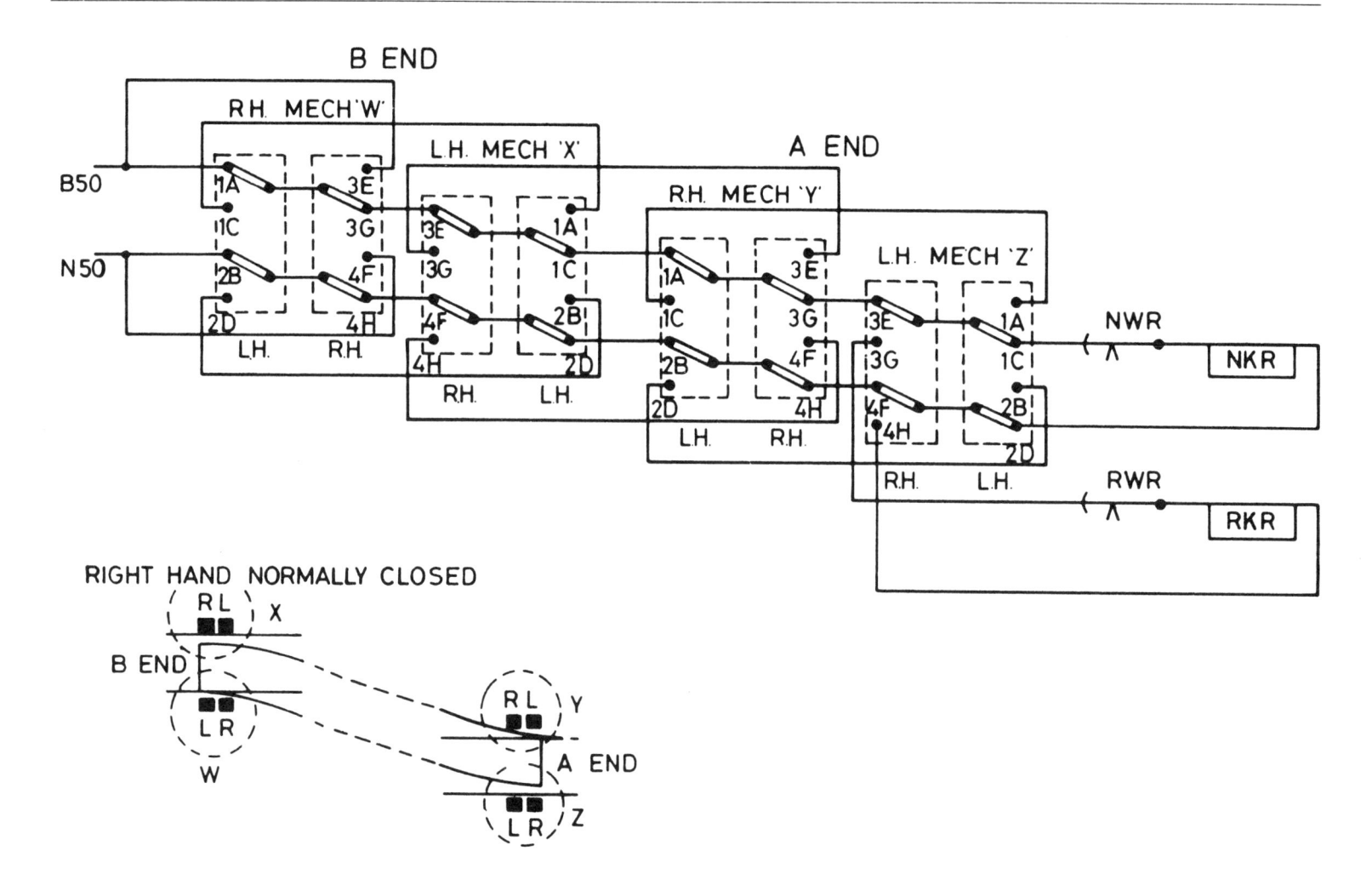

Fig. 7.5 Clamp lock standard detection circuit

would also be possible for fibre optic main signals, but with the present 'state of the art', degradation of the colour transmitted can occur unless the length of the light guides is relatively short.

Clamp Locks

The construction of the clamp lock, together with its electrical and hydraulic control circuits, are described in the first volume of the Textbook.

As a result of operating experience, a number of changes have taken place in constructional details which do not affect its method of operation.

Alterations have, however, been made to the detection circuits, the standardised version of which is now as shown in **Fig. 7.5**. It is now standard practice to detect the normally closed switch, which in the example illustrated is the right hand one. Also down-proving of the 'N' relay (NWKR) in the 'R' circuit, and vice versa, is no longer included and double-cutting of the circuits through pairs of NWR and RWR front contacts is replaced by single-cutting only since the control is in the same housing as the relay.

It will be apparent that this simplification does not affect the integrity of the detection arrangements.

Train-operated Points System

Despite the increasing sophistication of remote control systems, particularly for main lines, there are sections of lightly used rural railway where such expense cannot be justified, and where there remains a need for economy in operating costs. For these areas, methods of radio control have been developed, associated with unstaffed passing loops at which the points are operated by the trains themselves.

A simple passing loop has a single turnout at each end, the points being normally set in the facing direction for entry into the loop from the single line. When leaving the loop in the trailing direction, the points are reversed by the action of the wheel flanges, and restored to their normal position once the train has passed.

The system developed by British Railways for this purpose is shown in **Fig. 7.6**.

Each installation consists of:

- An energy pack, or accumulator unit, comprising a single-acting hydraulic actuator connected to the normally closed switch, a control valve block with pressure gauge, pressure detection switch and overload safety device and a hydro-pneumatic accumulator.
- A manual operation pack, comprising a second single-acting hydraulic actuator connected to the normally open switch, a hand pump and a manual control valve.
- An electric detector.

Ground connections are provided to second or third point drive stretcher bars as necessary from the leading drive stretcher bar.

The accumulator is pre-charged to the required pressure (75 bar) to generate a force in the extended actuator to hold the points closed in the normal position.

When the points are traversed in the trailing direction, the force generated by the wheel flanges exceeds the hydraulic holding force. This opens the closed switch and causes the actuator to retract. The hydraulic fluid thus displaced passes via a flow regulator control valve into the accumulator, so increasing the pneumatic pressure. As soon as the wheel flange generated force is removed or reduced, the potential energy in the accumulator is released into the system to reclose the points, in a controlled time of up to 20 s to minimise wear on the points as they are trailed.

The points can be moved to the reverse position manually by operating the hand pump, and can be secured in that position by a manual over-ride valve. Releasing this valve allows

196 EQUIPMENT

Fig. 7.6 Train-operated points

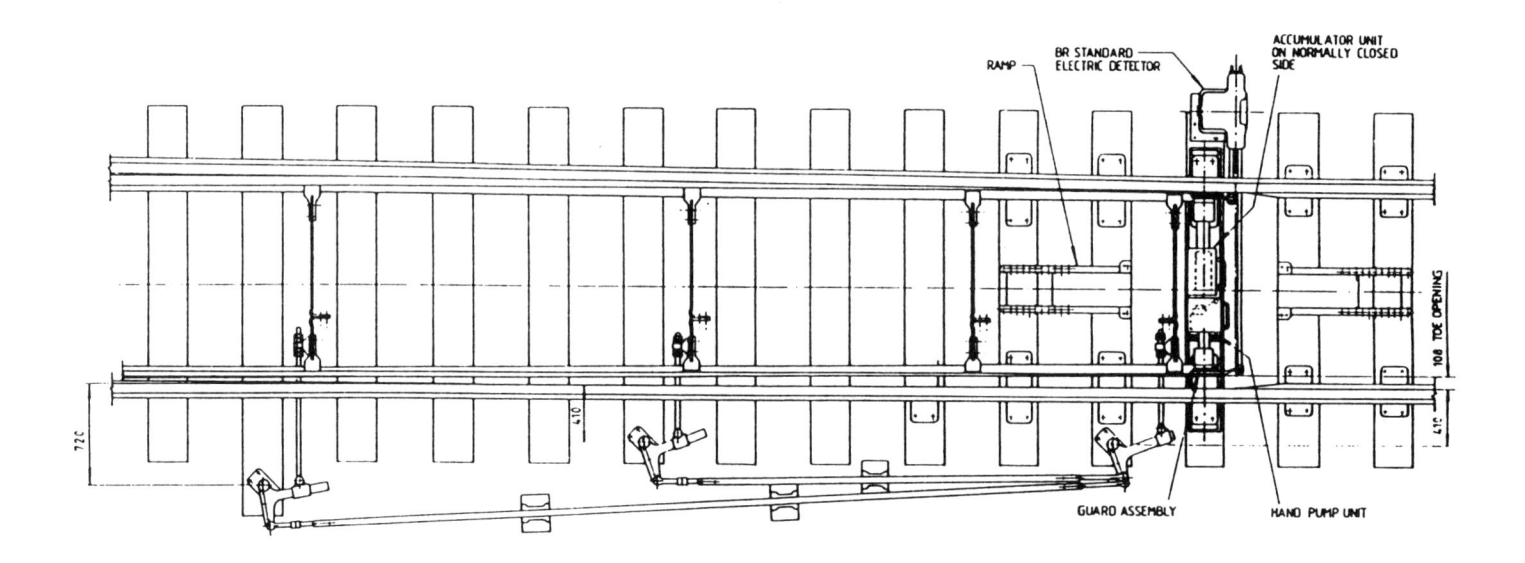

the points to be restored to normal under the action of the stored energy in the main system. Both energy pack and manual operating pack are self-contained and do not require hydraulic interconnections.

Because of the nature of the operating arrangements, the maximum permitted speed over such installations is restricted to 24 km/h. It is also important that to avoid possible derailment, trains passing over the points in the trailing direction must not stop immediately beyond the points and reverse.

Level Crossing Barriers

Originally, the law required that public highway level crossings in the United Kingdom should be protected by gates closing alternately over the railway and the road.

Under the British Transport Commission Act 1954 (s. 40), British Railways was generally empowered to replace these gates by lifting barriers similar to those in widespread use in continental Europe and elsewhere.

The basic operational and constructional requirements for lifting barriers are summarised below, whilst detailed discussion of controls and electrical circuits for different types of lifting barrier installations will be found in Chapter 6.

OPERATIONAL REQUIREMENTS

The requirements for such barriers are covered by BR Specification 843, together with the Department of Transport document 'Railway Construction and Operation Requirements for Level Crossings'.

There are two essential types of barrier installation:

- Automatic half barriers (AHBs) covering approximately half the width of the highway on the approach side to the railway. The barriers normally stand raised and after visual and audible warning are automatically lowered across the road on the approach of a train or trains.

- Manually controlled barriers (MCBs), generally of full width across the highway, pairs of barriers on each side of the railway being used where the width of the road makes this necessary. These may either be locally controlled, in conjunction with local railway signalling and road traffic lights, or remotely operated from the control centre with supervision by closed circuit television monitors (CCTV) at the crossing.

It is a requirement that the barriers shall be capable of descending against adverse wind conditions in both controlled operation or in the event of loss of power. The maximum wind speeds specified are 46 m/s for AHBs and 30 m/s for MCBs, a lower maximum being acceptable for the latter since they are usually longer and their operation is monitored. The most adverse condition occurs when the wind is blowing directly along the track against the direction of descent of the barrier.

The barriers must always descend reasonably rapidly, because an imprudent road user must be discouraged from any attempt to cross the railway once warning of closure has been given. This is of particular importance for automatic barriers since any serious delay could result in a train approaching dangerously close to a crossing before the road is closed. It is equally important for there to be as little delay as possible between closure and the arrival of the train, to discourage the foolhardy from 'zigzagging' round the lowered arms. It is also desirable that the barriers shall rise as quickly as possible once a train has passed, to avoid unnecessary delays to road users and possible traffic congestion.

British Railways specify a descent time through 85 degrees of 6–8 s for barriers up to 7.6 m long and 6–10 s for longer ones. The time taken for arms to rise through the same arc can vary from not more than 6 s for barriers up to 5.6 m long to not more than 10 s for those over 7.6 m.

CONSTRUCTION

An arrangement of a right hand barrier (which would be installed on the left hand side of the roadway) is shown in **Fig. 7.7**.

The equipment illustrated is typical of a manually controlled barrier installation, the principal components of which are:

- An electro-hydraulically operated power unit in a suitable housing, which also accommodates the electrical controls, and a means of manual operation. This is described in further detail below.
- Counterweighted side arms, mounted on the machine pivot shaft. A variety of balance weights may be used, either singly or in combination, to ensure that sufficient out-of-balance force due to the boom (and its skirt where provided) remains to allow the barrier to descend against the maximum wind speeds specified.
- The barrier boom itself mounted on one side arm. The boom is of a special extruded aluminium alloy section, having a tee slot at one side for mounting the barrier lights, and a second recess on the underside to accept the skirt pivot pins. The top surface is approximately semi-circular, minimising the risk of objects being left on the boom either inadvertently or deliberately and reducing the wind effect during operation.
- The boom is attached to its side arm by an adaptor. The fixing arrangements of the adaptor are such that in the event of impact from a road vehicle, the barrier, with its adaptor, will be released and drop out of its side arm. A device is included which will break an electrical circuit should this occur.
- The shortest standard barrier length (measured from pivot to tip) is 3.6 m and the longest 9.1 m, with intermediate variations in steps of 500 mm.
- The boom is coloured with alternate red and white vertical stripes, each 600 mm long, commencing with red at the tip, the red being retro-reflective.
- A support member mounted by a suitable adaptor on the second side arm, to stiffen the boom in the horizontal plane. This will likewise become detached on displacement of the boom itself.
- A straining wire to stiffen the boom in the vertical plane. The support member and straining wire are only provided for barriers over 6.6 m long, so that these features are not necessary on the shorter barriers.
- A skirt, pivoted from the boom, to close the space between the roadway and boom when lowered, and which folds back against the boom when raised. It consists of a number of equally spaced vertical members, restrained by a horizontal bottom rail. The skirt is not provided on AHBs since such barriers only partially close the road, but is usually fitted to manually controlled barriers.
- Red barrier lights placed along the boom at appropriate intervals. All barriers have one lamp located within 150 mm of the tip of the boom. Those up to 6 m long have a second lamp approximately at the centre of the boom and longer booms have additional lamps for every 3 m of boom length or part thereof, approximately equally spaced.

The lamp units provide front and rear illumination with provision for horizontal adjustment. Hoods are fitted to restrict the beam to 45 degrees either side of the lamp centre line, so that they are not visible to approaching trains, and as far as practicable, the lamp units are vandal-proof.

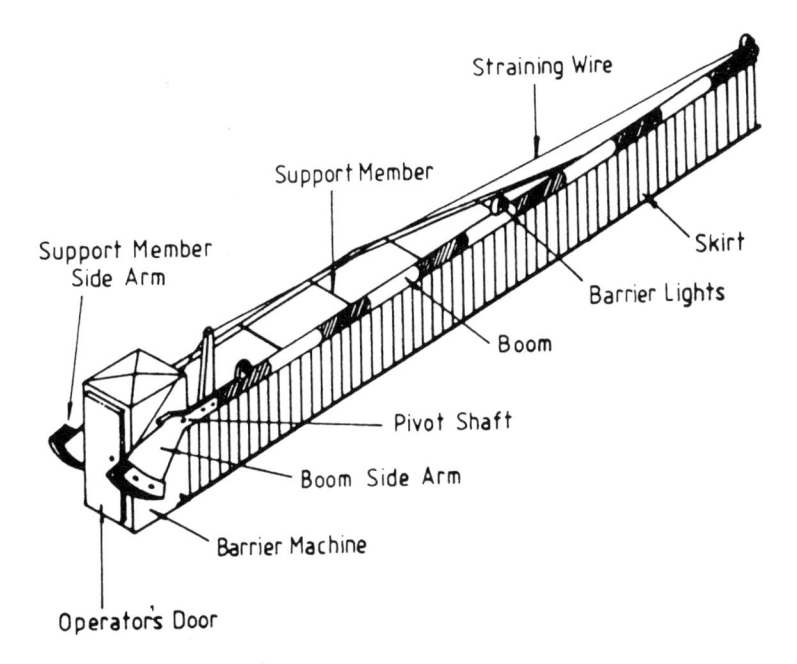

Fig. 7.7 Typical arrangement of lifting barrier unit

POWER UNIT

The BR requirements for the electro-hydraulic equipment forming the power unit for barrier machines are covered by BR Specification 985.

The arrangements are such that the barrier is driven to its raised position hydraulically where it is held by energising a solenoid valve. This, when de-energised, permits the barrier to return to its horizontal position by gravity. Re-energisation of the solenoid valve during descent of the barrier will immediately arrest the descent.

The principal features of the equipment are:

- A hydraulic linear actuator pivoted at its lower end on a trunnion secured to the supporting structure, and at its upper end on a second trunnion attached to the barrier operating arm.
- An electrically driven pump to deliver oil under pressure to extend the actuator and thus raise the barrier. The motor driving the pump to raise the barrier is energised through contacts remote from the unit.

- A normally open solenoid valve which, when energised, is closed to prevent the flow of oil from the actuator and thus maintain the barrier in its raised position.
- A pressure compensated flow regulator adjustable to control the rate of flow of oil from the actuator, and thus the speed of descent of the barrier. The arrangements are such that further control is exercised over the last 10–15 degrees of descent, to bring the barrier gently to rest.
- Means of selecting 'auto' or 'manual' mode of operation, for AHBs and MCBs respectively, the unit being locked and sealed in the selected mode.
- A hand lever operated pump system for local operation of the barrier, which can be operated at will to over-ride power operation. The lever is extended for hand pumping, and only permits the access door to be closed and power operation to be resumed when the lever is retracted and stowed in its out-of-use position. The lever has a 'hold' position which prevents the barrier from lowering and is capable of arresting its movement at any point while in motion.
- An oil reservoir forming an integral assembly with the actuator. Suitable oil filtration arrangements are provided to prevent foreign particles entering the hydraulic circuit. Pressure relief arrangements to prevent overloads damaging the system or the motor driving the pump are also provided.

The hydraulic circuit is a self-contained closed system such that the power unit may be installed and removed from the barrier machine housing without making any hydraulic disconnections. As far as practicable, elements of the hydraulic circuit are contained in a suitable manifold to minimise the number of hydraulic connections and the consequential risk of leakage.

When 'auto' operation is selected, the hydraulic circuit permits the barrier to be raised by an external force other than that generated by the power unit and offers minimum resistance to such operation. When the external force is removed, the barrier will descend at its controlled speed under gravity in the usual way.

When 'manual' operation is selected, the application of such a force is resisted, and the barrier may only be raised by the power unit. A relief valve is incorporated to relieve excess pressure generated by the application of an external force in these circumstances.

CHAPTER EIGHT

Operator Interface

The Integrated Electronic Control Centre (IECC)

The development of the solid state interlocking (SSI) was only the first stage in the application of microprocessor systems to railway signalling. In parallel with SSI development, rapid progress was also being made with the application of similar technology to signalman's control and display systems and to automatic route setting systems.

The IECC is essentially the combination of these three principal strands to give an integrated electronic signalling control and display system which also offers automation of timetabled train movements. An IECC must provide data to adjacent signalboxes and to other information systems concerned with the local area.

The IECC does not incorporate telecommunication facilities since to do so would jeopardise the availability of these systems in the event of a major failure of the IECC.

HARDWARE

Introduction

Designing new electronic systems is expensive in the first instance. There must therefore be a significant expectation of obtaining a reasonable system life and of reduced costs for later installations.

These parameters are most likely to be obtained by building systems around competitively supplied commercial equipment that is based on an acknowledged standard. This is the philosophy adopted for IECC.

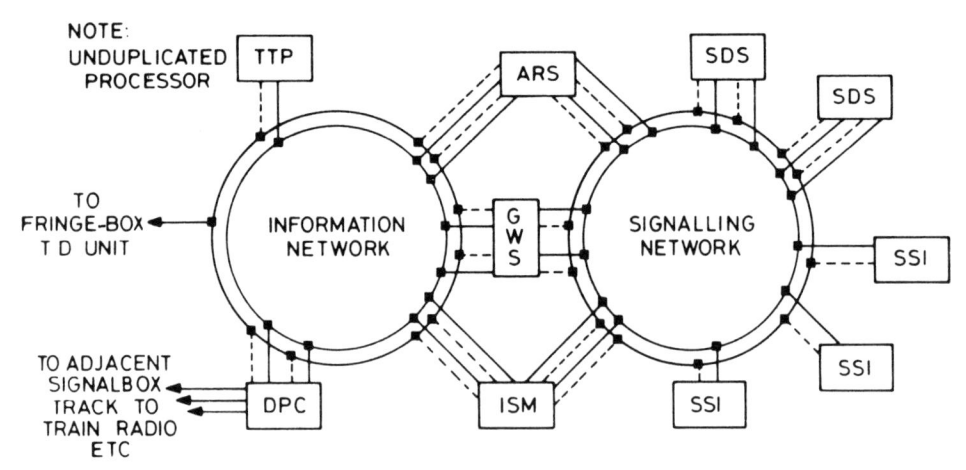

Fig. 8.1 The networks and systems of IECC

Systems
The basic networks and systems forming the IECC are shown in **Fig. 8.1**.

Signalman's Display System (SDS)	— provides the signalman's operating equipment
Automatic Route Setting (ARS)	— provides automatic route selection
Solid State Interlocking (SSI)	— provides the safety functions
Timetable Processor (TTP)	— provides timetable storage and manipulation
Gateway System (GWS)	— provides the bridge between the two networks
IECC System Monitor (ISM)	— provides the monitoring and technician's aids
Data Protocol Convertor (DPC)	— provides message protocol (message structure) conversion to the formats needed by connected systems
Fringe Box Systems	— fulfil the role defined for train describer fringe box systems

Modularity
Each of the systems described above (except for the fringe box) is duplicated and consists of two sets of processor boards, memory boards, and application-specific interface boards basically as shown in **Fig. 8.2**.

All systems are designed around the VME bus architecture, one of several systems allowing multi-processor configurations to be used.

Each system has an application processor card which provides the computing capacity needed to perform the specific

Fig. 8.2 IECC general subsystem configuration

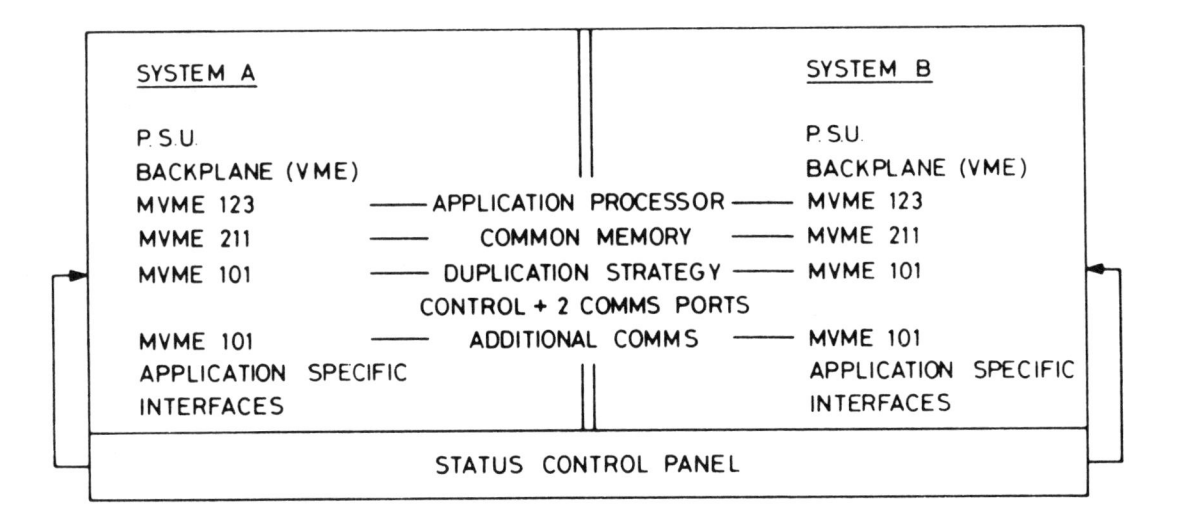

```
┌─────────────────────────────────────┬─────────────────────────────────────┐
│  SYSTEM A                           │  SYSTEM B                           │
│                                     │                                     │
│  P.S.U.                             │  P.S.U.                             │
│  BACKPLANE (VME)                    │  BACKPLANE (VME)                    │
│  MVME 123    ——— APPLICATION PROCESSOR ———    MVME 123                    │
│  MVME 211    ———    COMMON MEMORY      ———    MVME 211                    │
│  MVME 101    ——— DUPLICATION STRATEGY  ———    MVME 101                    │
│                  CONTROL + 2 COMMS PORTS                                  │
│  MVME 101    ———   ADDITIONAL COMMS    ———    MVME 101                    │
│  APPLICATION  SPECIFIC              │  APPLICATION  SPECIFIC              │
│  INTERFACES                         │  INTERFACES                         │
├─────────────────────────────────────┴─────────────────────────────────────┤
│                      STATUS  CONTROL  PANEL                               │
└───────────────────────────────────────────────────────────────────────────┘
```

duties of that system. There is a common memory board primarily used by the application processor, but accessible to the other processors to provide communication with each other.

In addition, each system has a second processor card for communication control and duplication strategy functions. It is this processor which, in conjunction with its counterpart in the duplicate subsystem, decides on the command status of the unit. This card also provides communication to the primary networks. Any additional communication links, be they to external equipment or to another network, are controlled by further processor boards which do not run the duplication strategy program.

Finally application-specific interface cards are used to access any peripheral equipment specific to the system. Examples include the display driver boards in the SDS or tape logging interfaces for ARS.

Networks
The IECC consists of two networks, both of which are duplicated for availability.

The signalling network provides all the communication facilities needed to control the movement of trains either manually or automatically. The information network provides data to or from adjacent signalbox systems or other information users.

The two networks are segregated to control the amount of information flowing in the signalling network where the response time, particularly to a signalman's actions, must be brisk and well-defined. Generally, systems on the information network can withstand delays measured in seconds and some variability of delay in the event of the network being busy.

The gateway provides a bridge between the two networks allowing relevant information to be exchanged. In addition, it provides filtering and buffering of exchanged data.

Both networks operate at 1.5 Mbaud with transmission between the network nodes and the systems being at 19.2 kbaud except for SSI which communicates at 9.6 kbaud.

Network Capacity

The prime determinant of the size of an IECC is the capacity of the signalling network to handle the data traffic between the control devices (SDS, ARS) and the interlockings.

Since the most frequent messages are those between the interlockings and the control devices, it is also most likely that this traffic will result in message collision.

As a result it was decided to operate a polling technique from SDS, ARS or GWS to each interlocking. Polling, however, results in additional message characters both to generate the poll, possibly when there is no data to exchange, and in setting up the network path. These additional characters represent an overhead to the message, and consequently mean that the message takes longer to transmit.

Polling has the advantage of quickly identifying the failure of a communication path, possibly before it would otherwise have been used.

Since SSI will only generate or respond to changes of state once during each major cycle, this can be defined as the minimum poll interval. The major cycle time of an SSI is about 600 ms.

The maximum poll interval is determined by the acceptable response time to an action. Manual route setting requires a prompt response if human frustration is to be avoided, hence a short interval. Automatic route setting is not quite so critical but still needs a reasonably short interval if events are not to appear in the wrong order. External information systems (via the GWS) are not usually time critical within a few seconds and hence can wait correspondingly longer for an update.

Each SDS will need to communicate with typically three or four interlockings but certainly not more than six. The actual exchange is limited by the 9.6 kbaud speed of the SSI link. However, six SSIs can be comfortably handled by one SDS even though the poll rate must be relatively brisk.

The ARS needs to poll all SSIs and although this can be at a lower rate than for SDS, analysis of the intervals required indicated that not more than 12 SSIs could be supported on one network.

The GWS must also poll all SSIs on the network, but as the speed is less critical than ARS, this is not a limiting factor.

A second design parameter, this time in SSI, is the number of change of state files held. SSI holds any changes of state that may occur in a file until the appropriate user device polls for the data. There are six such files in SSI representing $3 \times$ SDS, $1 \times$ ARS, $1 \times$ GWS and one available for the technician's display or other such user.

In summary, network capacity and file limitations result in any one IECC system having a maximum size of $3 \times$ SDS, $1 \times$ ARS, $1 \times$ GWS and $12 \times$ SSI or equivalent interlockings.

SIGNALMAN'S DISPLAY SYSTEM (SDS)

The SDS acts as the signalman's interface to both the signalling system and automatic route setting. The SDS is housed in a workstation which includes the telecommunication and other facilities, such as level crossing controls, required by that position.

Each SDS consists of a duplicated processor system driving typically four graphics screens through interface boards on the VME bus. In addition, there are interfaces to the trackerball controller and keyboard.

Two types of interface board have been designed and built for the SDS equipment. The 'control and display' board provides all the video timing signals and drives one display monitor. One is provided in each SDS computer unit. The second type is the 'display and display' board which drives two further monitors, receiving the relevant timing signals from the control and display board. Sufficient display and display boards are provided to ensure all monitors driven by the SDS can be operated.

The display drive boards are designed to generate the necessary high resolution (128 character by 48 line) pictures with the special character set required to depict a railway layout. Another important feature is the need to be able to change pictures on any of the display screens with a maximum delay of one screen refresh cycle and to have access to eight such pictures.

One important difference between SDS and the other systems is the need for an automatic changeover switch between the computer systems. Only one set of picture monitors is provided, with the consequent need to connect these to the currently on-line computer.

The Displays

The four graphics displays are normally configured to show two overviews, one detail view and a general purpose display. The symbols used on the overviews and detail views are mostly derived from those used on traditional panels.

The two overviews provide a complete view of the area controlled by the workstation, together with any necessary overlap with adjacent areas. The detail view may be one of six possible views giving close-ups of part of the control area. The general purpose display provides up to eight TD style map displays of adjacent areas, command input lines for confirmation of keyboard entry and output message plus alarm windows.

Main route setting can be accomplished by reference to the overviews or the details views. Shunt routes require reference to the relevant detail display, since it is only here that full track circuit, point indications and shunt signals are shown. It is also necessary to refer to the detail view whenever points are to be operated by any means other than route calling.

The eight graphic views are all fully interchangeable between three VDUs and can be configured to suit the current operating requirement. The general purpose display is dedicated during installation to a specific screen.

Overview Display

The overview display provides only information appropriate to route setting between main running signals. All details are shown at the minimum size compatible with legibility. Part of a typical overview layout is shown in **Fig. 8.3**.

The track layout is grey when no route is set or track circuit occupied. When a route is set, the relevant part of the diagram, including the overlap, is shown in white with point positions indicated. An occupied track circuit will be displayed red with, where relevant, the train description. Consecutive plain line track circuits may be displayed as a single section.

Full aspect information is displayed for controlled signals whilst automatic signals are indicated grey. Shunt signals are not shown.

Finally, ARS status indications together with other indications are given on the display, usually as a coloured symbol and an associated legend geographically positioned relative to their area of control.

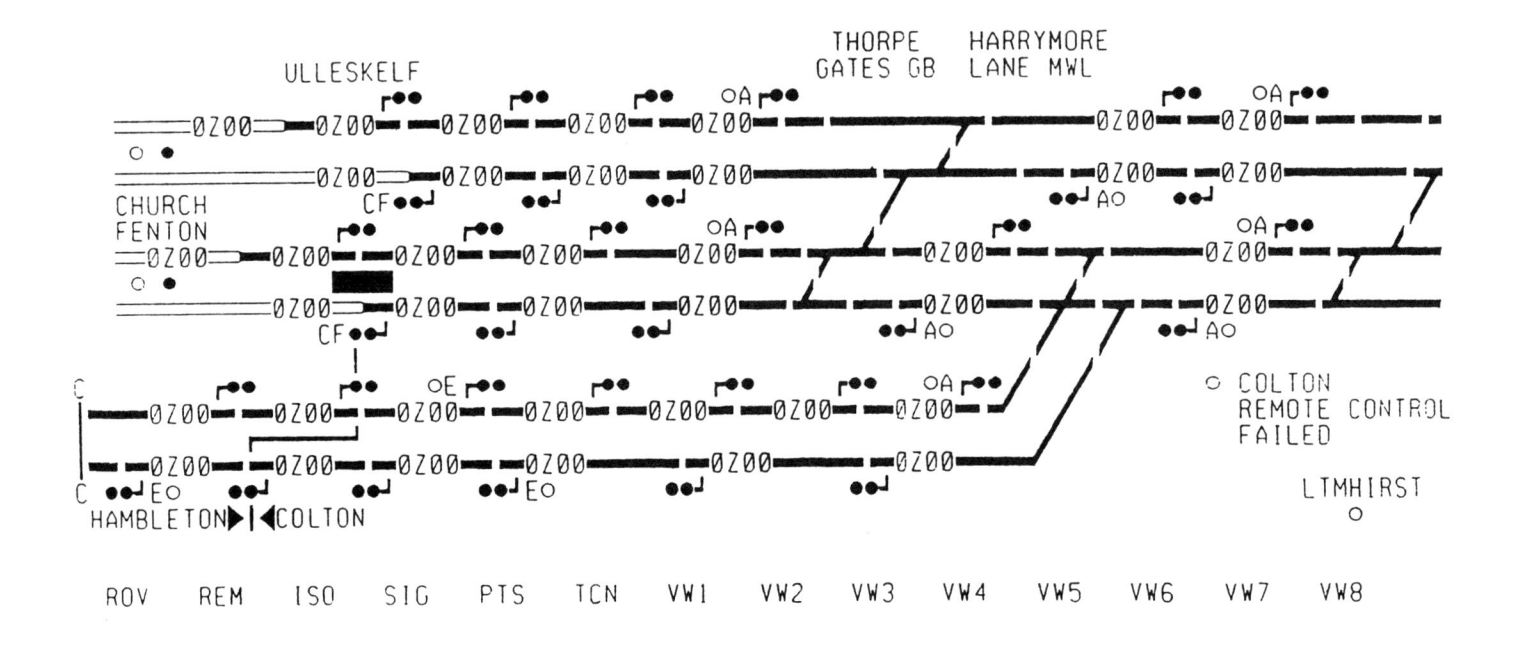

Fig. 8.3 IECC signalman's display — typical overview

Fig. 8.4 IECC signalman's display — typical detail view

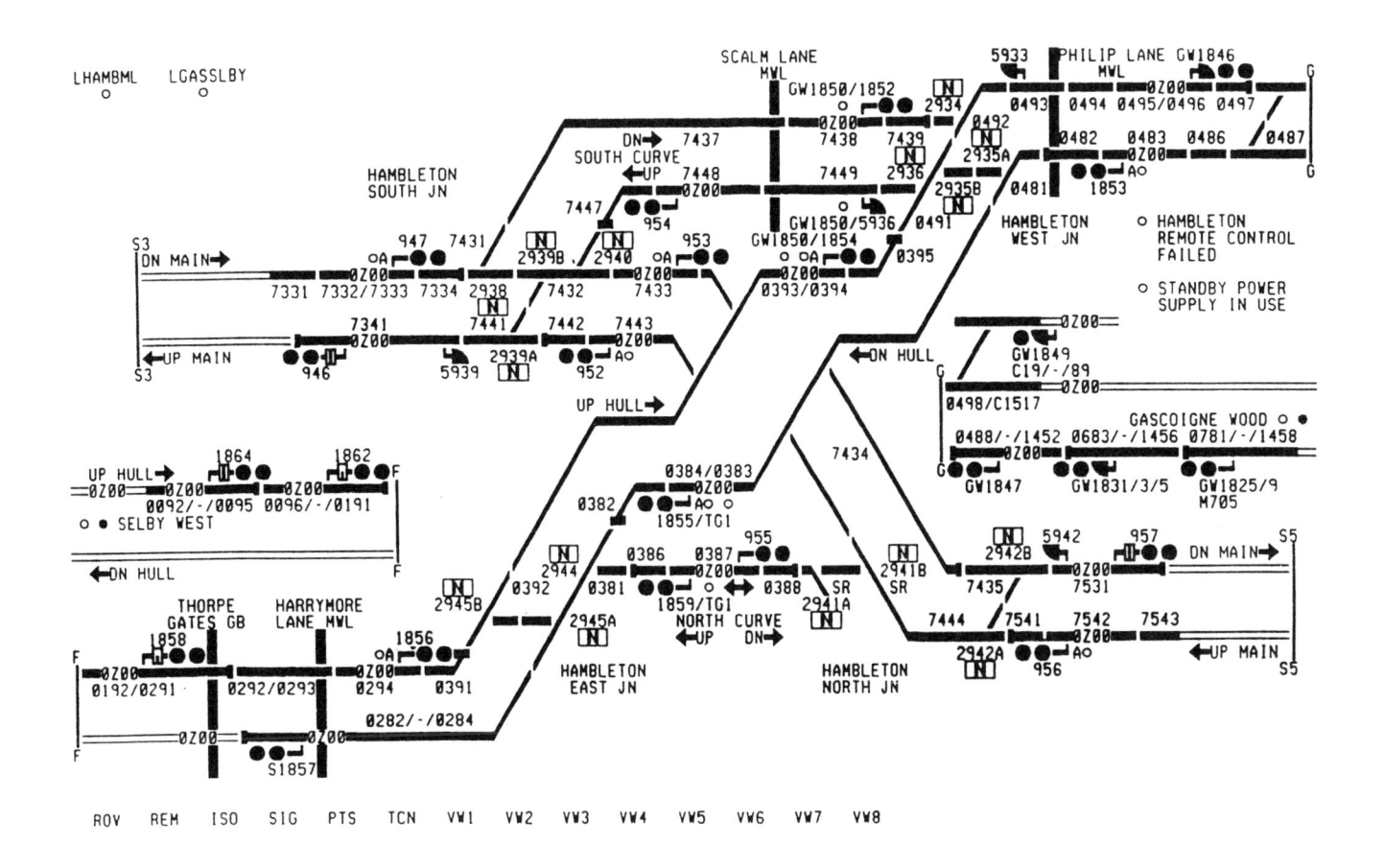

Detail View

The detail view characters are double the size of those used on the overview for most of the graphical elements. Part of a detail view screen is shown in **Fig. 8.4**.

In addition to the details shown on the overview, these pictures always depict the lie of the points together with an indication if the points are currently controlled by the key switch.

All track circuits are shown together with overlap limit markers where appropriate. All signals are displayed including shunt signals and notice boards.

The screen layout can display all signal, point and track circuit names, though the signalman can request any combination of these three sets to be displayed or suppressed at any time.

Reminder Display

Both overview and detail view screens can display reminders to the signalman. These take two forms, the 'memory jogger' and the 'operational inhibitor' or 'collar'.

In the first category are the possession and isolation area reminders. These are coloured backgrounds that can be displayed behind a length of track. Blue indicates a track possession while pink indicates an isolation of the electric traction. Combinations can be used. The presence of such reminders on the display does not inhibit the signalman from sending a train into the section. They resemble the overlays (eg traction isolation) used on hard panels.

In the second category are the reminder collars applied to signals and other route control functions. They also take the form of a blue or pink background with the same meaning as above, this time placed around the signal. When such a collar is applied, a special function is set in the interlocking which inhibits route setting. If necessary, the inhibit may be bypassed by use of an additional routine by the signalman during request for the route. The provision of a bypass facility was considered necessary to avoid the risk of a collar being temporarily removed and subsequently not replaced.

General Purpose Display

A screen having 80 × 24 characters, displays one of the eight train describer maps surrounded by various windows for data entry, output messages and alarms. A typical layout is shown in **Fig. 8.5**.

A review conducted prior to implementation indicated that significant improvements could be made to this presentation. Whilst some changes have already been incorporated, the layout and use will be further reviewed after operational experience has been gained.

Control

The primary means of manual signalling control is the trackerball, although some aspects such as train description entry require keyboard data.

TRACKERBALL FACILITIES The trackerball was chosen as a physically robust means of positioning a cursor that allows convenient operation for both right and left handed operators.

Around the ball are individual buttons representing the push and pull functions associated with a panel button and the three positions, normal, centre and reverse associated with a point switch. The 'push' and 'pull' (set and cancel) buttons are most frequently used and are duplicated to assist either handed operation.

The operator positions the cross hair cursor on an appropriate display symbol (eg signal) and presses a button. This follows the entrance–exit practice adopted on traditional panels. Typical symbols that may be targets for such control are signals, points (detail view only) ARS subarea controls, over-ride controls and alarm acknowledgements.

Fig. 8.5 IECC general purpose display

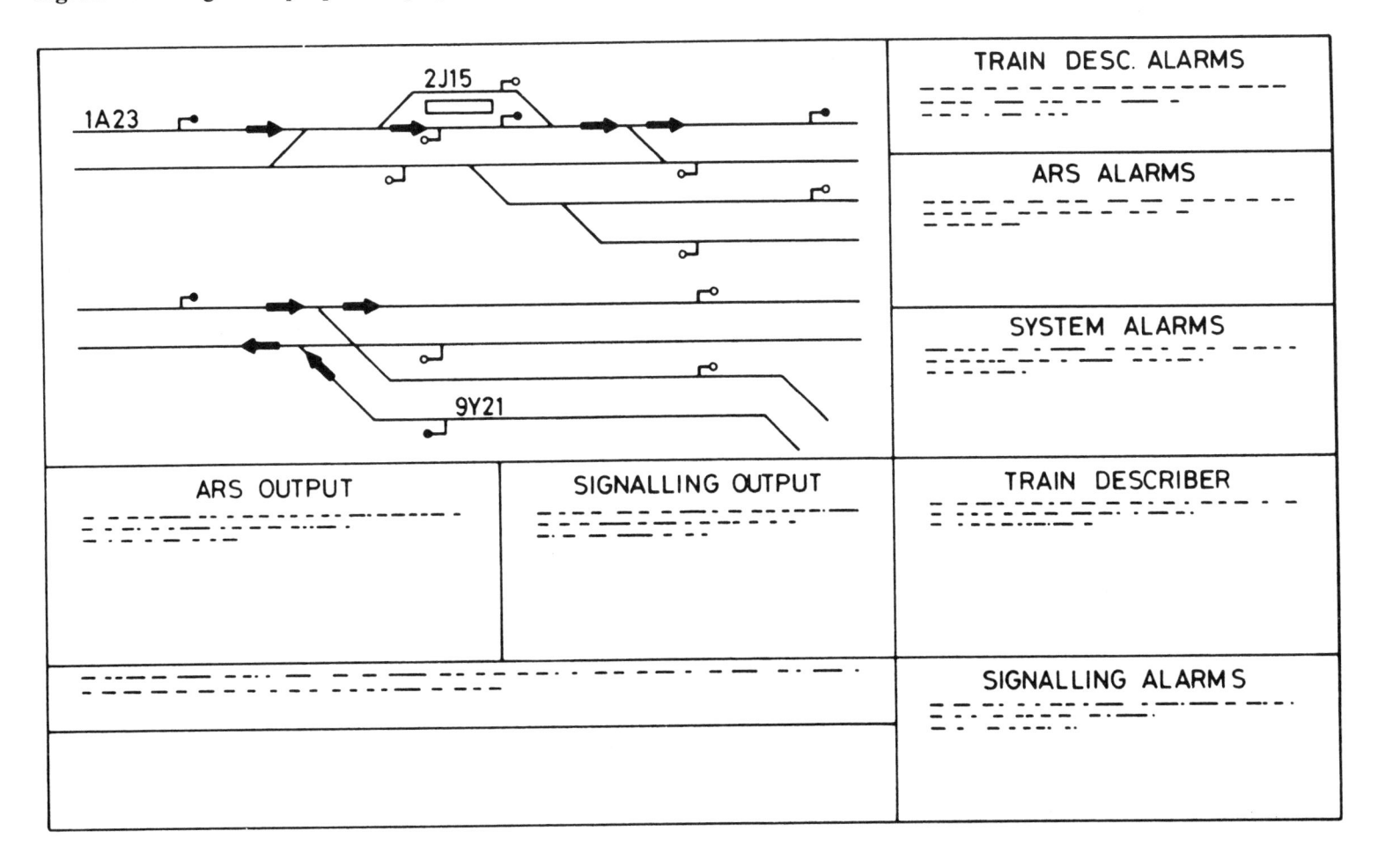

To ease cursor positioning, the relevant target is made as large as possible, covering, in the case of a signal, between two and six character positions. The hit area for any symbol also includes a half character margin all round. Movement of the cursor is aided by a software-controlled three-speed function whereby the distance moved by the cursor for a given angular movement of the trackerball is a function of the speed of the trackerball.

At the bottom of each graphical display is a set of ikons which enable trackerball control to be exercised over a set of additional facilities. The basic mode of operation is as described above, with the cursor being positioned on the ikon and the relevant function key being operated. In some cases it is then necessary to move to a signalling symbol on the diagram to complete the action.

Typical facilities operated in this way include the application of collars to signals and points, the temporary overriding of such collars, the selection for display or suppression of signal, point or track circuit names and the changing of one view for another on a given screen.

KEYBOARD FACILITIES A keyboard is provided for train description entry and to provide a means of interrogation of the ARS system. In addition, keyboard commands act as a fall-back method of operation in case of trackerball failure.

The keyboard has four separate colour coded key sets. The basic alpha-numeric keys used for data entry are white. There are specific function keys for signalling (red), automatic route setting (yellow) and train describer (green) arranged in independent groups as shown in **Fig. 8.6**.

All commands are executed by entering the required alpha-numeric data and then using the appropriate function key. In some cases, an initial function key operation is required to identify the subsequently entered data.

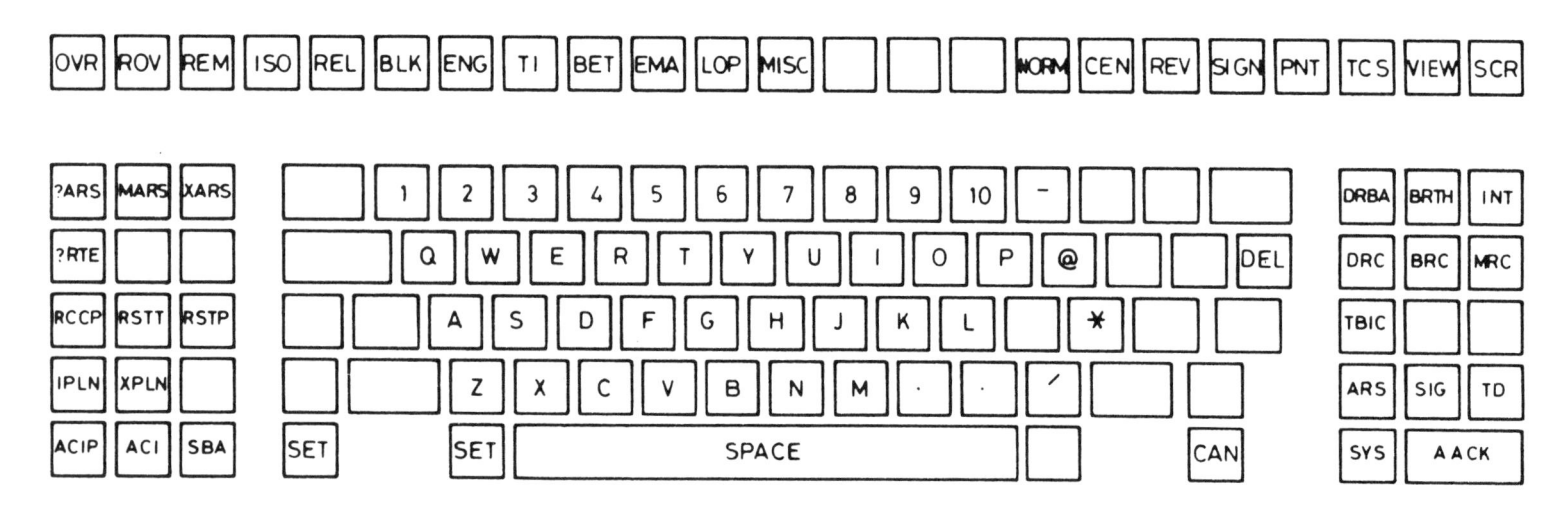

Fig. 8.6 IECC keyboard layout and legends

During data entry, each key stroke is echoed to a command entry line on the general purpose (GP) screen in white. If an initial function key is operated, a mnemonic code is displayed. When the execution function key is operated, the command is actioned and confirmation is given by moving the entered data up one line and changing its colour. Should the data be invalid, an error message is displayed beneath the entered command and where possible, the field in error is flashed.

The train describer commands involve the typical interpose, cancel and interrogation functions of a train describer system together with TD map recall commands associated with the general purpose display.

The automatic route setting commands involve train specific commands or interrogations such as 'remove train from ARS working' or 'why has route not set for train?'. In addition there are commands to invoke special contingency timetable sequences and to revoke such timetables.

Display Control
The SDS receives information from SSI as individual state changes for separately identified functions within the SSI.

The display required is often a combination of individual states such as 'route set' and 'track circuit not occupied'. It is an SDS responsibility to undertake the relevant processing to generate the correct display.

Similarly all the train describer stepping controls which are fundamentally driven by the same SSI source data, are part of the SDS. In fact, the whole TD function is a subset of SDS.

SSI responds to specific requests to operate a facility. However as route setting is based on the traditional entrance – exit system used by British Railways, SDS has to pre-process the data to generate the right command for SSI. Facilities are also provided to allow strings of signals to be set by a single entrance – exit combination provided these strings are defined during implementation. This is known as 'long route' setting.

The above requirements are best met by a databased software structure such that an alteration can be accommodated without recourse to specialist software.

Within SDS there are two major subdivisions of the data. The first defines the screen images and the cursor hit areas. It is usually referred to as the 'create' data. The second defines the relationship between the individual screen elements or train describer functions and SSI formats together with the keyboard command names that may be used. This is referred to as the 'relational' data.

Overall, this represents a significant volume of data which needs to be generated as part of the design process and subsequently tested prior to installation.

Some Basic Rules
The SDS picture layouts are based on screens able to show 48 lines of 128 characters each on a high definition display. As railways tend to be long and thin, some cut sections are inevitable. Consideration also has to be given to the ability of the signalman to operate the depicted layout safely and efficiently.

This gives rise to a number of restrictions on the screen presentation. The most important limitations are:

- Not more than two lengths of four-track railway should be displayed on one screen (short sidings and loops excluded).
- Not more than three lengths of two-track railway should be displayed on one screen (short sidings and loops excluded).
- Interfaces between screen layouts should take account of the relative position of entrance and exit functions and either avoid crossing the boundary or duplicate the functions.
- Unless unavoidable, cut sections on detail view screens should not occur in the middle of a controlled signal section. This is particularly important if the layout is complex. In such circumstances, the signalman may be led to a wrong conclusion during a failure.

- Abutted screen layouts should wherever possible, share a common track configuration at the boundary to avoid misinterpretation.

In addition to these essentially ergonomic rules, there is one significant technical limit that needs to be remembered.

The hit area of a symbol extends half a character outside the area of the symbol itself. It is therefore not possible to place two symbols representing hit areas for different functions adjacent to each other. There must always be a one character gap. Symbols used for display only can abut either to each other or to hit area symbols.

Ergonomics

Investigations into criticisms about working with VDUs have almost exclusively shown that it is not the VDU that is causing the problem, but the way in which it is worked and the environment in which it is installed.

One of the fundamental difficulties of working with VDUs is that of reflection. The relative brightness of the screen tends to be low, with the result that any sharp changes in reflected light cause obliteration of an area of the picture. These phenomena are controlled by making sure that the glare (caused by sharp contrast changes) does not exceed a specified index value. The sort of actions needed to achieve this are:

- Never have windows behind the operator.
- Aim for uniformly lit ceilings, preferably by use of uplighters.
- Avoid ceiling lighting fittings.
- Avoid reflective surfaces behind the operator.
- Graduate the workstation colour scheme so that there are no sharp contrasts at the edges of the screen or likely to be reflected in the screen.

These requirements have an appreciable effect on the colour scheme of the workstation and the room, together with the lighting and orientation of equipment in the room.

A second important issue is that of accommodating the essential viewing requirements of the average person. The relaxed state of the eye muscles results in a focal point at infinity. In practical terms, this is anything more than 3.6 m away. At the other end of the scale, maximum muscle tension is required to focus on close objects with the increased risk of viewing difficulties. At viewing distances of less than 0.9 m, different people have differing accommodation of preferred ranges of view; it is thus almost impossible to design a piece of equipment that suits every operator. Finally, the size of the characters depicted on the screen defines a maximum reading distance. The mix of these factors results in a low relief desk to permit a long focus eye in the rested position, with screens placed about 1.2 m in front of the operator in an arced layout.

A third issue of importance concerns the posture of the operator. Good posture avoids muscle fatigue and its associated aches and pains. It helps alertness and facilitates right actions being taken at the right time. The fundamental issues are determined by the quality and style of seating, together with the dimensional relationship between the seated operator and the various controls he has to reach. To obtain the best fit for a range of users as is essential in any facility used by different operators, it is wise to allow as much freedom of movement as possible and to take account of both left and right handed people. This has been achieved by allowing the keyboard to be free standing, by providing relevant duplicate trackerball buttons and by ensuring the seat is both mobile and fully adjustable.

In addition to the main physical conditions associated with VDUs, there is a range of more general physical and psychological issues that need to be considered. Is the working environment appropriate to the job? Is the air at the appropriate humidity? Can the real world outside be seen? These issues relate to the design, decoration and ventilation of the room and thus determine the parameters that need to be input to the architect's specification.

Failures and Fall-back Systems

It is essential with any signalling system that failures which impact on the operation of the railway are rare and that when they do occur, trains can still be moved under special procedures.

The safe operation of trains is controlled by the fail-safe design of the interlockings, be they solid state or relay types. The console with indications and controls is nonetheless an integral part of the operational system.

The early consoles had very few pieces of common equipment, in many cases only the power supply, and hence widespread failures were rare. The widening areas of control led to more complex systems driving the console with eventually panel processors being used. The risk of wide area failures increased and duplication of equipment was adopted to minimise this risk.

The IECC system is duplicated to minimise failure. The screen views presented on each VDU (except GP) can all be interchanged, thus allowing continued operation if one screen fails. All functions such as train describer entry needing the GP screen can be performed at another workstation if the need arises. Most major controls can be achieved by either trackerball or keyboard. It is however conceivable that a complete workstation failure will occur. Design calculations estimate that this will happen once in 20 years as a statistical average.

A traditional panel still indicates the layout of the track when it fails. However, a VDU-based workstation is very likely to go blank. To assist operation under these conditions, colour photographs of the normal track layouts have been encapsulated in plastic film and are held in the workstation. In addition, a hard copy diagram of the area has been prepared, encapsulated and mounted on ferromagnetic backing so that magnetic reminders can be used. These facilities will, together with the telphone/radio systems, be used to enable trains to be moved under special arrangements in the event of failure.

AUTOMATIC ROUTE SETTING (ARS)

The automatic route setting system is designed to operate main running signal routes, and a limited set of shunting routes, automatically in response to the approach of trains. The system is designed to be able to fulfil this role even when the trains arrive in a sequence other than that defined in the timetable. The primary objective of ARS is to select an operating sequence which will minimise the resultant deviation from timetable whenever a conflict occurs.

ARS always allows the signalman to remain in charge. The signalman may therefore set routes ahead of ARS; he may restrict the area over which ARS works, or restrict the set of trains for which it works. Finally the signalman can ask ARS to explain what is happening at the current time.

This section will first describe the methods used within ARS to enable route setting to occur and then describe how the signalman interacts with the system.

ARS Models

To enable ARS to operate, a considerable volume of data has to be available. Some of the data is static over significant periods of time, some is dynamic over the medium term and some is dynamic in the short term. Static data is typified by geographical or operating strategy data. Timetable data can be viewed as medium term dynamic, while route and track condition data is clearly short term. Most of this data is contained in three primary models, though the models do not correlate to the length of validity described above.

TIMETABLE MODEL   ARS needs access to timetable data to set routes that will minimise deviation from timetable.

Timetable data is needed over an area larger than the ARS control area. Consider the impact of a slow train proceeding ahead of an express along a line without passing places. Clearly ARS, which may control only one end, must estimate the effect at the far end (see **Fig. 8.7**).

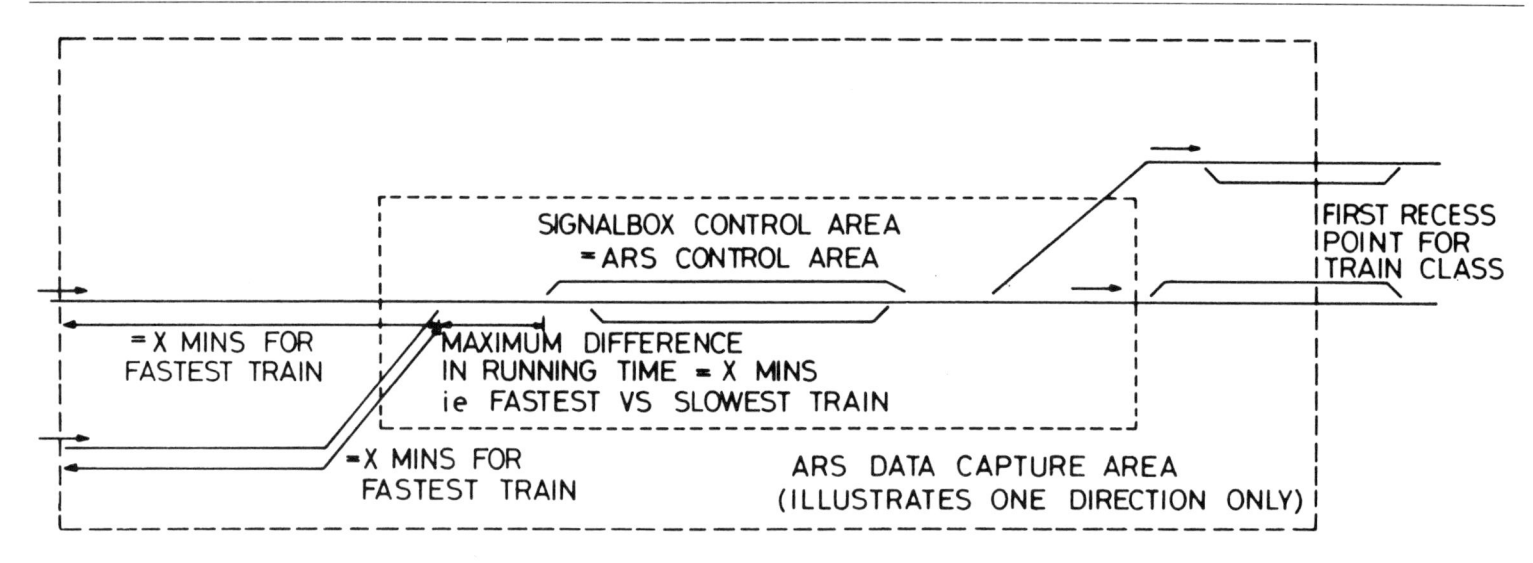

Fig. 8.7 ARS — illustration of principle

Within this area of influence, ARS needs comprehensive timetable data. It requires all scheduled passing, arrival and/or departure times. It needs details of any recovery time in the timetable together with any timing allowances for pathing or other operational reasons. The line (fast or slow, etc) to be used by a train must be specified, as must the platform at a station. This data is all contained within the working timetable, though in a few isolated cases it has been necessary to insert additional line references where ambiguity was possible.

Once ARS has the basic timetable data, it prepares for each train a sectional running time between signals. Whilst these sectional timings are only intended to be estimates, a reasonable level of accuracy is needed and the existence of recovery or pathing times must be excluded from this process. The sectional running times are primarily calculated on the basis of relative distances, though special allowance can be made if the length includes a major change in speed.

SIGNALLING MODEL The signalling model describes the routes available between the timing points. It thus describes the essential geography of the area concerned.

Additionally, all the features of each route are effectively described within the model, such as what points exist in the route, which subroutes and track circuits are present, and how many aspects can be displayed by the signal.

One of the design parameters built into ARS is that it should not constantly test an interlocking by requesting a route that is unavailable. The signalling model therefore contains the basic interlocking definitions, so that it can compute when the route should be available in accordance with the data being received from the interlocking.

The majority of the data held in the signalling model can be derived from the scheme plan and the control tables. As most IECC schemes interface to SSIs, it is preferable to use the 'points free to move' (PFM) and 'panel route request' (PRR) files from SSI as sources of data in place of the control tables.

TRAIN DESCRIBER MODEL ARS must know the identity of any train. For this reason, all ARS activity is initially triggered by a train describer step. It is therefore necessary for ARS to contain data areas defining each train describer berth and identifying all the steps that may take place between berths.

Auxiliary Data

Whilst the main data areas are largely contained in the models described above, a significant volume of auxiliary data is necessary.

Much of this describes particular operating characteristics such as a weighting of the relative importance of different trains. It will also describe threshold limits, which are used to maintain particular sequences or to create alarms for the signalman.

Basic Operating Algorithms

The original operational remit was that 'ARS is to be as efficient as a good signalman'. Experience so far indicates that ARS meets, and in many ways exceeds this target, particularly as it does not forget to set an occasional route during busy times. The basic techniques used by ARS in reaching a route setting decision are outlined below.

The first point to note is that ARS is regularly updating a predictive image of train movements. Whenever a train describer step occurs, ARS notes the time and then, using the sectional timings for that train held in the timetable model, predicts its arrival times at subsequent points on the journey.

It is not necessary for this estimate to be very accurate, especially over accumulated lengths, since the estimate will be refined when the next step occurs. A reasonable degree of accuracy is necessary over the medium distance (say five signal sections) since this will influence route setting.

As a result of the routing data held in the timetable, ARS is able to identify possible areas of conflict. These can take three basic forms: following moves on the same line, opposing moves on the same line and moves on lines which cross. The last can be regarded as the minimum case of either of the previous forms.

The following explanation is based on a following move conflict, though minor adjustment to the formula results in the same basic functions being true for opposing movements.

Fig. 8.8 depicts a track layout with an opportunity for conflict. The limit point is defined as the point beyond which one train cannot proceed without obstructing the progress of the other train. The clearing point is the point at which the train clears the conflict section.

Fig. 8.9 depicts the distance/time graph for trains A and B according to the timetable and as predicted by ARS during actual approach. It will be noticed that the predicted slopes are identical to the timetable slopes because the predicted timings are derived from the timetable.

According to the timetable the trains will arrive at the limit points at times $T(LA)$ and $T(LB)$, and at the clearing points at $T(CA)$ and $T(CB)$. The timetable is designed to ensure that $T(LA) < T(LB)$ and $T(CA) < T(CB)$ indicating that train A will arrive and depart before train B.

The predicted timings on the other hand indicate a conflict, because $T(LA) < T(LB)$ but $T(CA) > T(CB)$ showing that although train A arrives first train B would depart first.

The third option in which train B arrives first is also shown in **Fig. 8.9**. Mathematically depicted as $T(LA) > T(LB)$ and $T(CA) > T(CB)$, it is again clear that no conflict occurs.

A brief analysis of the above options shows that conflict occurs when the signs relating $T(LA)$ to $T(LB)$ and $T(CA)$ to

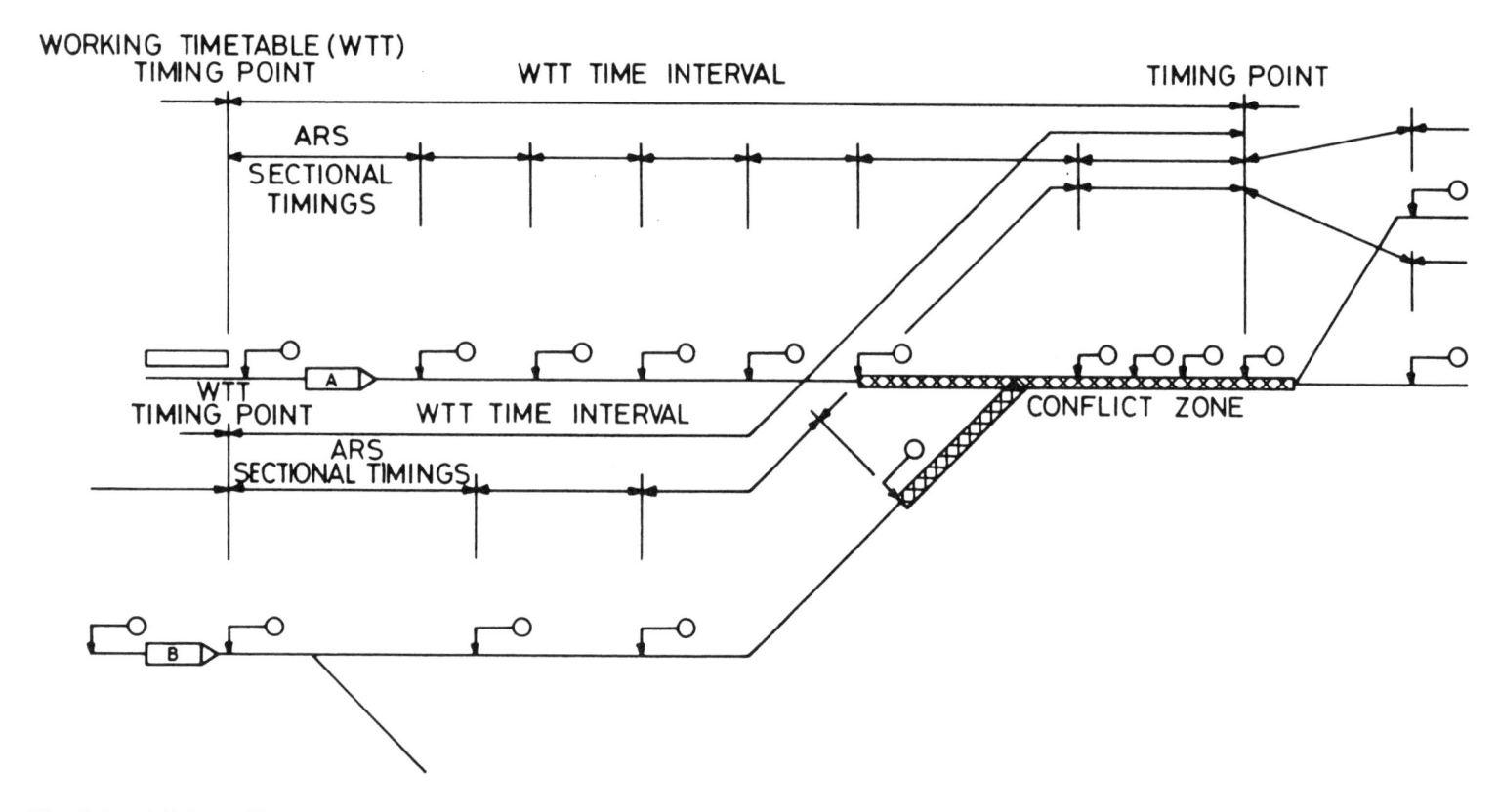

Fig. 8.8 ARS conflict assessment

OPERATOR INTERFACE 217

Fig. 8.9 ARS — timetable and predicted train graphs

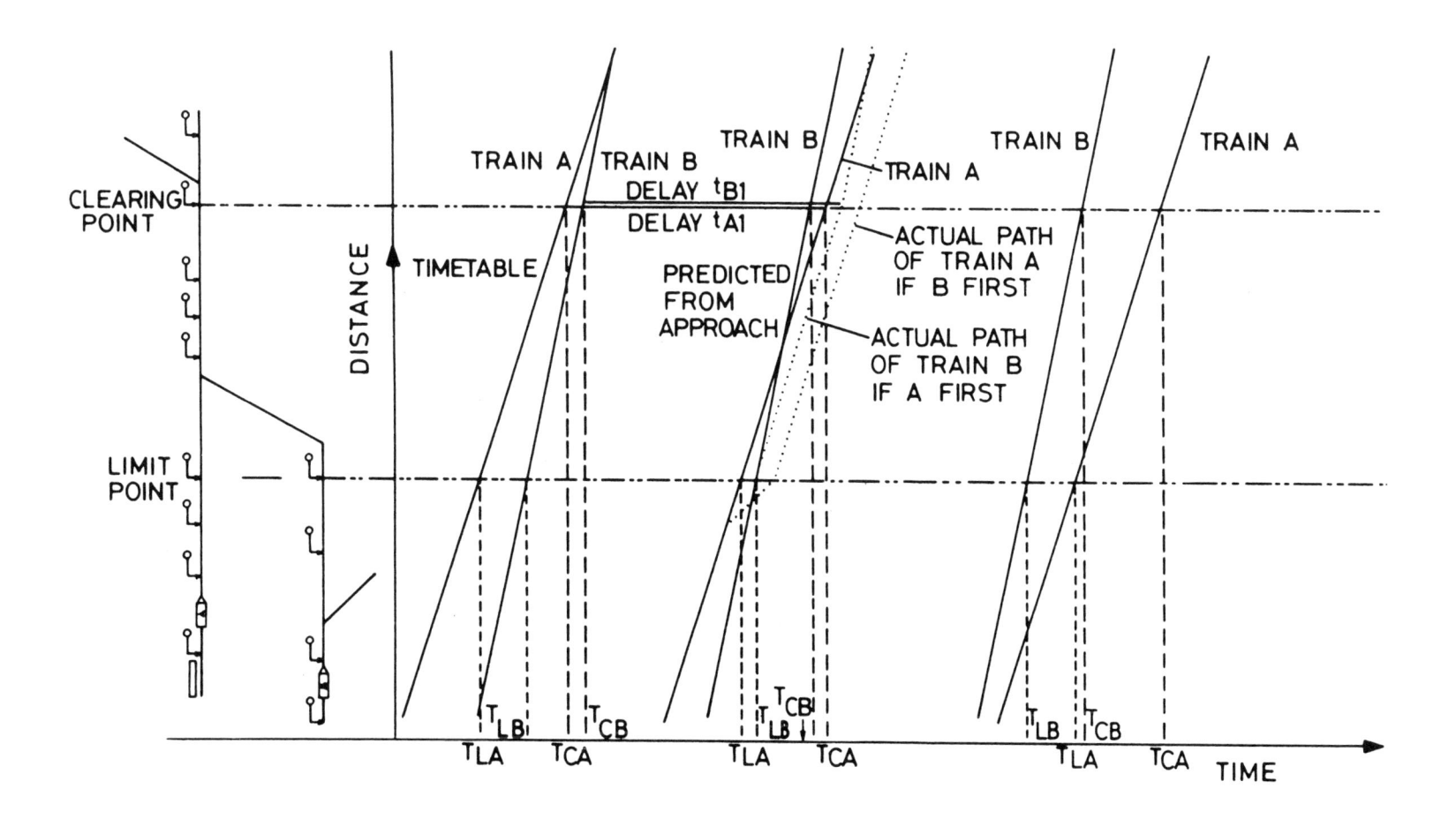

T(CB) are different. Analysis of such mathematical signs is easily performed by the computer.

In practice, considerations of headway and train performance need to be included in the calculations, but these do not change the principles outlined above.

If an opposing move conflict is considered in the same manner, it is still necessary to take into account the timings of the two trains at common geographical points. Geographically, the limit point for a train in one direction is the clearing point for a train in the opposing direction. Therefore T(LA) is compared against T(CB) and T(CA) with T(LB). Otherwise the equations are identical with identical interpretations.

Having identified a conflict, it then becomes necessary to calculate the timetable deviations for the two options of allowing train A or train B to proceed first. The relevant timing is at the clearing point since this gives the delay following the movement.

If train A proceeds first, the delays will be t(A1) for train A and t(B1) for train B. If train B proceeds first the delay will be t(A2) for train A and t(B2) to train B.

If train A is a Class 7 freight train and train B is a Class 1 express passenger train, then operating data will, typically, have weighted the delay to train B at 10 and train A at 2. The weighted train delay is given by the estimated real time delay multiplied by the weighting.

Therefore the weighted delay if train A goes first is:

$$[t(B1) \times 10] + [t(A1) \times 2] = X$$

If train B goes first the weighted delay is:

$$[t(B2) \times 10] + [t(A2) \times 2] = Y$$

The option will be chosen from the sign of the sum $Y - X$. If the answer is positive, train A proceeds first. If it is negative, train B goes first. In practice, if the alternative options generate delays that are almost equal, there may be other advantages in remaining in timetable sequence. A test is carried out against an operator-determined threshold and decisions are based on the results.

All conflicts are assessed in this manner, but the system recognises that a solution to the first conflict may create subsequent conflicts. It has recovery routines to check and, if necessary, resolve these.

The basic ARS operating algorithms include a number of other features. ARS may decide to request a reduced overlap (warning) approach either because the overlap is currently occupied or is about to be. ARS may also select calling-on routes if the timetable permits a second train at the platform, and it may re-platform trains according to a pre-defined list of alternative platforms prepared in priority order, or use alternative lines where these are available.

Additional algorithms are developed as the need is identified within a scheme. It is therefore part of the design process to select the relevant algorithms for the scheme and identify any that may need development.

Resetting the Route

The ARS system within a signalbox is divided into a number of subareas. A subarea can be turned on or off by the signalman so that an operating problem not visible to ARS (eg point failure, special train) can be managed by the signalman. ARS will still collect data from a subarea that is switched off and route setting can thus be resumed outside the area, or within the area immediately it is turned back on.

Using data from the signalling model, ARS will identify when a train has only one green signal ahead (ie restrictive aspects will be encountered after the next signal). When this point is reached and provided there is no conflict with another train, ARS will check the state of the route required by examining its SSI state data, and if the route is apparently available and provided the relevant ARS subareas are switched on, a route request will be issued direct to SSI via the signalling network.

If a second train is now within the route setting zone, ARS will continually monitor the availability of the route and as soon as it is available, ARS will issue the route request direct to SSI.

Signalman Interactions

The routes are set when required by direct interaction between ARS and SSI. The signalman observes the result of these route requests as the appearance of route set indications and changes to the signal aspects on his SDS screen.

The signalman can at any time issue his own route setting requests in the usual manner. He must of course do this before the relevant ARS route is set, otherwise the interlocking will ignore his request.

If the signalman needs to take full control of a particular area, he can turn off the relevant subarea either by trackerball operation of the subarea control on the SDS screen or by keyboard command. If at any time the signalman cancels a route, then ARS will itself turn off the relevant subareas. This is a safety precaution in case the route cancellation was an emergency measure and prevents ARS setting routes for other trains into that area. ARS cannot, by design, cancel a route which has been set.

The signalman may also decide to take control of a specific train. This can be done by entering to SDS a keyboard command including the train description. A message relayed to ARS via the network results in the train being declared 'non-ARS'. Equally, this command can be rescinded using an alternative command.

Sometimes ARS may not respond in the manner or at the time the signalman would expect. This will largely occur during the initial period of operation. To enable the signalman to understand what ARS is doing, a small set of interrogations is available. ARS will generate the relevant reply from a set of responses. For example, the question 'why has ARS not set route for train ABC?' may result in the reply 'awaiting departure time XYZ for train ABC' or 'waiting arrival of connection DEF'. The reply is as far as possible in English text, but the length is limited to 40 characters.

There will be an inevitable reduction in the vigilance of the signalman in areas usually worked by ARS. This may result in him failing to observe a track circuit changing state unexpectedly. He may thus overlook the early indications of a mishap. ARS therefore monitors the occupation and clearance of track circuits and raises alarms if they become unexpectedly occupied or clear. Similar alarms are raised when a train remains an exceptional time in any one TD berth, after allowing for signal aspects or timetable requirements.

ARS Impact on the Interlocking

The ARS system will attempt to set a required route immediately it is detected as free.

Although interlockings have considerable protection against loss of route holding, the basic design allows the route to be released if a track circuit should become clear under a train.

Normally the signalman would see the train disappear from his display and reappear an instant later. He would not attempt to set a conflicting route under these conditions. Indeed it is most unlikely that he would be poised to set such a route in the first place.

ARS on the other hand may be waiting to set a conflicting route and would issue the relevant command as soon as the conflicting route is released. It would not suspect that the route has been released in error.

In consequence additional protection, which is to some extent valuable even without ARS, has been built into the interlocking. This prevents a section of route from being released unless the following section is occupied. To allow recovery from failure situations, release will occur if the condition lasts for more than 15 s.

Additional ARS Functions

As well as the prime function of setting routes for the signalman, ARS provides comprehensive train monitoring against the timetabled plan. It thus has knowledge of where a train is and if the train has completed its journey. It also knows the next trip for that set of rolling stock.

The ARS system is thus able to generate the next train description of the rolling stock and to interpose this into the train describer, further relieving the signalman from routine tasks.

The insertion of new train codes (automatic code insertion or ACI) is of course more complex than briefly outlined above, due to the need to cope with the splitting and joining of trains at many locations. All the necessary data is available to ARS to enable this to be comprehensively achieved.

TIMETABLE PROCESSOR

One of the prominent features of ARS is its dependence on timetable data. In most schemes to which ARS is applicable, the volume of timetable data is substantial. It is unacceptable in terms of both the resources necessary and the error risk, for such large amounts of data to be entered locally.

BR has a train service database (TSDB) which contains the timetable for all scheduled trains. TSDB provides data for the public timetable, working timetable, TOPS and many other functions. ARS obtains the majority of its timetable data from this source via a timetable processor provided as part of each IECC.

The timetable processor actually consists of two independent computer systems as shown in **Fig. 8.10**. An ordinary IBM compatible personal computer with communication facilities provides the special program functions necessary to accept file to file data transfers from the mainframe computer via the high speed data ring. Having received these files, the communication format is altered and the file passed on to the main timetable processor.

The main timetable processor (TTP) is a standard commercial unit again utilising VME equipment. However the timetable processor configuration does not follow the standard pattern defined earlier, primarily because it is not configured as a duplicate system.

Duplication was seen as unnecessary for this application because under normal circumstances, the ARS computer will hold between nine and 21 hours of timetable data and thus a reasonable repair and reload time is always available.

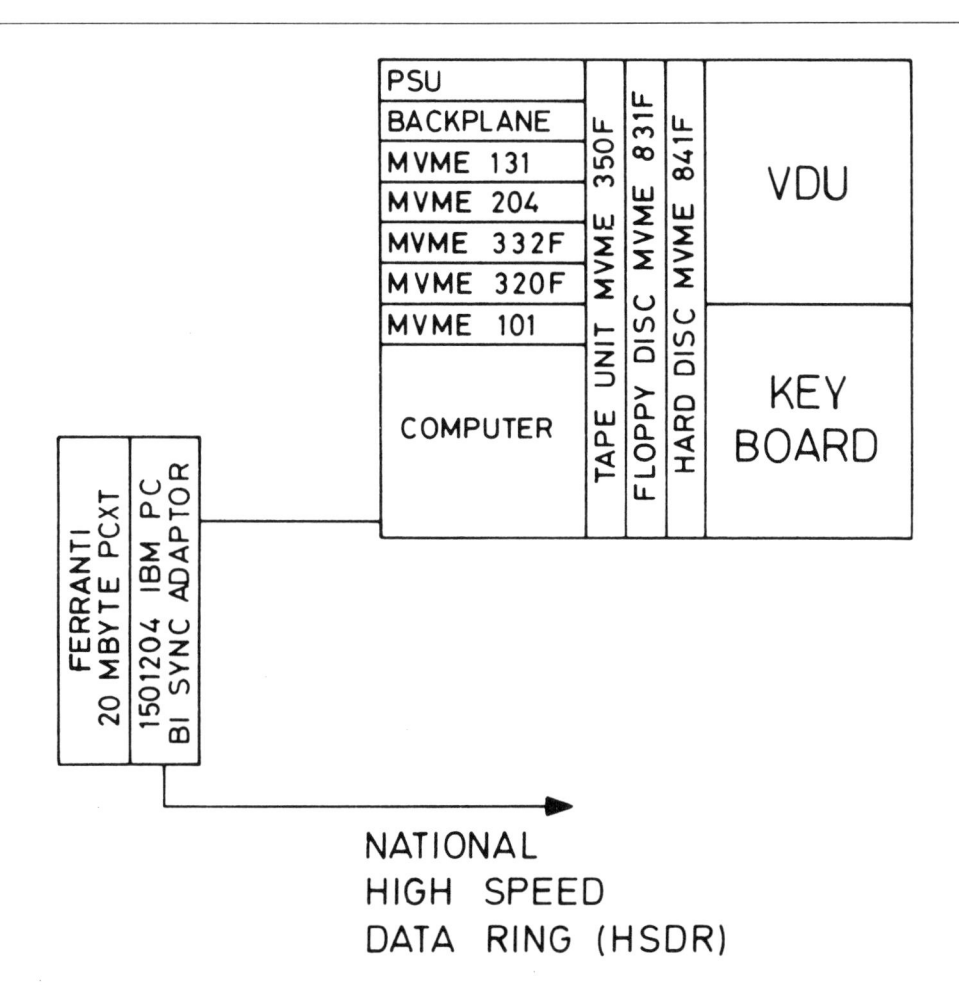

Fig. 8.10 Timetable processor

The TTP provides bulk storage of timetable data downloaded from the mainframe initially at six monthly intervals corresponding to the major timetable updates. This data is stored on a 70 Mbyte hard disc in the form of weekday, Saturday and Sunday timetables. It is anticipated that the update frequency will be increased as more short-term amendment facilities are added to the train service database (TSDB).

In addition, certain platforming and stock diagram details are not available from the TSDB, and terminals are provided so that these and short-term alterations can be added to the timetable used by ARS. Typically, one terminal will be provided in the Area Manager's office for longer-term data additions, while a second terminal in the signalbox is used for current amendments.

The timetable processor is also capable of storing and bringing into use (on request from the signalman via ARS) up to 99 different sets of contingency timetable data. These could range from train specific contingency plans to complete timetable recasts applied as a result of some major operational problem. These contingency plans all have to be locally entered, since no provision currently exists on the national database.

Clearly with such substantial volumes, it is essential that the data is not lost in the event of a failure. In addition to the run time storage on hard disc, a tape unit is also provided to permit regular archive recording of the data.

The role of the timetable processor is to reformat the data stored so that it is compatible with ARS requirements, and then to divide the timetable into the 12 hour chunks that ARS can hold in its solid state memory. ARS will request from the TTP the next section of timetable about three hours after an earlier section has lapsed. The delay enables route setting to continue even if the trains are late. The timetable processor ensures that it supplies the relevant section for the correct day of the week.

The timetable processor has been specifically designed to make its timetable data available on the information network to any system requiring it. There is thus an expanding role in both passenger and management information systems for such data, but this is outside the scope of this chapter.

GATEWAY SYSTEM
It was recognised at an early stage in the design of the IECC, that there would be a need to regulate the traffic on the network linking the safety-critical signalling systems carefully, in order to achieve consistent response times. This led to the adoption of the two networks, the signalling and the information networks with the gateway system (GWS) between them.

The signalling systems are effectively isolated from the variable levels of demand for information from the systems on the information network. These systems may include fringe signalboxes, adjacent signalboxes with train describers, track-to-track radio systems, adjacent IECCs and passenger and management information systems which rely on real-time signalling and train describer data.

The GWS maintains an up-to-date picture of the IECC area in terms of current signalling states, train describer berth occupations and contingency plans in operation. It polls the SSIs for signalling state data, receives train describer messages (steps, interposes and cancels) unsolicited from SDS and receives contingency plan data from ARS. The GWS is configured to supply a selected subset of the data making up its current picture of the IECC area to each system on the information network. The GWS updates these systems in response to data changes received from the systems on the signalling network, and in reply to requests from the systems themselves.

The GWS also passes on information from outside the IECC area to systems on the signalling network. For example, TD map updates are passed to the SDSs allowing the IECC signalmen to see approaching trains which are yet to reach their area.

IECC SYSTEM MONITOR (ISM)
The ISM provides the technician's fault recording and interrogation facilities together with an ability to control the operational configuration. In addition, the ISM provides a central real-time facility via a Rugby clock so that any messages containing time data or time dependent actions have a common reference standard.

As discussed above, the individual subsystems have their own network duplication strategy (NDS) process, and remote control and monitoring (RCM) process. The NDS process decides what status each element of the subsystem will adopt and reports changes to the ISM. The RCM process is essentially controlled by the ISM and allows monitoring conditions to be set up or controls from the ISM to be achieved. It can be regarded as an outbased portion of the ISM program, resident in every subsystem on the network.

By utilising its own representation of the installation together with data received from the outbased programs, the ISM provides extensive monitoring of the system status and failure reporting, in most cases to printed circuit board level. Reports can be presented on a local printer and displayed on a VDU. Remote monitoring facilities are also being provided to the Area Engineer's office, for example.

The time facility is provided by a duplicate set of Rugby clock radio systems interfaced to the ISM. Each system has a free running internal clock, but this is synchronised with the time standard by messages from the ISM. All time sensitive events can thus be harmonised but in the event of a failure, continued operation is ensured albeit with a very small time drift.

PROGRAM, DATA AND DATA PREPARATION

Almost all software systems can be divided into four distinct parts; the operating system, the program, the fixed data parameters and the run-time dynamic variables.

The design, coding and testing of the operating system and program is a time consuming and expensive process. It also requires a specific set of skills and knowledge. These skills must include the use of the particular language and the overall structure of the system being built. Such skills or knowledge may well be lost or eroded by time. It is therefore important that changes to the program should be minimal once a system has been successfully tested and placed into service.

This can be achieved by ensuring that the program is very general in nature. All the detail facilities are controlled by data that can be readily understood by people familiar with the application. Such programs are said to be data driven. It is then only necessary to find people familiar with the application when an amendment is needed, rather than people who are familiar with both the application and the program techniques.

In may ways, this is precisely the same as the design of any signalling scheme. It is not usual to redesign the point operating mechanism, the signal head or even the interlocking principles for every scheme. Normally these elements are accepted and the way they are interconnected is adjusted to achieve the required result. Data driven programs offer the same facility to software systems.

The IECC has been designed as a data driven system. The basic code for the program has been written once by specialist engineers and computer programmers and will only be modified if a major shortcoming or new development is identified. Individual schemes are configured by signal engineers familiar with data preparation, IECC concepts and traditional principles of signalling and safety.

In consequence, a considerable amount of effort has been devoted to the design of general purpose program functions that can be data driven. It is also true that such data driven packages require substantial resources to be applied to the task of data identification, and then data preparation, as part of the design process.

The map data and relational data referred to earlier are stored in the SDS. This describes the picture layout and its relationship to the trackside railway so that SDS can present the appropriate signalman's display. ARS contains several levels of data, but those which are fixed at the time of design relate to the trackside functions and their interlocking, information originated by SSI, and some parameters to optimise the operating characteristics to the local geography. Similarly the gateway data is primarily concerned with trackside functions, train descriptions from SDS and contingency and other data from ARS.

Much of this data is repetitive, a feature of a distributed processing system. Many of these data elements either originate from, or are echoed in, the data preparation required for the SSI. Although this is not the place to discuss distributed processing against single unit computing in detail, one may wish to consider the relative merits of multiple data preparation against the difficulties of debugging large and complex programs in single machine environments. It was decided that data preparation can be mechanistic and repetitive so a distributed system was chosen.

Having decided that a distributed processing environment with interlinked data processing is the right solution, one then has to face the project management task. It is probably true to say, paraphrasing an ancient proverb, 'look after the data preparation and the hardware will look after itself'. New all-electronic control centres will depend very heavily on the right data being available in the correct form at the right time. Any time saved from the development period will be won by improved sequences or methods of data preparation. The hardware will be delivered a well-defined period after ordering.

It would be unrealistic to expect that no further program development would be needed by adopting data driven techniques after the first system has been introduced. Systems in general, and thus the data, can only be designed within the vision of the original design team and by taking account of the financial limitations. Invariably, there is either an aspect of future connections that are outside the experience of the designers, or following installations, the user sees a new opportunity.

Both are likely to result in additional development as each generation of equipment is introduced. Program updates will not be completely banished by data driven techniques, though they can be reduced.

TESTING

The testing of IECC can be divided into three elements:

- Testing the program.
- Testing the data.
- Correspondence testing.

Testing of the program has been done once as part of the IECC development and need only be done again when program alterations are made. The testing of data and correspondence check must, however, be performed for every installation.

Testing the Program
Whilst rigorous quality assurance checks were carried out as the software was developed, substantial testing was still necessary. These tests were performed on individual program modules, complete suites of programs having a particular function (ie subsystems) and at the system level (eg SDS, ARS).

Finally a complete IECC, configured to the layout of Liverpool Street/Bethnal Green, was set up and connected to SSI simulators. These simulators ran the true SSI program and appropriate interlocking data. In addition they ran programs that simulated train movements which generated the appropriate inputs from the trackside modules.

This configuration was initially run to test pre-defined routines, and thus ensure that the principal paths through the program were tested. It is, however, virtually impossible to test all possible branches in such a complex system, so an extended period of operation, to a version of the Liverpool Street timetable, was used to verify as many branches of the program as practical.

Testing the Data
The data used in an IECC to configure the system to the particular location needs to be tested. All the data needs some testing, but it is particularly important that the data controlling the SDS screen displays and the route requests from the SDS is tested.

The method adopted uses what is known as a type B simulator. This runs the panel processor end of the SSI program, but allows a keyboard or similar input device to set all the bits in the SSI memory area. It therefore appears as an SSI to the IECC but in fact, the interlocking is driven by the testing staff.

Data files were prepared on floppy disc so that each item of SSI identity files was changed through all possible states. The data files included prompts so that the testing staff were able to check that the correct display change took place. In addition, they ensured that no unexpected change occurred. When satisfied with the results, the testing staff initiated the next test. Once all display item tests had been completed, further data files provided prompts requesting particular routes to be set, and then provided the relevant panel request identity for cross-check purposes.

In principle, similar data files could be set up to test the train describer and ARS functions, but such files could not be mechanistically created. This tends to limit the testing as the file preparation becomes time consuming and prone to error.

Correspondence Testing
Having performed the above tests, and accepting that even more exhaustive testing is performed in parallel in the interlocking, it then becomes necessary to connect the two together. Functional and correspondence tests in the usual manner must then be performed prior to final commissioning.

THE FUTURE
The IECC is a new technology. It is therefore at the beginning of a development cycle and no doubt has several as yet unexplored opportunities.

The major areas of development that can be visualised are:

- SDS.
- ARS.
- Management and passenger information.

SDS
The ability to increase the number of maps available to one workstation above eight is probably a necessary extension. This, together with a relaxation of the requirement to be able to observe continuously all parts of a control area, would allow one operator with the help of comprehensive ARS to control a significantly larger area. An abnormal event would automatically cause the relevant picture to be displayed.

A more immediate development is likely to be in the form of a comprehensive system of map selection to back-desk supervisors in place of the currently restricted subset of pictures available.

ARS
As discussed earlier, additional algorithms will be prepared for ARS as operating sequences are identified which need them.

In addition to such new features, it may be necessary to provide ARS with some expert learning features. In particular, ARS at present relies on the timetable speed profiles of a train when making forward projections. The performance of a train when running may be different (eg it may be underpowered) and as speeds rise, such differentials become more critical. ARS may need to adapt to such situations. Similarly the impact of TSRs will be train dependent, or at least train type dependent and this may need further consideration in due course.

Management and Passenger Information
The existence of comprehensive timetable data in the timetable processor, together with comprehensive signalling data in the IECC and the route projections of ARS, all in electronic form and all easily exchanged, makes the development possibilities very wide indeed.

Of immediate interest is the ability to provide to operating staff, signal by signal indications of the position of a train and its deviation from timetable. Relevant parts of this data could easily be passed to passenger information systems.

Automation of the passenger information system, particular delay notification and public address announcements, must be the next feature.

The prediction of future delays is also possible together with advance warning of re-platforming, if the full set of ARS functions were accessed. Such facilities could be linked to other systems to provide 'what if' analysis of the options available to recover after an incident.

Experience so far indicates that an integrated system of data exchange is the fundamental building block on which the future of train control and information systems will rest.

Panel Processors

Before the introduction of microprocessor-based remote control systems, it was common practice to carry the minimum number of indications back from a remote interlocking in order to economise in the size of the system. As each output from the receiver had only the equivalent of one relay contact, the coming of train describers requiring stepping controls caused problems. What would now be called a non-vital interface was used to generate the required indications and stepping controls.

Considering the requirements of the interface with respect to plain track circuits and signals, very little conversion was needed to generate the required indications. For tracks there would have been two functions transmitted — track clear (TK) and route set (UK); from these the interface would have to generate the route set (UKE) and track occupied (TKE) indications. Again for signals, the conversion would be very simple. From the approach locked (ALK), on (RGK) and off (DGK) functions, the flashing or steady red (RGKE) and the green (DGKE) indications would be produced. However, for points, the conversion was far more complex. For single ended points, it was necessary to transmit five functions in the remote control system — normal and reverse detection (NWK and RWK), route set over points normal and reverse (NUK and RUK), and point track clear (TK). From these as many as 11 different panel indications would be generated. For instance, from the NWK and RWK functions, the basic normal (NWKE) and reverse (RWKE) indications were easily obtained, but in addition the out-of-correspondence indication (OCKE) could be generated if neither NWK or RWK were present. Similarly the full range of route and track lights could be provided.

Over a period of time, these became standardised into the requirement for common, normal and reverse track lights (CTKE, NTKE and RTKE respectively), and five different route light indications, again common, normal and reverse (CUKE, NUKE and RUKE), and normal and reverse incorporating the appropriate point detection (sometimes somewhat clumsily designated NKUKE and RKUKE). The conversion also ensured that if the point track circuit became occupied without a route being set, or if a train ran past a signal into an overlap, then the track circuit indications were 'flooded', ie CTKE, NTKE and RTKE all illuminated simultaneously.

In addition to the generation of panel indications, any train describer stepping controls would have been constructed as required.

As no safety is involved in the interface, it was not necessary to use BR 930 type relays. In addition, few contacts were generally needed on each relay, so it was found practical to use standard Post Office, or even smaller types, although reliability is of prime importance when selecting a relay type.

The basic interface for every signal is similar, as it is for every point end. This standard concept is reinforced by the fact that geographical interlocking systems generate a consistent set of indications for feeding into the remote control system. Using similar arguments to those for the introduction of geographical circuitry, it therefore became the practice to use pre-wired packages for these interfaces. The relays might either be housed on a small metal chassis, or be carried on a printed circuit board. The former would be mounted on relay racks with a jack-in arrangement, while the latter would be housed in cubicles. In either case, the amount of floor space taken up by the remote control equipment and its associated interface was significant. This was most noticeable with the introduction of the major centralised schemes in the late 1970s, where the buildings required incorporated large relay rooms. The central relay room, whilst housing perhaps a quite small local interlocking, also needed to accommodate the remote control systems and interfaces for a large number of associated remote interlockings. An additional problem was the practicality of cabling all the indications, numbering many thousands, from the relay room to the panel, which was usually on another floor of the building.

With the introduction of microprocessor-based remote control systems, it was realised that the microprocessor was not fully utilised, and could additionally undertake the interface conversion. This, coupled with the reduction in physical size brought about by the use of microprocessors, resulted in the possibility of all the remote control and interface equipment being mounted within the body of a normal sized control panel.

The immediate benefits included a considerable reduction in building size and a significant saving in cable and installation costs.

The technique of using the remote control system microprocessor for this additional task was initially known as 'panel processing', but at present is more accurately described as 'indications processing'. The introduction of indications processing was perhaps the first occasion that signal application engineers became involved with the software side of electronics, and the first principle of the design was that no computing knowledge should be needed to apply indications processing to a scheme. In signalling, every interlocking is different and it would in any case be quite impractical to write a unique program from scratch for each scheme. It followed that the necessary software should fall into two parts; firstly the basic program, which is identical for every installation, and secondly the data which engineers the system for a particular installation. The basic program is written by computer programmers in an appropriate computer language and is not accessible for alteration by the application engineer. The data, however, is written by the signal application engineer in terms that he understands, ie in terms of signalling functions. Such systems are termed 'data driven', and subsequently, both SSI and IECC followed the same principle.

The program is held in a programmable memory chip (EPROM) which is plugged into a socket on the microprocessor printed circuit board. The data is similarly held in an EPROM, usually separately from the program EPROM. In order to convert the data into the appropriate format and coding, a data preparation program (DPP) is required. This program is only used when preparing, or editing, data and is not connected in any way with the remote control/panel processor program running when the system is operating. The main task of the DPP is to take the various files of information typed in by the engineer and compile the data, ie convert it from the 'high level' language that the engineer understands into the machine code appropriate to the type of microprocessor being used. The DPP also combines the various files of information into a co-ordinated whole and formats everything into the state required for 'blowing' into an EPROM.

However, the DPP may also be used to take a lot of the drudgery away from the engineer. A well thought out DPP allows the engineer to produce the data with the minimum of effort, by undertaking a large amount of the repetitive work. This feature is described below.

Systems produced by different manufacturers will vary in detail. The system to be described is typical and will show the basis of operation.

The indications processor converts a set of functions transmitted in the remote control system into another set of functions required by the panel. The required logic is broken down into a collection of standard packages, or logic modules, which may be likened to the pre-wired relay packages described above.

The data requirements fall into four files of information:

- Details of the incoming functions (input file).
- Details of the logic modules required (logic module file).
- Details of how the logic modules are to be used to transform the inputs into the outputs (function file).
- Details of the outgoing panel functions (output file).

INPUT FILE

The logic state of each input is stored in the cells of a memory chip. The logic states are described as '0' if the input is de-energised, and '1' if the input is energised. The system has to know which incoming signalling function is stored in which memory cell, so the input file consists of a list of the functions in the same order as they are input into the remote control system at the remote relay room. The DPP then allocates each one a cell in the memory. Each cell has a numerical addresss which the indications processor uses, but which the applications engineer does not need to know. The remote control process keeps the contents of the input memory updated, so at any moment the memory contents reflect the state of the incoming inputs.

The majority of functions in the input file fall into standard blocks, for example all shunt signals have the same set of input functions. Because of this repetitive nature of the file, it is possible to define a set of shorthand expressions, the use of which saves a lot of effort by the engineer. Each expression, termed a macro, enables several file lines to be replaced by just one, resulting in a typical reduction in file length of about two-thirds. The DPP expands each macro into its full form as part of its compilation process.

LOGIC MODULE FILE

This file defines the logic required within each module to set the logic state of its outputs each time the inputs change. Each module has a name defined which is meaningful to the engineer.

Some modules are for general purposes, such as AND and OR gates. Others are for specific purposes such as driving the route and track lights of a track circuit. It is possible to incorporate default function names in the logic module definitions if required, and this is especially beneficial for the specific modules.

FUNCTION FILE

This data defines how the logic modules are used to construct the required panel indications. Relay packages were engineered for a particular scheme by the way they were wired between the remote control system and the panel. An important difference however is that a signal relay package was needed for every signal, whereas in an indications processor the same logic module is used in turn for each signal, the results of each use being stored in the output memory.

The file consists of a list of instructions for each output, detailing which inputs affect it, and which logic module is to be used. If a module definition incorporates default function names, then it is only necessary to detail the logic module and the number of, for instance, the signal being considered, rather than listing out all the input and output names in full.

In early systems, this list was continuously processed so that all outputs were continuously updated every second or so. Later systems work on a 'change of state' principle so that when one of the inputs is detected as having changed state, those outputs depending on it are updated immediately.

OUTPUT FILE

The output file is very similar in principle to the input file. Again the state of each output is stored in a cell in a memory chip. Logic state '0' represents no output required (ie lamp not lit), whilst logic state '1' represents an output required (ie lamp lit). The file consists of a list of the outputs, and again the DPP allocates an addressable cell for each. The contents of these cells are kept updated by the function process, and in turn are used to control triacs which actually drive the outputs to the panel. The output printed circuit boards have a triac for each output, controlled by the logic state of the equivalent cell in the output memory. If a flashing output is required, then the triac is switched on and off at the appropriate rate, as long as its memory cell is in logic state '1'.

The output file is similar to the input file in as much that it also contains standard blocks of functions. It is again beneficial to define a set of macros to reduce the work of the engineer.

From the above description, it can be seen how a well-devised DPP can reduce the effort required by the engineer and enable him, with no particular computing knowledge, to draw up with a minimum of effort, the data for an indications processor system.

Controls processing may also be carried out. This would be particularly useful if a non-route setting interlocking was to be controlled from an NX panel. The controls process examines the pairs of button pushes and converts them into individual point and signal controls. The point controls resulting from route setting would be combined with those from the individual point keys. The normal features of button rings would be built in, but additional features may also be incorporated. The new features include the detection of 'stuck' buttons (ie buttons appearing to be pressed for longer than say 10 s) and a time period within which the exit button has to be pushed following the entrance push.

No extra hardware is normally required to add the facilities of indications and controls processing to a remote control system, but the savings in space and cabling are considerable when the equipment may be mounted in the panel itself.

Centralised Traffic Control

INTRODUCTION

The term centralised traffic control (CTC) can be applied to any form of railway operation where the control of a section of line is exercised remotely from a central location. It has however, by virtue of its original applications, assumed a more specific meaning. Those applications mainly involved the consolidation into one central control room of the supervision of stretches of single line railway with passing loops. However, many CTC installations comprised multiple tracks and varied in size from one remote interlocking to hundreds.

ORIGIN

Prior to the introduction of CTC, the predominant means of signalling was by the issue of written train orders or the use of electric staff apparatus. Both of these systems are inherently slow.

The first CTC system was brought into service in 1927 over a distance of 40 miles between Stanley and Berwick on the Toledo and Ohio Central Railroad in the USA. Direct wire control was employed between the control machine at Fostoria and the various relay interlockings.

Since then, CTC has been applied in many countries throughout the world and has today reached a high level of sophistication, amply demonstrating, in a number of situations, an increase in traffic density. The basic concept comprises a non-vital control machine which acts as the operator interface and which communicates via some medium to the various remote stations where fail-safe interlockings operate the field equipment.

TIME CODING

Direct wire control is naturally costly in the use of line wires. In order to control one or more stations by only one pair of wires therefore, a system of time coding was developed. This used a series of pulses selected over switches and relays to transmit coded control messages and likewise employed return pulses over indication relays at the field to transmit indications back to the office control machine. The codes contained the station identity followed by the controls or indications. The capacity of a relay time code system was typically 35 stations, but where necessary this could be increased by employing carrier techniques to allow the addition of further field stations.

Solid state coding using transistor circuitry, high level logic and CMOS have followed on in later apparatus although operation remains similar. Now computers form an integral part of the control machine, handling communication with both the field and the operator. Modern time division multiplex systems operate over voice frequency channels carried on suitable bearers, ranging from copper wire to PCM optical fibre systems and even radio transmission, to communicate between the office and field stations. Speed of operation is greatly enhanced by the telemetry available today and also by the computing power which interfaces between the operator and the telemetry system. The computer is also used to give train describer facilities.

THE OFFICE CONTROL MACHINE

Early control machines were constructed as combined control and indication boards, very often as a series of modules, known as storage panel sections, joined together to form a continuous console.

At the top of each panel was a representation of the station layout containing track circuit and block indications. Below this were the signal, point and ground frame switches, together with their respective indications.

Operation consisted of turning the appropriate thumb switches for a station and then pressing a start button. The controls as selected were then transmitted by a series of impulses to the relevant field unit. Indications were returned from the field for observance by the operator. Should a number of start buttons be pressed in close succession, controls were stored by the machine and were sent out in order. To reduce the number of switches and to aid selection, each group of opposing signals was controlled by a three-position switch. For example, referring to **Fig. 8.11** turning switch no. 4 to the left would call to clear either 4LA or 4LB signal depending upon the lie of no. 3 points, whilst turning the same switch to the right would call to clear 4R signal.

The indications for points and signals were displayed above the relevant switches. N and R stencil indications were normally provided for points with red and green indications for signals.

Associated with the control machine was usually a traingraph recorder which automatically recorded track occupation against time on a slow moving paper chart; this could additionally record such things as point position and signal clearance if desired.

PUSH BUTTON OPERATION

In order to reduce the size of the control machine, simple push button control panels evolved on which the train controller could select the station to be communicated with, followed by the desired operations. This was provided as separate from the mimic diagram which contained the track layouts and all indications in their relevant geographic positions. The operator was therefore able to remain seated and did not have to move around a large machine.

The designation number of the equipment at each station followed a standard pattern, thereby ensuring a uniformity in operation. At the simpler stations those controls not appropriate were incapable of transmission.

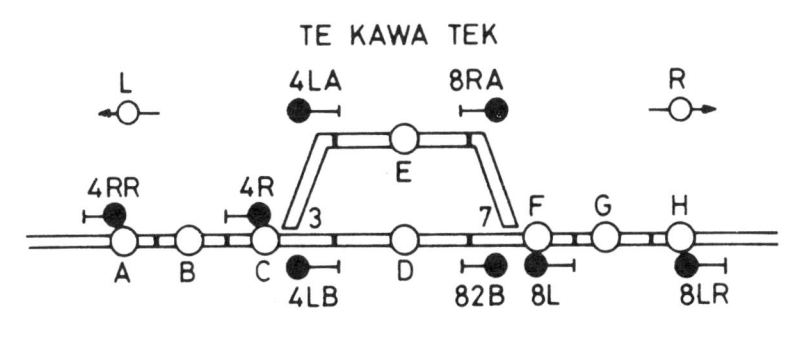

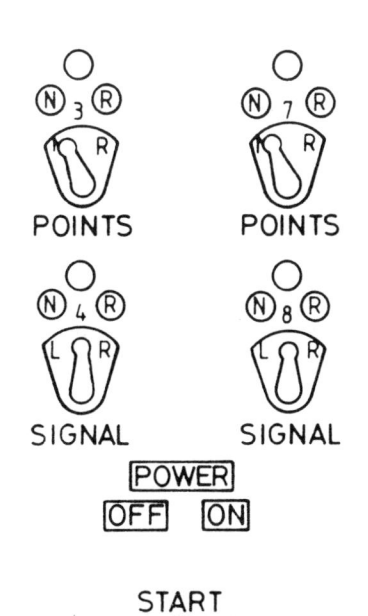

Fig. 8.11 CTC office control machine face layout

COMPUTER CONTROL

With the introduction of computers into the control office, the operator has been relieved of many of the routine operations, allowing him to increase his productivity by controlling traffic over a wider area. Dual minicomputers are normally provided for availability with one being in hot standby mode. Controls are entered via a keyboard rather than a dedicated panel. Simple commands to move a set of points or to clear a signal are available and the software is organised to check that these commands are valid before being output to the appropriate field. The true benefit is derived from the use of complex commands which the software will arrange automatically into the required controls in the correct sequence and at the correct time.

Commands are prefixed by station name in order to select the correct field station. A simple command such as TEK/C4LA would clear signal 4LA at TeKawa (see **Fig. 8.12**). The software checks that no. 3 points are normal, opposing signals are all red and that no train is approaching in the opposite direction, before issuing the control to the field station. If the points were reverse, then provided they were free to move they would automatically be called normal first by the software. If any of the checks revealed that the signal should not be cleared, the command is stored awaiting availability, this being shown by the flashing of the green signal indication.

An example of a more complex command would be TEK/XFL. This is a request to cross two trains, with the first train to arrive taking the loop. The required sequence of point and signal operation for the particular station would be determined by the software with each event being initiated as the correct sequence of indications are returned from the field.

A range of commands which allow for holding, passing, departing and crossing moves are available, for example:

    DUL — depart the up train from the loop
    PUM — pass the up train through the main
    XUL — cross up loop
    PDL — pass the down train through the loop
    MG   — enter the main line and then go

The office computer is further used to allow access by the maintenance technician to interrogate the complete system. Event recording and train run recording are integral features, whilst the blocking of individual tracks or pieces of equipment can be applied by separate commands. This blocking is the software equivalent of placing a reminder appliance over a lever or switch in other systems.

AUTOMATIC ROUTING
By adding timetable and train running information to the computer, automatic routing can be achieved. By also adding decision criteria to this package, a certain amount of intelligence can be built in. These criteria include rules for such things as routing empty freight trains into the loop whilst running loaded trains non-stop on the main.

INTERACTIVE TRAIN GRAPH
Basic to all forms of train control is the use of a train graph to depict the timetable service and to plan and plot stops and crosses. The controller, having decided the next moves with pencil and straight edge on his graph, will implement these by entering the desired commands through the control machines.

Further enhancement to operator interface is now available using an interactive train graph. All keyboard commands are provided as described so that the controller has complete freedom of control but in addition, the train graph is displayed on a colour VDU. Timetable, actual and projected train movements are depicted. The operator, using a mouse or trackerball can adjust the projected movements to plan the service for a period ahead. The computer then routes traffic in accordance with the information contained on the train graph.

Associated with the keyboard is a VDU which echoes the operator inputs and which also displays meaningful error messages, alarms and prompts.

MEANINGFUL WARNING MESSAGES
As the system constantly monitors the status of all equipment and train movements, it is able to generate warning messages when irregular or abnormal conditions are detected. These include such things as train passed signal at red or loss of point detection.

OPERATOR PROMPTS
A prompt is an actionable advice message generated to attract the attention of the operator by audible means. It advises such action that may be necessary by a text message on the VDU. For instance, a train may be approaching a signal at red or may have cleared a set of points when making a shunt move. The prompt message will invite the operator to issue the next command.

OPERATOR DISPLAYS
Display of the state of the railway and indications of the field equipment is carried out either on mimic diagram and/or VDU displays. Nowadays where a diagram is used, this is usually simplified as an overview, detail station pictures being available on the VDUs. Where a diagram is not provided, then both overview and detail pictures can be displayed on the VDU. A recent development has been the use of video projection to display an image of the area under control, rather than by means of a mimic diagram.

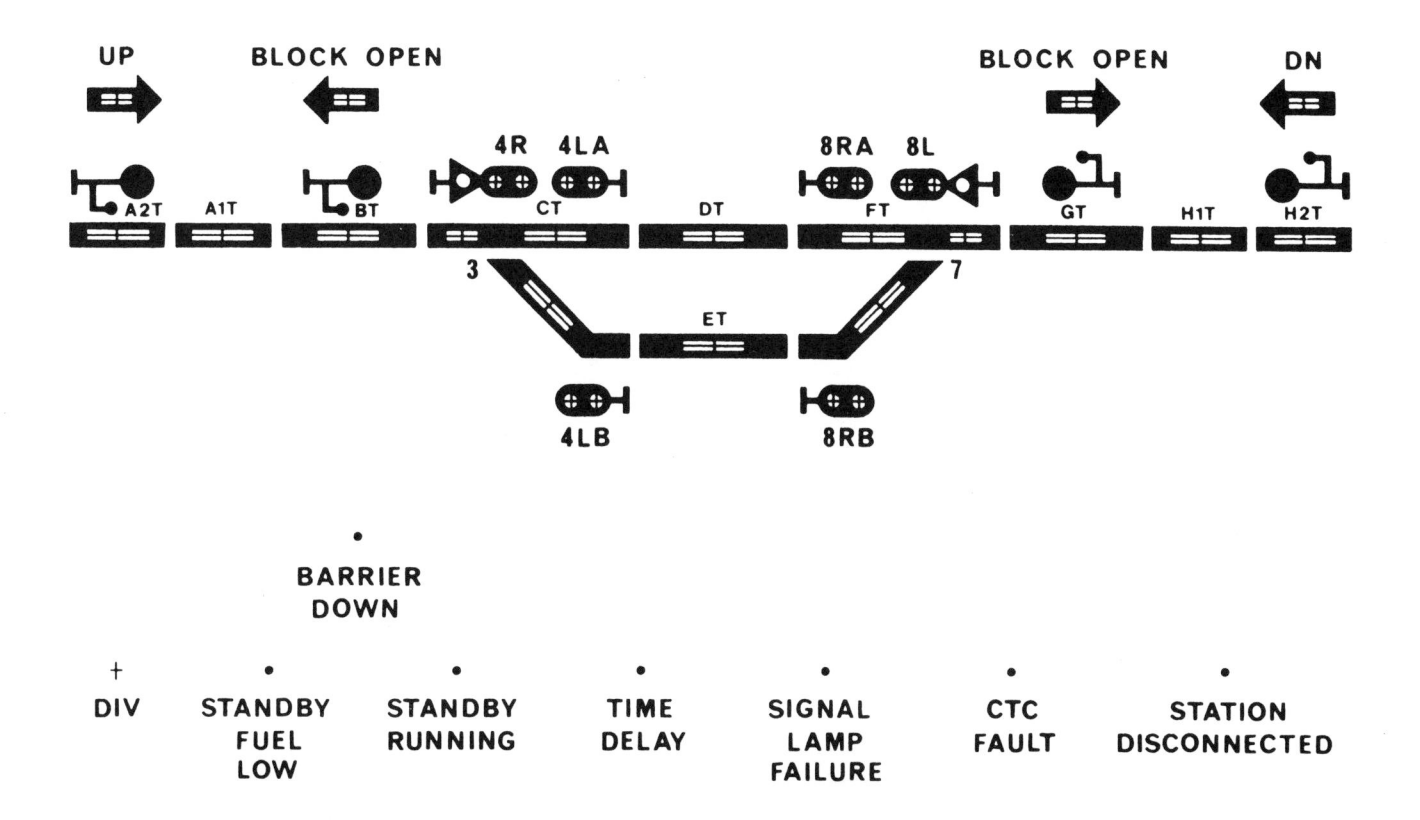

Fig. 8.12 CTC typical field station indications in control office

THE FUTURE
With the undoubted advantages it offers in operating efficiency, CTC is seen as continuing to form the basis of the control system for many lines in the future. Techniques in computer control and transmission continue to be developed. Processors are now being used for the vital interlocking at stations, whilst in-cab displays are beginning to replace the wayside signal. With the use of data transmission over radio, two-way communication between the CTC office computer and train-borne computers is possible, enabling such enhancements as fuel economy and better train handling by intelligent driving based upon information of activity ahead.

CHAPTER NINE
Signalling the Passenger

Introduction

In this chapter, various techniques to provide information are described. Starting with visual and audible ways of presenting information, various control systems are then dealt with, finishing with an explanation of the way in which systems are applied to various categories of station and what may lie in store in the future. The techniques described are current at the time of writing but technological advances are so fast that improvements will almost certainly have occurred by the time of publication.

The need to provide rail travellers with train running information has been recognised since the early days of trains. When considering how best to fulfil this need it is first necessary to consider its cause. Rail travel places most passengers in an unfamiliar environment. It removes the responsibility of the journey from them, their only remaining decision being to ensure they board the correct train. In the more complex suburban areas, the decision is often which is the best train to board because of alternative routes or trains running with different calling patterns. Passengers are therefore either seeking reassurance and information concerning their particular train or the best train to board for their particular journey. The information provided must be clear, concise and above all correct, since once credibility is lost in an information source, it ceases to serve its purpose.

Some of the earlier attempts to meet the need for information included providing:

- Station staff to relay the information to customers. This is the most basic form of train information system to provide and is only as good as the staff, and the information they are given. It does have an advantage in that some passengers prefer to have human reassurance. However it is clearly impractical for large stations to rely on this as the sole source of information.
- Information boards, which consisted of some form of frame with slots for slide-in pieces of hardboard which had been stencilled with train information. This form of train information board is still in use today at a number of stations as barrier end indicators to supplement a more modern platform indicator system. This type of device can be used to communicate with large groups of passengers although it is very inflexible in the information it can display.
- Chalkboards are still used to display train information. They are very flexible and cheap to provide. Their disadvantages are that they rely on the writing ability of the operator and cannot be read from a distance; they also require local staff to keep them updated. They are however invaluable for the dissemination of emergency information should other more sophisticated systems not be available.
- Public address systems still provide in some stations the only source of current train running information. They are easy to use, flexible and comparatively inexpensive. Their main disadvantage is that the information is not accessible to the passengers when they require it, but is presented to all at the same time. The quality of reproduction is very much dependent upon the announcer, the system design and the subsequent maintenance.

In the early 1980s, British Railways introduced the concept of sector management, the chief aim of which was to make the railway become more business conscious. This entailed the division of the railway network into the separate businesses of InterCity, London and Southeast (subsequently renamed Network SouthEast), Provincial, Freight and Parcels, with each group being held responsible for its own performance. This caused a more businesslike approach to all matters,

Fig. 9.1 Average ratings of attributes of importance scale

Rank	Attribute	Average Rating
1	Information about delays and cancellations	4.74
2	Information about train times	4.65
3	Help and advice from station staff	4.52
4	Shelter from weather while waiting	4.43
5	Cleanliness of toilets	4.37
6	Ease of buying a ticket	4.34
7	Availability of toilets	4.19
8	Maintenance of stations	4.11
9	Time for ticket collection and inspection	4.09
10	General cleanliness of station	4.09
11	Ease of getting to platforms	4.04
12	General look of station	3.82
13	Heating in the waiting room	3.80
14	Number of seats in waiting room	3.54
15	Somewhere to buy newspapers and books	3.54
16	Number of seats on platform	3.53
17	Ease of walking round on platform	3.49
18	Somewhere to buy drinks and snacks	2.71
19	Somewhere to buy cooked meals	1.90

Note: Average ratings are based on a range of 1 to 5, from least to most important.

including the provision of information to the customer. One of the features of this change has been the regular use of internal and external studies to gain a measure of customer reaction to the services provided and to new initiatives. A report was produced by Martin Vorhees Associates Consultancy in November 1985 entitled 'Station Modernisation Priority and Pay-offs'. This was based on earlier studies at BR Network SouthEast stations, and produced the results shown in **Fig. 9.1**, extracted from the report. This emphasises the importance of information to customers, which has been reflected in the investment undertaken by BR in information systems in recent years.

Display Techniques — Mechanical

FLAP INDICATORS

Split flap indicators have been the most widely used mechanised display to date. Because of the inherent contrast ratio, they produce the clearest display under all ambient light conditions, although supplementary lighting is needed at low light levels. All combination of characters or colour can be silk screen printed on to the flap and any typeface or symbol can be produced with very clear definition. Flap units are available in various sizes up to a maximum of 80-way units. The individual flaps are made of plastic (PVC); where there is likelihood of direct sunlight, metal or plastic thermo-resistant material is used, since PVC flaps are liable to buckle. When used as individual platform indicators, up to three units can be installed, dependent upon the local requirements. The units are mounted within a weatherpoof housing, protected by a vandal-proof transparent screen. If used as a main departure board, any number of units can be mounted vertically as required. At intermediate size stations, repeat indicators of each platform are sometimes provided in ticket halls, on overbridges or in subways. A typical example is shown in **Fig. 9.2**.

The drive mechanism, although electromechanical, requires the minimum of maintenance. It consists of a step-by-step motor driven by pulses of alternating polarity. The rotor of the motor comprises a ferromagnetic cylinder with alternating poles, and an anti-reversal wheel ensures the flaps rotate in one direction only. Two detectors are used to ensure correct synchronisation of the rotor to the controller. One detects the position of the rotor during flap setting, and the other detects the blank position. The flap control can be a simple thumbwheel switch used on a small to medium size station, or part of a more complex control system at a large station. The main disadvantage of the flap indicator is the need for the re-scripting of the unit when any alteration occurs to the pattern of train services. This is expensive both in the reprinting of the flaps and in the manual replacement of the altered flaps at each indicator.

ELECTROMECHANICAL DOT MATRIX
Nearly all reflective dot matrix display devices use electromechanical techniques whereby one half of a hinged disc is flipped or a cylindrical drum is partially rotated either to expose, or blank off, a painted pixel, under the control of a switched magnetic field. Each pixel represents one part of a character. Thus it can be seen that an electromechanical dot matrix display of even modest proportion represents a highly complex assembly consisting of a considerable number of individual moving parts. In general most displays of this type utilise a 9 × 5 dot matrix arrangement which limits the available character set.

A transmissive display of this type consists of moving metal segments contained within a thin-walled plastic membrane. A bank of fluorescent tubes is located behind this membrane to provide the light source. A traversing magnet passes over each segment during the display refresh cycle and the segment is moved out of, or in line with, a transparent section in the membrane depending upon whether the dot is to be displayed or not.

Fig. 9.2 Split flap indicator

Display Techniques — Electronic

As the name suggests, display of characters is achieved by purely electronic means, and as such no moving parts are used. At the time of writing, the types of electronic display available are:

- Closed circuit television (CCTV) monitors.
- Light emitting diodes (LED).
- Liquid crystal displays (LCD).
- Electrochemical displays (ECD).

CLOSED CIRCUIT TELEVISION MONITOR (CCTV)

Probably the most prolific display device used in modern systems, it has a part to play from the simplest customer information system to the most complex. At present the most common type of monitor used is monochrome with a number of phosphor colours available — white, yellow and green being the most common; but recent reductions in cost and better availability have now made the colour monitor a viable option. A typical monochrome monitor is shown in **Fig. 9.3**.

A purely television-based display system in its most basic form consists of a television/monitor connected to some form of video drive unit. The video drive unit will address pixels and lines on the monitors to form whatever information it has been programmed to display. The video drive unit will refresh the screen periodically (typical rates are 50–64 Hz) such that the screen information appears static. Typical values for a monochrome drive unit would be 1,024 pixels per line and 286 lines, with individual characters formed of 16 pixels width and 24 pixels depth. This enables rounded clear characters to be formed and facilitates the use of graphics. Connection between the monitor and video drive unit is by a single 75 ohm co-axial cable. When colour is used, the definition is often reduced slightly because of the need to address the three colour guns, whose combinations form the one colour pixel, with approximately 600 pixels per line available. Connection between the monitor and video drive unit usually consists of four 75 ohm co-axial cables, because of the need to drive each of the red, blue and green colour guns separately and to supply a synchronising signal to control them.

Fig. 9.3 CCTV monitor

The basic requirements of CCTV monitors are as follows:

- 75 ohm unbalanced isolated inputs (isolated inputs provided to improve interference performance).
- 625 lines per frame.
- 50 frames per second (non-interlaced).
- Switchable terminated and high impedence inputs.
- Where colour monitors are used, separate red, green and blue (RGB) and synchronisation inputs. In some cases the synchronisation signal may be superimposed on the green signal input.

The CCTV monitor offers a very versatile means of information display for high resolution characters. The major limiting factor is the amount of information that can be displayed at any one time. However when used with a well designed and constructed housing, monitors provide an aesthetically acceptable and robust display device which can be located virtually anywhere, usually showing departure and arrival summaries or individual displays for each platform. The CCTV monitor is unique in the field of dot matrix display devices, in that the monitor itself does not have any built-in intelligence in the form of a display controller, as do all other dot matrix display devices. The display controller receives incoming display instruction from the system controller or computer and arranges for the correct display of the information required, using a character set pre-programmed within the display controller. In the case of CCTV monitors, this function is provided by either a character generator or a digital-to-video converter.

Character generators (CG) are exclusively used on monochrome CCTV display systems. These are multi-channel, multi-page devices using either a vertical or horizontal wipe technique to signify the transition from one display page to the next. A number of display attributes may also be available such as flash, inverse video, underline and halftone to highlight part of the information displayed. Character generators are usually centrally located adjacent to the system computer from which they receive all the necessary display instructions. The video signal outputs are then distributed to the required monitors on any given display channel by means of a co-axial

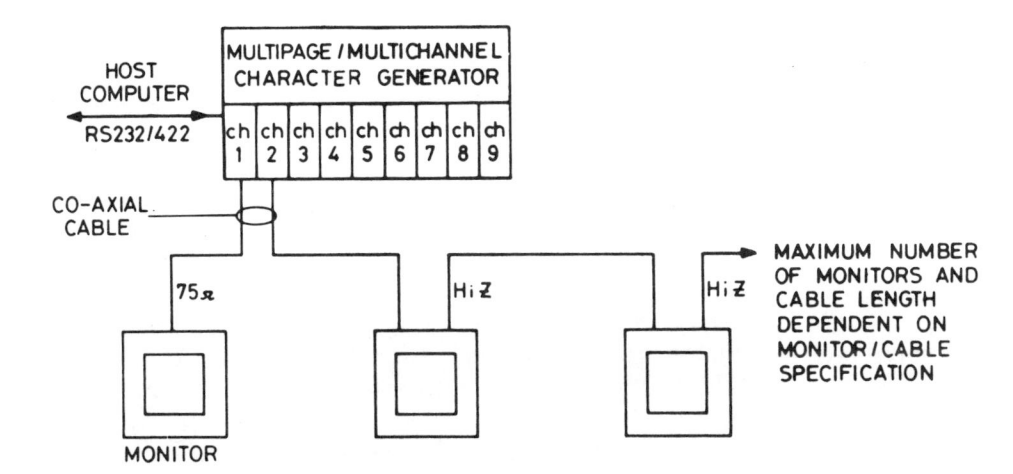

Fig. 9.4 Character generator-based system schematic

Fig. 9.5 Digital-to-video convertor-based system schematic

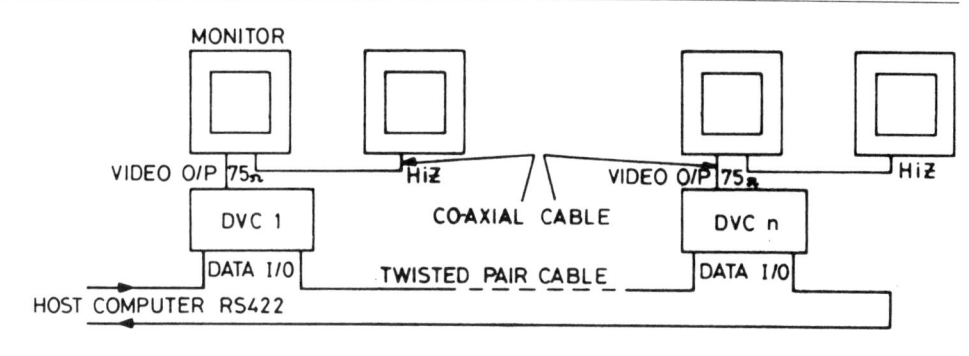

a. MONOCHROME SYSTEM

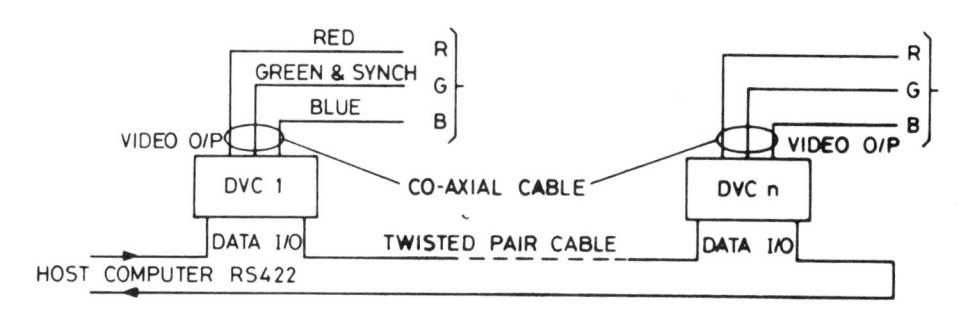

b COLOUR SYSTEM

cable (see **Fig. 9.4**). This technique has a number of disadvantages:

- The video distribution network can be susceptible to varying forms of interference with the video signal.
- Cable length can be a limiting factor, dependent upon video signal output level and pre-emphasis and co-axial cable characteristics.
- The number of monitors which can be multi-dropped off any one video circuit is determined by monitor input sensitivity and CG output and cable characteristics as described above.

However despite these disadvantages, it is a technique widely used for the provision of information systems of varying sizes.

Digital-to-video converters (DVC) in essence provide the same function as individual channels of a centrally located character generator, the main difference being that DVCs are generally located adjacent to, or in, the monitor. Each DVC is usually connected to the host computer or controller by means of an omnibus data ring as shown in **Fig. 9.5**, from which it receives addressed display instructions. It has been the DVC, along with the decreasing costs of colour monitors, that has heralded the introduction of full colour information systems. The DVC offers the system designer a number of advantages, making them the most common devices being installed at present in the UK on main line railway information systems:

- Monitor-based system less prone to video interference.
- Data distribution networks are usually less restricted in length, for example the RS442 protocol is specified for up to 1,000 m cable length at 1 MHz in a noisy electrical environment.
- Slave monitors can be fed by simple video distribution from the master DVC/monitor location or from another DVC.

LIGHT EMITTING DIODE (LED)

Displays using these devices are made up of individual LEDs which are available in a limited range of colours. A typical display as used on the London Underground network and shown in **Fig. 9.6**, consists of a matrix of 24 high × 256 across LEDs at a pitch of 8 mm. This allows two rows of characters to be displayed with a character height of approximately 75 mm. LED display devices have a number of disadvantages which have impeded their use as displays on BR for information display systems, the main use being in booking halls for providing a scrolling display for sales information:

Fig. 9.6 Dot matrix LED display

- Viewing performance deteriorates markedly in high ambient light levels since there is little or no control available of overall display brilliance.
- Viewing angle is restricted to approximately 30–50 degrees.

These disadvantages do not apply in any significant degree on underground mass transit railways and LEDs have been used extensively by London Underground to replace older, light-box displays.

LIQUID CRYSTAL DISPLAY (LCD)
Liquid crystal display systems were introduced in the UK in 1985 and provide a serious contender to the more conventional electromechanical dot matrix and flap devices which are presently extensively used. Liquid crystal display technology has been well known for many years and is basically derived from the 'Kerr Cell'. These devices have become a more viable option due to two main factors — improved contrast ratio and better and more reliable manufacturing techniques.

The majority of LCD displays installed to date on railways in the UK are of the transmissive type as shown in **Fig. 9.7** and as such require a light source mounted behind the display. In larger displays this light source requires a light-sensing device controlling dimmer circuits in the light source to adjust the brightness of the display in variable ambient lighting conditions. Current production techniques in LCD technology can now produce dot matrix displays of 38 mm and 50 mm height with a 12 × 6 matrix providing a character set of upper and lower case with descenders and limited graphics.

Future development in LCD technology lies in the areas of improved contrast ratio so that reflective displays become viable in all lighting conditions, thus eliminating the light source requirements for transmissive displays, and in smaller characters with increased pixel population leading in its ultimate form to LCDs becoming a replacement device for CCTV monitors.

ELECTROCHEMICAL DISPLAY (ECD)
This relatively new display exploits the principle whereby an electric field can induce a chemical change and a consequent colour change in a liquid between two electrodes. All devices available at present are reflective and suffer the major drawback of poor contrast ratio, but continuing development is likely to improve performance. Electrochemical display devices do offer a major advantage over reflective LCDs in regard to display refresh cycles. Once a given display is set on an ECD, it only requires refreshing once every few hours, whereas LCDs require refreshing several times a second. Thus theoretically the drive electronics for an ECD may be very much simpler than for a comparable LCD.

Audio Techniques

What the passenger hears is important. Audible messages reinforce information given by visual signs and can supplement it by the provision of more detail than can be conveniently displayed on a sign, and in a form which passengers, particularly occasional passengers, can more readily understand. Public address systems consist of various types depending upon the size and complexity of stations, and in turn comprise a number of subsystems, described below. Many individual designs are possible, and the details given are typical of the systems available.

LOCAL PUBLIC ADDRESS (LPA)
Local public address systems include announcing points fitted with selection controls and microphones, amplifiers and switching modules, and speaker circuits. Microphones can be of the hand held type or can be desk mounted if installed as part of an information desk. They have a load impedance of

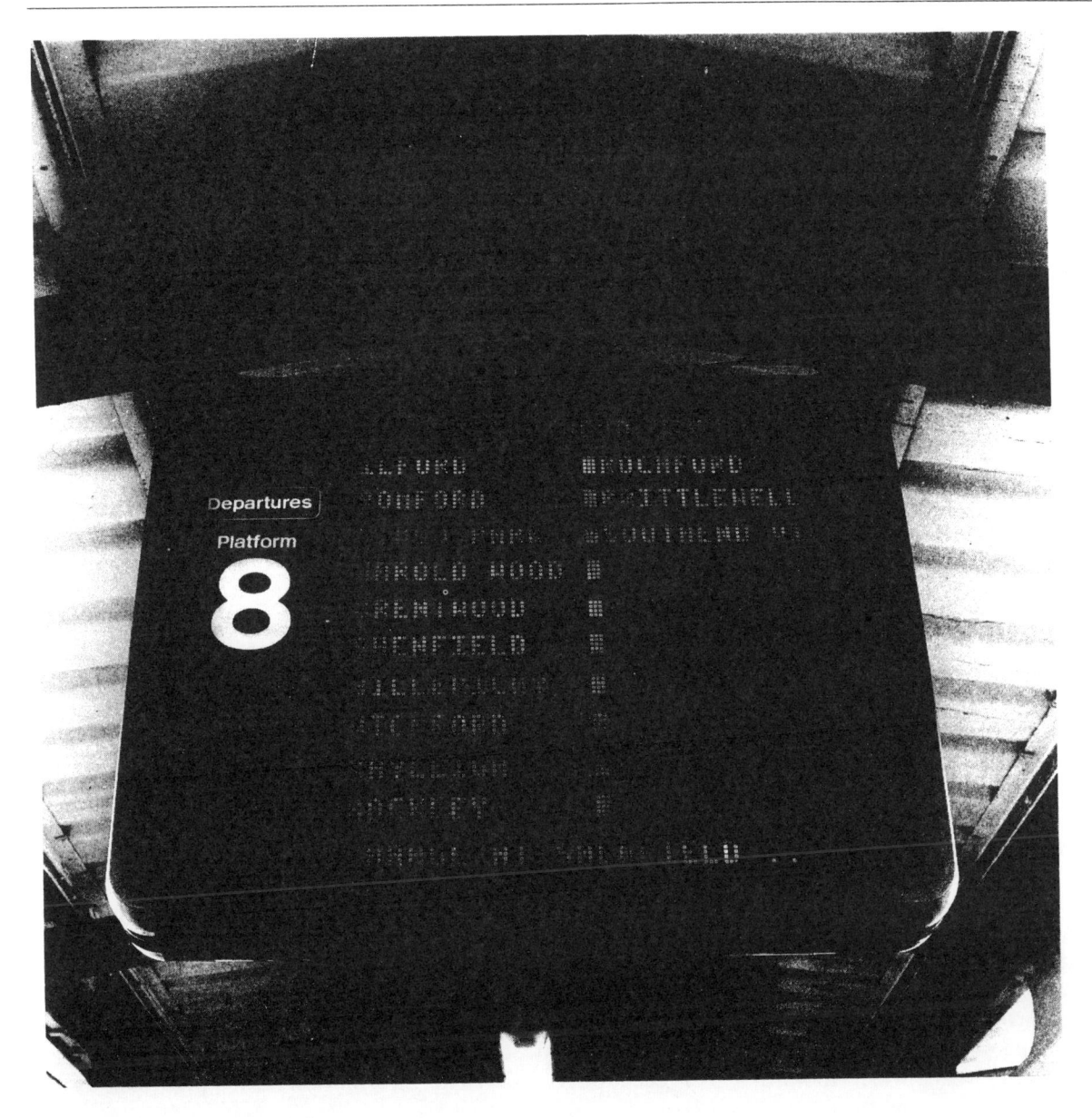

Fig. 9.7 Liquid crystal display

200 ohms and are connected to the amplifier by a screened twisted pair cable. The selection controls, where fitted, allow the announcer to select the required geographic zone on the station for the announcement by operating appropriate locking buttons. The buttons can be mounted on a standard size modular control unit (MCU) when situated in the station supervisor's office or at various platform announcing points on the station.

There are separate pre-amplifiers for each microphone, together with a pre-amplifier for other subsystems of the PA installation, such as recorded announcements (RA) and long line public address (LLPA). The RA and LLPA pre-amplifiers are designed for a balanced 600 ohms input. The input level can be ±3 dB over a frequency range of 50 Hz to 15 kHz. Each main amplifier has an output voltage of 100 volts rms with a regulation better than 1.5 dB. Provision can be made for reduced output during night hours, controlled by a local timeswitch or by an instruction from the LLPA system if connected.

The switching module uses electronic switching techniques to connect the main amplifier to the required platform speaker circuit. Each platform may have a 'long' and 'short' output circuit in order to reduce environmental noise at off-peak times when trains are shorter. The short output circuit covers the central part of the platform, and the long circuit extends to the whole length of the platform. In the case of island platforms, the short output circuits are on each side of the station building which usually occupies the central part of the platform, whilst the long circuit is common to, and selected with, either of the two short circuits. Selection circuits are also provided for the ticket halls and overbridge/subway areas where appropriate. At stations where there is more than one announcing point, there may be a requirement to make simultaneous announcements to two or more platforms independently, for which the appropriate number of main amplifiers are provided. A miscellaneous unit is also provided with the amplifiers to provide alarm circuits which monitor power supplies, and provide a visual and audio alarm at the manual control unit if a fault occurs.

Loudspeaker designs consist of the horn, beam and column type and are used as appropriate for each situation. Normal practice is to install a speaker every 15–20 m on platforms. The provision of speakers within large covered stations requires special consideration with regard to the poor acoustic environment. Two approaches have been adopted, which both give a satisfactory result. One uses low power column loudspeakers placed throughout the station affording an even coverage, with minimised interference and overlapping of sound from different speakers. In addition, ambient noise-sensing is provided using a small microphone within the loudspeakers to sample the local noise level, and to minimise reverberation in low ambient noise conditions. The second approach is to use delay line circuits with speakers grouped in zones, with a delay of some 60 ms between sounds emitted from successive loudspeakers. This ensures that at any position, the sound waves arrive at that point at the same time, from whichever loudspeaker the sound was emitted. These two techniques are mixed on some systems. Special headphones are issued to the announcer in order to provide feedback, and ensure that the announcer speaks at the correct speed and level.

RECORDED ANNOUNCEMENTS (RA)

In order to reduce the workload for operators making individual announcements, and to ensure consistency of presentation, RA systems were developed some years ago using tape machines. Now recorded announcement systems comprise some form of solid state memory and microcomputers to store and process the speech messages. Current practice is to use digitally stored and regenerated recorded speech rather than synthesised speech, since the objective is to have sound quality as near to that of a human announcer as possible. The words and phrases are first recorded on a studio quality analogue tape recorder. This information is then processed using pulse code modulation techniques by sampling the speech signal at regular intervals, quantising the resulting samples and storing them in binary form in the memory. Superfluous information is removed to reduce memory capacity whilst ensuring there is no effect on the intelligibility of the speech.

A central processing unit (CPU) controls the processing of the binary representation of the words and phrases, forming them into the required pattern for the particular announcement. This binary information is then output through a digital-to-analogue converter to re-create the analogue signal. The analogue signal is then input to the LPA amplifier. There are two methods of creating the messages and controlling the output. The first method is to preform the messages with an index number and store them on eraseable programmable read-only-memory integrated circuits (EPROMs), allowing a simple operator control panel to be used adjacent to the LPA controls. This is shown in **Fig. 9.8** and consists of selections for pre-announcements made before the train arrives (eg *the next train at platform 4 ...*) and the main announcement when the train arrives at the platform (eg *the train now standing ...*). The disadvantage of this method is the requirement to re-create the messages when timetable amendments are made, necessitating the reprogramming of the EPROMs. The second method which is more appropriate to the larger station, is to have a VDU-based controller and a light pen to enable new messages to be formed from the stored vocabulary and to keep them in memory or on disc. This allows the operator when there is an alteration to the planned working, to create or amend any message required and store it for subsequent output.

LONG LINE PUBLIC ADDRESS (LLPA)
Long line public address systems allow a central point to issue announcements to remote stations either independently or in groups. Any station may contain a number of addressable zones to which announcements can be made. The LLPA system interfaces with the LPA system at each station. A 'staff circuit' may be provided with output loudspeakers in the supervisor's or ticket office and is driven directly from the LLPA system to ensure an announcement can be made on this circuit even when the LPA is in use. The LLPA can only access the LPA system if the latter is not in use, and it is

Fig. 9.8 Manual control unit

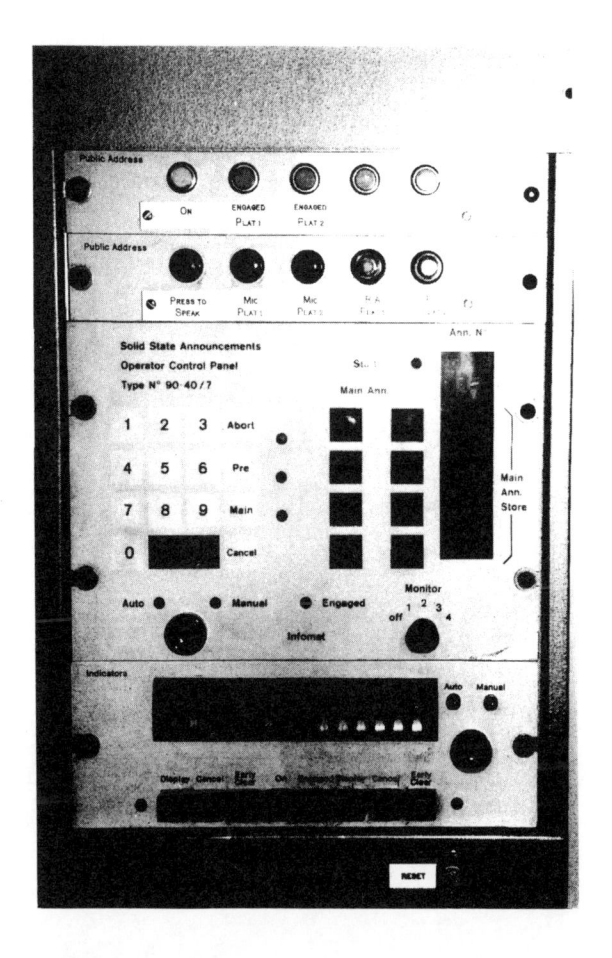

normal to give an indication to the central point if this is the case. Fault indications can also be given to the central announcer and a technician's fault logging printer can be provided. If any outstation fails due to a local fault or power supply failure, the line circuit is switched through to enable the remaining outstations on the circuit to operate. The speech signal is frequency modulated on to a carrier where physical line circuits are used, or can be directly fed into a PCM transmission channel when available. The outstation consists of a demodulator, switching unit and modulator.

There are variations in the operational design of LLPA systems depending upon local operating requirements. One method is the 'key station' approach, where the central announcer makes announcements to staff at a small number of key stations. At these stations, local staff repeat those messages which are relevant to their own station and to the group of stations served by that key station. A different concept is to have all outstations served directly from the central point with all long line announcements made by the one announcer. **Fig. 9.9** illustrates the general principles.

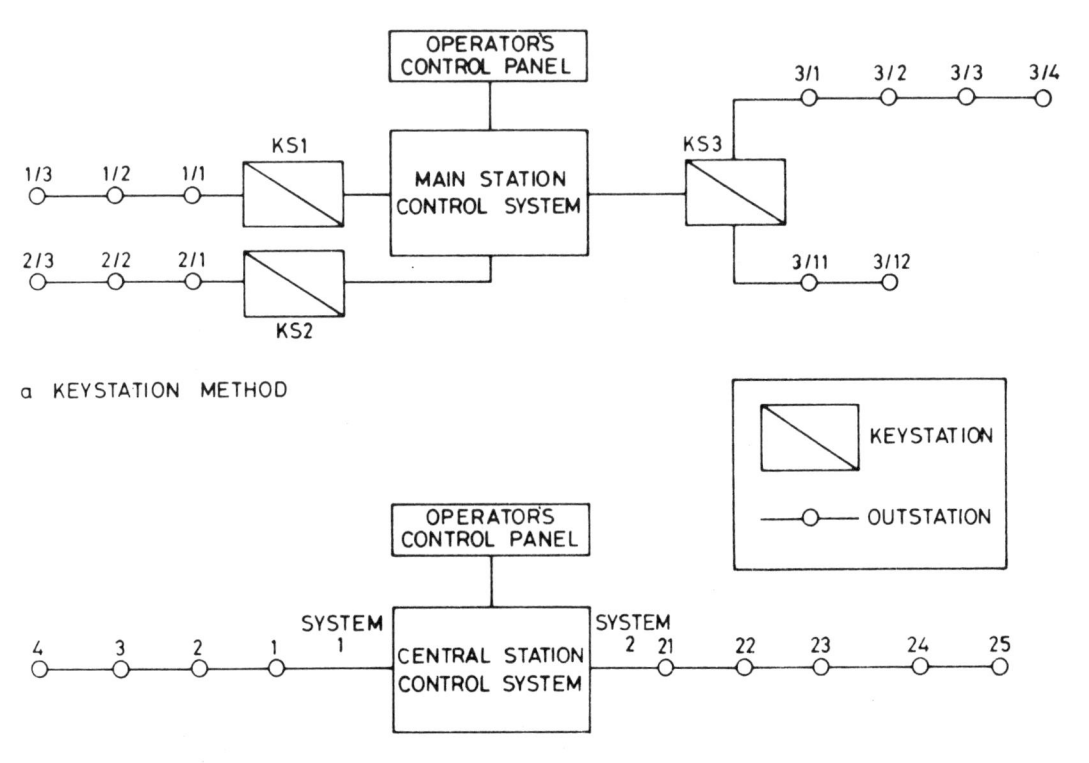

Fig. 9.9 Schematic of long line public address system

Control Techniques

CONTROL AND AUTOMATION

Any information system be it large or small, must be made up of a number of common elements:

- Display devices consisting of CCTV monitors, flap boards or dot matrix displays, etc.
- Display controllers such as character generators, digital-to-video convertors, flap or dot matrix board controllers.
- System controllers. Computers with sufficient processing and storage capacity must be provided to encompass present and future requirements.
- Operator terminal. Usually today some form of VDU and keyboard is provided to enable efficient monitoring of overall system performance and status, and to allow effective and speedy operational amendments to be made as required.

Additional features that may be required are:

- Digital input facilities to enable relay contacts to be monitored to receive additional external inputs such as track circuits, right-away plungers, etc.
- Additional input/output serial ports to allow connection and control of other devices such as additional operator VDUs and keyboards for use at an alternative operator's position or for off-line database preparation, and a technician's terminal for fault reporting and the running of diagnostic software. Other uses are for other information systems such as train describers, recorded announcement systems and other information systems as part of an area control or long line system.

Fig. 9.10 illustrates how a more complex information system may be configured. It is not meant to show a definitive answer as each system will require individual tailoring in order to meet specific engineering and operational requirements. The heart of any customer information system is the system controller or processor. The size and capacity of this device will depend on the size of the system involved and will range from a desk-top microcomputer to a powerful minicomputer. However they all show some common features:

- Sufficient processing power to achieve fast, effective response to system inputs and outputs: for example, displays are cleared as soon as possible after a train has departed from a platform.
- Sufficient storage capacity of both volatile memory and disc. In many systems, data storage will be a combination of random access memory (RAM), hard and floppy discs with floppy disc used for the purposes of initial data input or amendment on to the hard disc. In larger systems, in order to reduce floppy disc requirements, a tape streamer back-up storage system may be utilised.
- Effective and user friendly system operating software to enable fast input of display parameters such as platform changes, delay, etc. The success of any information system will depend to a significant degree on the facilities available to the system operator for amending data.

SOFTWARE

Most software packages used in information systems consist of a number of basic modules. These are system operating, database, off-line data preparation, engineering diagnostics and system configuration software modules.

The system operating software, in conjunction with the system database, provides the second-by-second control of any information system, basically arranging for the timely display and clearance of information to the displays or other devices. It also arranges for the processing of information and

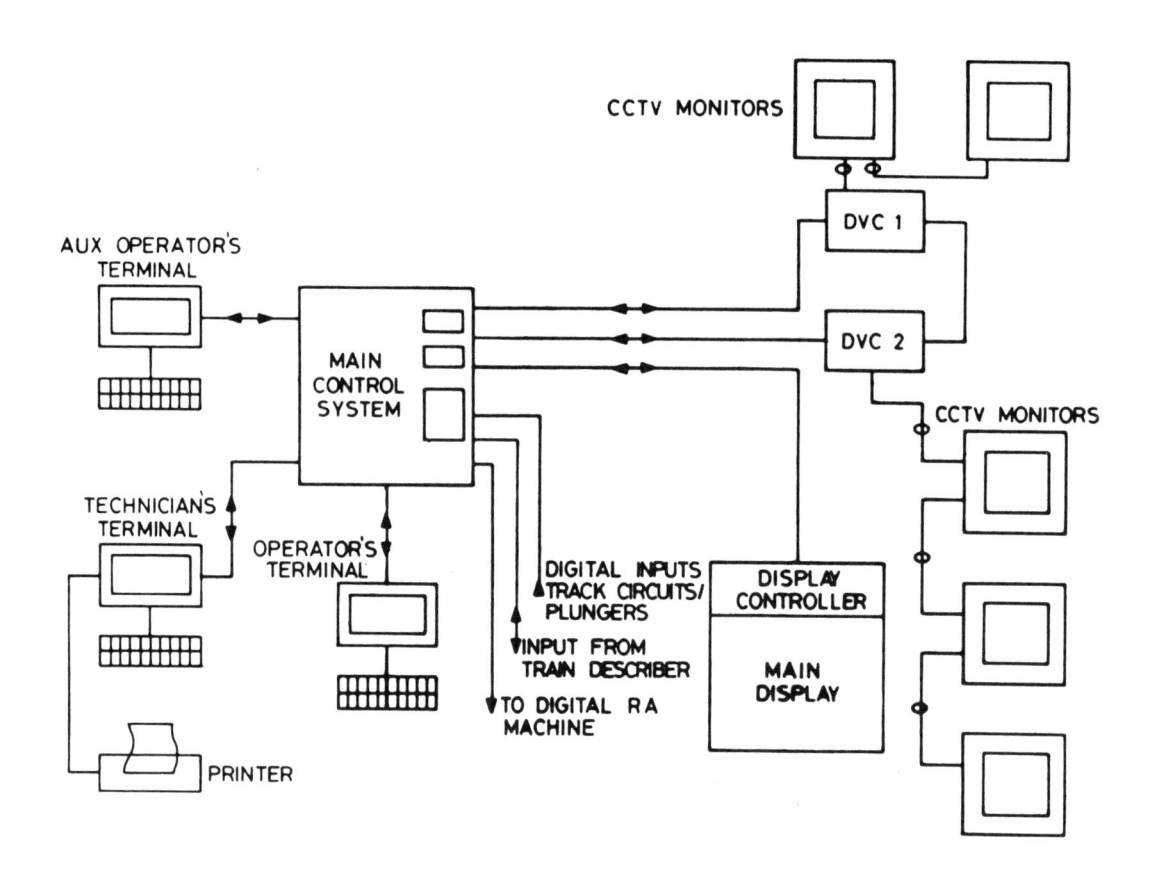

Fig. 9.10 Schematic of complex information system

instructions received from the system operator, and carries out such system checks as are possible at regular intervals and reports all detected faults to the system operator and technician's terminal where installed. Facilities provided within the software may include the following:

- Real time clock and calendar with British summer time changeover dates.
- System status page showing current status of the display system, for example the trains displayed, or queuing to be displayed.
- Audio announcements.
- Train delay information.
- Special notice creation, storage and display.
- Creation of special 'today only' trains.
- Deletion, cancellation and possible retrieval of any given train.
- Bank Holiday dates.

This list is not meant to be definitive but serves to illustrate the facilities and functions available to the system operator. The final list will obviously depend on the overall functional and operational requirements for any individual system.

The database represents the reference computer file from which the system operating software gathers the details of trains eligible for display to both operator and public at that time. Information contained within each train entry may include:

- Origin or destination.
- Train type, for example, arrival, departure, passing train.
- Calling pattern.
- Run days including Bank Holidays.
- Start and stop dates, for example, timetable changes, special trains, etc.
- Special remarks.
- Expected time of arrival and/or departure.
- Recorded announcement reference.
- Train describer references used to aid further automation of displays as trains move.

The off-line data preparation package is used for the preparation or amendment of the database to permit timetable changes and special trains, etc to be input into the system database at an appropriate time. It may be possible to use an extract from the main railway central train database, such as the train service database (TSDB) which holds details of all timetabled trains on BR, to minimise the amount of manual input necessary. A strict disc management system has to be adopted for data preparation to ensure separate back-up copies are available in case of failure.

Additional software packages may be included to enable engineers to configure the system by customising a standard information system package for a particular location. In addition, a diagnostic package may be provided to enable engineers effectively to maintain and find faults in the system.

AREA CONTROL OR LONG LINE PASSENGER INFORMATION SYSTEMS

The system illustrated in **Fig. 9.10** is typical of a large system. It could be further expanded to include the control of smaller, generally CCTV-based systems, provided at outlying stations. The outlying station systems would in essence be autonomous, self-contained systems as already described and would generally consist of a suitable desk-top microcomputer controlling a number of DVCs or CGs driving appropriately positioned CCTV monitors or other display devices. The only communications required between the central control and any outlying station systems would be:

- Regular polling to ensure integrity of communications link and time synchronisation.
- Confirmation of outlying display status.

- Confirmation of receipt and processing of display amendments input at the control point.
- Down-loading of information regarding a new train not held in the outlying station database.

The central system operating software for this type of system would need to incorporate facilities for displaying the status of outlying stations and the transmission of data over longer communication links: in order to minimise operator workload, connection to a train describer covering the area is considered essential. The database would need to be expanded to provide relevant information for the outlying station systems. Database preparation should be carried out centrally on an area basis, with individual databases then being produced for all outlying station systems to minimise keying in of information and to ensure that data is consistent throughout the area.

Applications

The techniques used at stations to communicate to customers are influenced by the factors of cost, whether the station is manned, the revenue or potential revenue earnings of the station, and what, if any, financial support is available from outside bodies. Consider the following five categories of station:

1. The 'smaller unmanned station'. These are usually low revenue earners and may often be subsidised by the local Passenger Transport Executive or local authority. The most commonly used technique in this type of station is long line public address (LLPA). Control of the system is usually vested either with the signalman who controls the area, or someone with access to train running information. Announcements are usually limited to out-of-course running information. Closed circuit television security systems may also be installed to provide some form of security for passengers and buildings. Whilst long line controlled platform indicator systems are sometimes used, the costs are usually prohibitive for this size of station.

2. The 'smaller manned station'. Because of the revenue earnings and customer use of this type of station, some form of information system is normally provided. Historically, public address systems were used to convey information to passengers, but more recently both long line platform indicator systems and locally controlled PIS have been installed to supplement the LPA system. Many systems are now obtainable which are specifically designed to operate in situations where specialised operators are not always available, and which can be operated by station staff as part of their normal duties. These systems are often based on the use of commercially available microprocessors with software capable of posting and clearing train information automatically, dependent on time, or as more commonly used, automatic posting of information with a simple keyboard operation to clear train information. The timetable data is produced/collated centrally, usually within the Area Manager's office using a down-loaded data extract from the train service database file held in the BR mainframe computer systems.

3. The 'busy suburban station'. These are stations with high revenue used by commuters accustomed to modern technology who, as a result of regular rail travel, become very familiar with train operations. Historically, some form of visual platform indicator system was employed, the indicators being large to allow for easy and quick dissemination of information to large groups of passengers. The two most commonly used types of indicators are flap boards and light boxes. The flap indicator itself is normally worked from a simple thumbwheel device which controls its stepping motor. The light boxes are operated by simple switches by the station staff. One of the more modern techniques used at this type of station is an automatic long line platform indicator system. This is controlled from the train describer provided as part of the main signalling control function, with manual intervention provided for out-of-course events. Systems associated solely with the train describer are limited, however, in that they do

not display train departure time and are not set more than a few minutes in advance of the train arrival, both of which have recently been identified by business sectors as fundamental requirements.

A link is therefore necessary to a timetable, and this is described further below. Liquid crystal display indicators operated from stand-alone personal computers which contain timetable data are now available. Control of the indicators in this type of system is vested with the station staff who can use a thumbwheel switch unit to enter train time and platform to the processor which then causes the indicators to display the time, destination and calling information obtaining the necessary additional information from the database. Unscheduled information can also be displayed by being selected by station staff from a pre-determined library of messages.

4. The 'major terminal station' can be a large interchange or a terminus. The information system is required to direct all passengers at these major stations since train platforms are liable to be altered at short notice. The control point for the information systems is usually manned continuously and there may be a requirement for additional operator positions for use during the peak hours.

The indicator system normally consists of a main departure board using split flaps, electromechanical dot matrix or LCD display devices with a portion reserved for special announcements. An arrivals board may also be provided. Each platform will be equipped with an individual display device at the entrance to the platform, and sometimes at an intermediate point along the platform. Summary displays using CCTVs are distributed around the station. The control system operates in a manual, semi-automatic or fully automatic mode. In the manual mode, the operator is presented with a list of trains for a particular period from which the trains to be displayed are selected. In the semi-automatic mode, the operator is prompted to display or clear the indicator depending upon the time; whilst in the automatic mode, the system displays the requisite information at a pre-determined time and clears it at the appropriate departure time.

The public address system is divided into zones for different areas. Because of the size of the station, the PA system would normally use one of the techniques already described to limit sound reverberation. The volume of announcements made may necessitate the use of recorded announcements to ensure a high standard of speech quality for every announcement and to limit the workload on the operator.

5. The 'urban metro network'. The conventional train displays on London Underground are being replaced by LED dot matrix indicators to provide a standard and flexible method of displaying train information to passengers. The order of arrival, destination and time before arrival of up to the next three trains due at a station, are displayed. The standard train arrival information is interspersed with text messages containing train service or other information, which may, for example, inform passengers that the next train through to a certain branch will arrive within 10 min. On platform signs, this is normally displayed on the lower line while information concerning the first train continues to be displayed on the upper line. The sign controller has outputs for three sets of dot matrix signs, one for each direction and an additional 'concourse' output which can only be used to display text messages. Each output can drive up to four signs in parallel.

The system can operate in normal or on-line computer mode (OLC) and back-up or train description (TD) mode. Each of these modes acquires its train information from an independent source. In the on-line computer mode, the full range of train information and text messages is available and is transmitted to the sign controller over a 2,400 baud serial data link. There are two systems used to generate this data.

(a) *Centralised computer control.* If the line is equipped with centralised computer control, for example the Jubilee Line, the central computer itself generates the serial data. By referencing internally held timetables, train destination and train movement information received from on-site equipment, it can determine the positions and destinations of trains on the line, and thereby generate the necessary train destination and time before arrival information. Some text messages are generated automatically by the computer depending on the position of trains and the state of the service, while others are generated by an information assistant on a workstation. More than one sign controller may be connected to a single data link as each serial data packet can be addressed to one or more sign controllers and signs.

(b) *Stand alone interface control.* If the line is not equipped with centralised computer control, for example the Bakerloo Line, a stand alone interface generates the serial data from the existing train describer systems and from inputs from track circuits referred to as timing markers. As each train passes a certain point on the railway, its train description is transmitted and this is used by the interface to generate destination and an initial time before arrival for that train. The time before arrival for each train is counted down by an internal clock and corrected as the train reaches a timing marker. In addition to their timing function, track circuit inputs may be used to trigger text messages. The internal clock may be used either to trigger text messages directly or to enable them to be triggered at certain times.

To ensure that incorrect information is not displayed, train order, destination and time before arrival information is normally updated every 30 s. If updates are not received by the sign controller for 90 s, it will revert to the TD mode until updates are restored. In this mode, inputs from the existing train describer system are connected directly to the sign controller to provide train order and destination information. No time before arrival information is displayed, no text messages are displayed and trains are normally much closer to the station when they first appear on a sign. A simplified schematic of the system is shown in **Fig. 9.11**.

SIGNALLING THE PASSENGER   253

Fig. 9.11 Schematic of London Underground dot matrix LED information system

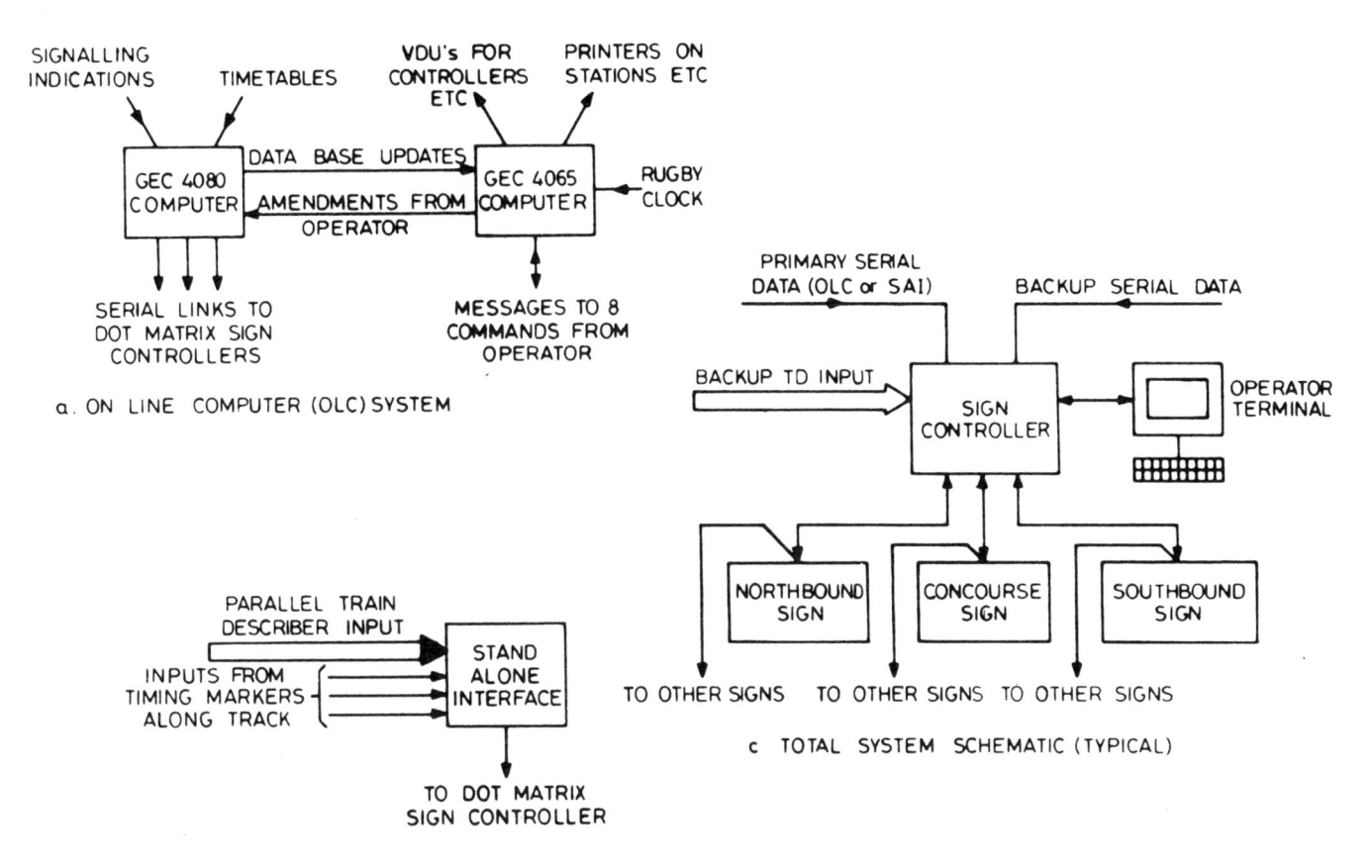

The Future

The advent of the Integrated Electronic Control Centre (IECC) has heralded the start of a new age in automatic train running information systems. To appreciate the effect the IECC has, the main problems faced in operating the earlier automatic passenger information systems must first be considered.

- Providing and keeping updated the train service timetable. Because the then national train service database (TSDB) did not contain short-term planned amendments, this entailed a considerable workload in separately inputting short-term alterations to the working timetable.
- Modifying the timetable information to accommodate non-planned, short-term changes. These include cancellations, changes to train calling patterns, etc. Again this involved a considerable workload and relied upon good communication between the train operating staff and customer information staff.

For these reasons, it could be argued that prior to the advent of the IECC, British Railways did not have operational a fully comprehensive, automated, timetable-based passenger information system. Various degrees of automation had been achieved, but in each case a considerable labour element was still required, which had effectively been transferred from one area of operations to another. A fully automated PIS requires from another source, a train timetable which is updated automatically, current train running information, and updated information concerning cancellations, changes to calling pattern, additional trains, etc.

In an IECC, automatic route setting is usually provided which necessitates the provision of a timetable processor (TTP) associated with the IECC. This processor must always contain all current timetable information. Modifications have also been carried out on the TSDB system to allow for short-term planned changes to be incorporated in the database, with non-planned, short-term changes being input direct to the TTP. The IECC is used to provide all the current train running information whilst the TTP gives all timetable information. This data is also required for other purposes such as station statistical records, staff information, and for interfacing to other systems, for example Prestel. In order to provide the interface to the IECC, a further processor, the information generator (IG) which is on the IECC information ring, is used to carry out any processing tasks common to the above system as shown in **Fig. 9.12**. The main tasks of the IG are to transfer to the other information systems:

- Timetable information from the IECC.
- Train running information from the IECC.
- To calculate delays and transfer delay information.

Because the IECC does not contain train cancellation information, this will be input at key locations into the IG by staff using the staff information system (SIS).

One option is for the TTP to send 12-hour timetables to the IG. A 12-hour timetable will cover either the am or pm periods and will be sent approximately 8 hours before the timetable is effective. Any short-term non-planned changes made to the already posted timetable will be transferred on receipt to the IG. The IG would then send filtered 12-hour timetables to each information system approximately 2–3 hours following its receipt from the TTP, supplemented by short-term changes. The real-time movement information from the IG to the information systems will consist of two categories — train journey information and train station information. The train journey information will include estimated delays, cancellations, and changes to train formation. It will be possible to insert a 'reasons for' code with each of these messages, which apart from delays, will need to be input to the IG via an operator or the SIS. Train station messages

SIGNALLING THE PASSENGER

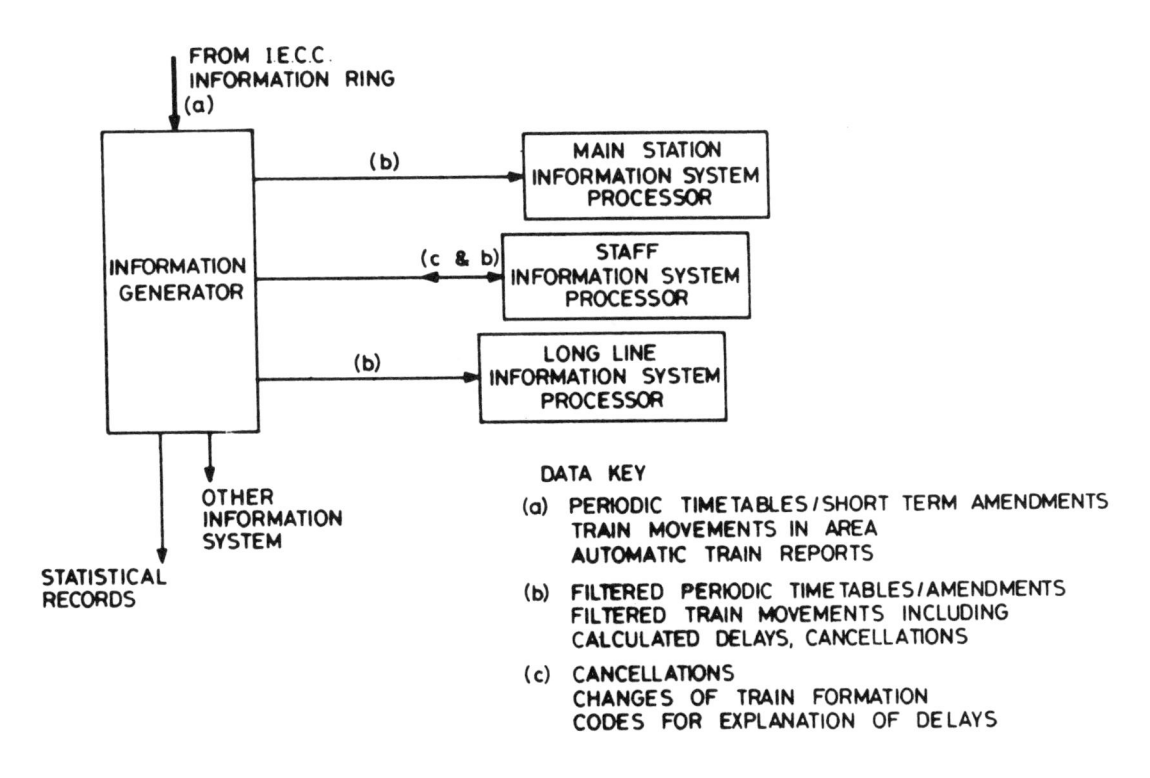

```
FROM I.E.C.C.
INFORMATION RING
(a)
        │
        ▼
┌──────────────┐   (b)    ┌──────────────────┐
│              │─────────▶│   MAIN STATION   │
│              │          │ INFORMATION SYSTEM│
│              │          │    PROCESSOR     │
│              │          └──────────────────┘
│ INFORMATION  │  (c & b) ┌──────────────────┐
│  GENERATOR   │◀────────▶│      STAFF       │
│              │          │ INFORMATION SYSTEM│
│              │          │    PROCESSOR     │
│              │   (b)    ┌──────────────────┐
│              │─────────▶│    LONG LINE     │
│              │          │ INFORMATION SYSTEM│
│              │          │    PROCESSOR     │
└──────┬───┬───┘          └──────────────────┘
       │   │
       │   ▼
       │  OTHER
       │  INFORMATION      DATA KEY
       │  SYSTEM
       ▼                (a) PERIODIC TIMETABLES/SHORT TERM AMENDMENTS
   STATISTICAL              TRAIN MOVEMENTS IN AREA
   RECORDS                  AUTOMATIC TRAIN REPORTS

                        (b) FILTERED PERIODIC TIMETABLES/AMENDMENTS
                            FILTERED TRAIN MOVEMENTS INCLUDING
                            CALCULATED DELAYS, CANCELLATIONS

                        (c) CANCELLATIONS
                            CHANGES OF TRAIN FORMATION
                            CODES FOR EXPLANATION OF DELAYS
```

Fig. 9.12 Proposed schematic of information generator for IECC systems

will include *train will arrive, train will depart, train has arrived, train has departed, non-stop train is approaching* and *non-stop train has cleared platform* messages.

In order to allow automatic control of the recorded announcements, the present policy of making a pre and main announcement will cease. Instead, one announcement will be made which will not be dependent upon the exact train position, for example 'Platform 1 for the ...'. This is necessary because of the difficulty in obtaining information on the precise position of a train within a station area.

These enhancements are already being planned. With the engineering techniques now available, it is possible to envisage facilities for customers to be able to gain direct access to information so that they can make their own enquiries without manual intervention by an operator. Viewdata and Teletext systems such as Prestel and the broadcast television based Ceefax and Oracle systems can be used to provide information about train timetables and fares. Further extension of the services that can be provided on these systems are likely in the future, and it is possible to envisage the presentation of real-time information on individual trains using these methods.

On some railway administrations, it is already the practice to provide a telephone on station platforms, linked to a manned control centre which can be used not only in an emergency for contacting the control centre but also for the seeking out of information about train services. The trend is to reduce the manning of station platforms, and such enquiry systems provide a vital method of reassuring the customer when the normal service is disrupted, particularly for those customers not familiar with the service. A logical extension to the present train enquiry bureaux, where customers telephone for information and are connected to operators who provide the information using either printed timetables or more automatic means, may be for speech recognition systems to be utilised so that routine requests of the type 'when is the next train to ...' can be recognised and the required information provided from a databank.

Trials have been carried out with customer operated information systems where the customer usually pressed two buttons for the originating and destination stations and a pictorial representation of the required journey showing stations where connections needed to be made, was shown on a visual display unit. However, a variety of reasons, probably varying from the lack of awareness in the use of such aids by customers to the difficulty of keeping the presented information updated in systems not linked to a master database, have limited the use of such systems to date. The developments described above will make the provision of these systems easier and more reliable.

A principal aim of any system must be to provide real-time information to customers, particularly when the normal pattern of timetable working is disrupted. It is, of course, worse to present misleading or incorrect information than no information at all, since this will give rise to a false sense of security. To avoid this presents considerable difficulties, since it becomes a requirement to advise the system of all disruptions to any particular train that will affect its journey. These disruptions can vary from the addition of station calling points perhaps because of the cancellation of other trains, to a change in the departure platform at a station for operational reasons. Whilst it is possible to design software routines to encompass many of the normal changes that become necessary during the average day in the life of a railway, effecting these changes by the system operator needs to be as straightforward as is possible. In particular, the logic must be able to cope with major disruptions to the train service whilst minimising the amount of operator intervention required. It is at such times of major disruption that the needs of the customer are greatest and the distraction of the system operator is also likely to be the greatest.

This is an area where insufficient thought has been given to the design of systems: the easier option of dealing with the normal routine operation of the information system has perhaps been concentrated upon, at the expense of the more difficult areas of system robustness. Automatic operation of systems reduces the likelihood of errors being caused by operator error and allows the consistent presentation of information, providing the problems of dealing with disruption can be overcome. The use of artificial intelligence techniques in control systems holds considerable promise that these difficulties can be overcome. Such systems use powerful computers and sophisticated software techniques to simulate the logic used by humans in deciding a course of action.

Quite apart from the aspect of information reliability discussed above, equipment reliability obviously has a major effect on the total reliability of the package. The improvement in equipment reliability now achieved is due to the widespread use of electronic systems with digital processing of information. The use of redundancy techniques for major customer information control systems has increased the mean time between failure of the complete system. Full diagnostic aids are now incorporated into complex systems to aid both the operator and the maintenance technician in recognising and being able to repair failures. Apart from the customary visual indications provided, such as power supply indications or transmission line indications, test sequences can be written into the software to test or exercise the various component parts of the system such as TV monitors, large displays and so on.

Conclusion

Whilst the methods used to provide information to the passenger have changed over the years, and will continue to change as technology advances, the fundamental need for clear concise information remains the same. Signal and telecommunications engineers have developed a methodology supported by principles and standards for signalling the train driver: the science of signalling the passenger must not be ignored, since without the passenger there is no need for passenger trains.

CHAPTER TEN

Automatic Train Protection

Many features of railway signalling systems are provided as a protection against errors on the part of the signalman. Indeed, a major part of the history of railway signalling can be seen as a story of progressive additions and refinements to signal interlocking and controls often prompted by serious accidents which resulted from error or lack of attention on the part of signalmen. Comparatively little has been done to protect against driver error, apart from provision of the automatic warning system (AWS), whose origins go back to the beginning of this century.

It is not surprising that a driver will occasionally make a mistake if he has not only to remember the last signal aspect, but also has to be constantly aware of the road ahead, with its gradient and curves, permanent and temporary speed restrictions, the acceleration and braking characteristics of his particular train under good and bad weather conditions, and the timetable.

The extensive provision of colour light signalling with its highly visible signal aspects reinforced by an AWS system has, over the years, largely reduced incidents of driver error. However, in spite of modernisation programmes which have generally replaced semaphore signalling by multi-aspect colour light signals, there has in recent years been an increasing trend in the number of signals passed at danger. This phenomenon which appears to have its roots in a complex of technical, psychological and sociological factors, has been experienced not only by BR but also by other European railways. The risks engendered, and accidents which have occurred in a number of countries, have prompted the decisions of the respective railways to adopt systems of automatic train protection to prevent the passing of signals at danger and to enforce speed limits positively.

The AWS System

The AWS system has permanent and electromagnet inductors installed at the approach to all running signals, and permanent magnet inductors at the approach to many fixed speed restrictions. Portable permanent magnet inductors are provided on the approach to temporary speed restrictions. The train driver will be given the same warning, make the same acknowledgement, and see the same visual reminder of the 'sunflower' display for all cautionary signal aspects and speed restrictions. The driver must then decide what action to take on the basis of his route knowledge and observation of lineside signals and warning boards. The system is incapable of checking his subsequent actions.

AWS makes an important contribution to the safety of the railway and has undoubtedly prevented many accidents since its introduction on BR. However, it is quite credible that a driver receiving a succession of AWS warnings may forget the aspect of the last signal passed, and approach a signal at red believing his last signal was at double yellow, or even green followed by a speed restriction. In such circumstances, overrunning the red signal, with a serious risk of a subsequent collision, is almost inevitable.

Train Stops

It is possible to protect against the consequences of a driver failing to observe signals by providing a train stop device to impose a full brake application on any train passing a red signal. Such systems are common on urban railways, but in order to ensure that an over-run will not result in a collision, it is necessary to provide an overlap beyond the red signal equal to full braking distance. For a main line railway, with long braking distances required by freight trains and high speed passenger trains, this would produce an unacceptable reduction in line capacity. In addition, train stops do not protect against trains exceeding speed restrictions.

Speed Supervision

The key to automatic protection on mixed traffic main line railways is speed supervision.

If, after passing a caution signal, the speed of the train is repeatedly checked it is possible to compute its braking rate. If this is insufficient to stop the train at the red signal, the speed supervision system can impose a full brake application to bring the train safely to a stand. This is illustrated in **Fig. 10.1**, which shows a speed/distance curve for a train which initially brakes at a rate which would cause it to run by the red signal. When the actual braking curve of the train intersects the service braking curve, a full service brake application is imposed and the train brought to a stand at the signal.

Similar arrangements can ensure that trains brake adequately to comply with speed restrictions, and continuous

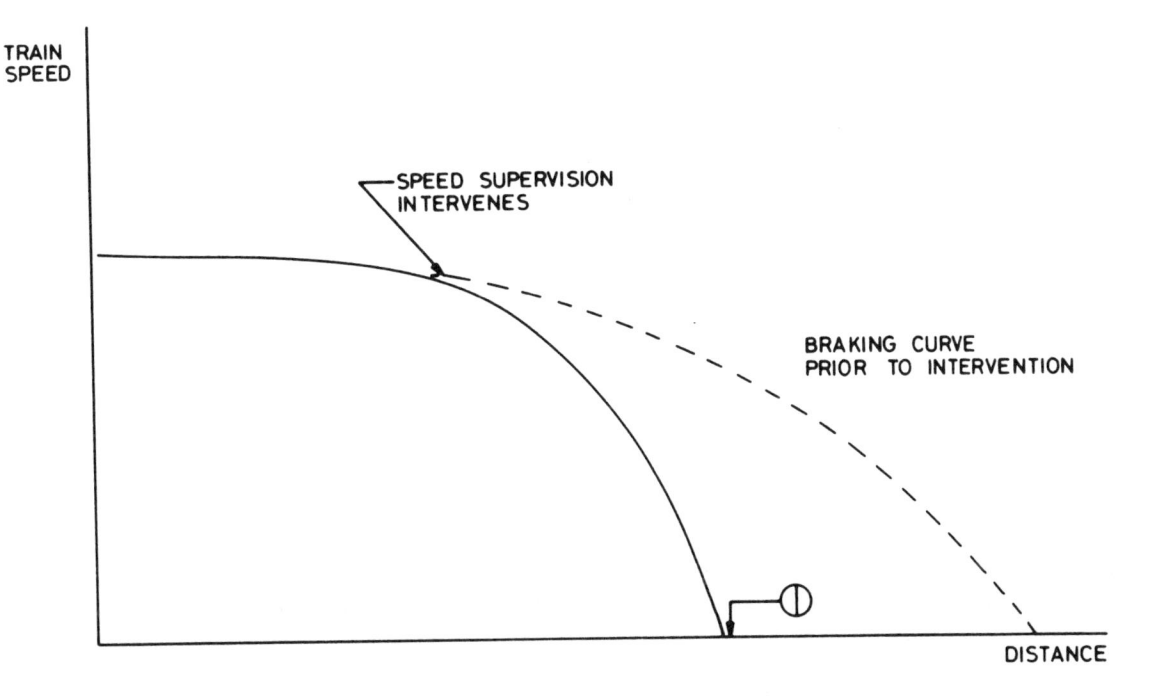

Fig. 10.1 Speed supervision to prevent run-by

speed checking can provide brake applications if speed restrictions or line speed limits are ever exceeded.

A system which enforces obedience to signals and speed restrictions by speed supervision is known as an automatic train protection (ATP) system.

Data Requirements of ATP

In order for an ATP system to make the correct decisions when supervising the speed of a train, the following data must be available:

- Distance to next signal at danger.
- Current maximum permitted speed (line speed, permanent or temporary speed restriction).
- Distance to next speed restriction.
- Value of next speed restriction.
- Distance to termination of speed restriction or uplift in permitted speed.
- Gradient.
- Actual speed of train.
- Braking ability of train.
- Maximum permitted speed of train.
- Length of train.

Some of this information has to be transmitted to the train from the ground, some may have to be entered by the driver at the start of the journey, some may be permanently programmed in, and some may be calculated by the train equipment.

ATP System Architecture

An ATP system consists of track and train subsystems (see **Fig. 10.2**). The main information processing and decision making takes place in the train subsystem, which is programmed with the train characteristics, some of which may have to be entered by the driver. The train subsystem also continuously measures the train speed and position.

The track subsystem transmits to the train signal aspect information, speed restriction data appropriate to the route over which the train is signalled, track parameters such as gradient, and provides position data at frequent intervals to re-align the train subsystem.

AUTOMATIC TRAIN PROTECTION 261

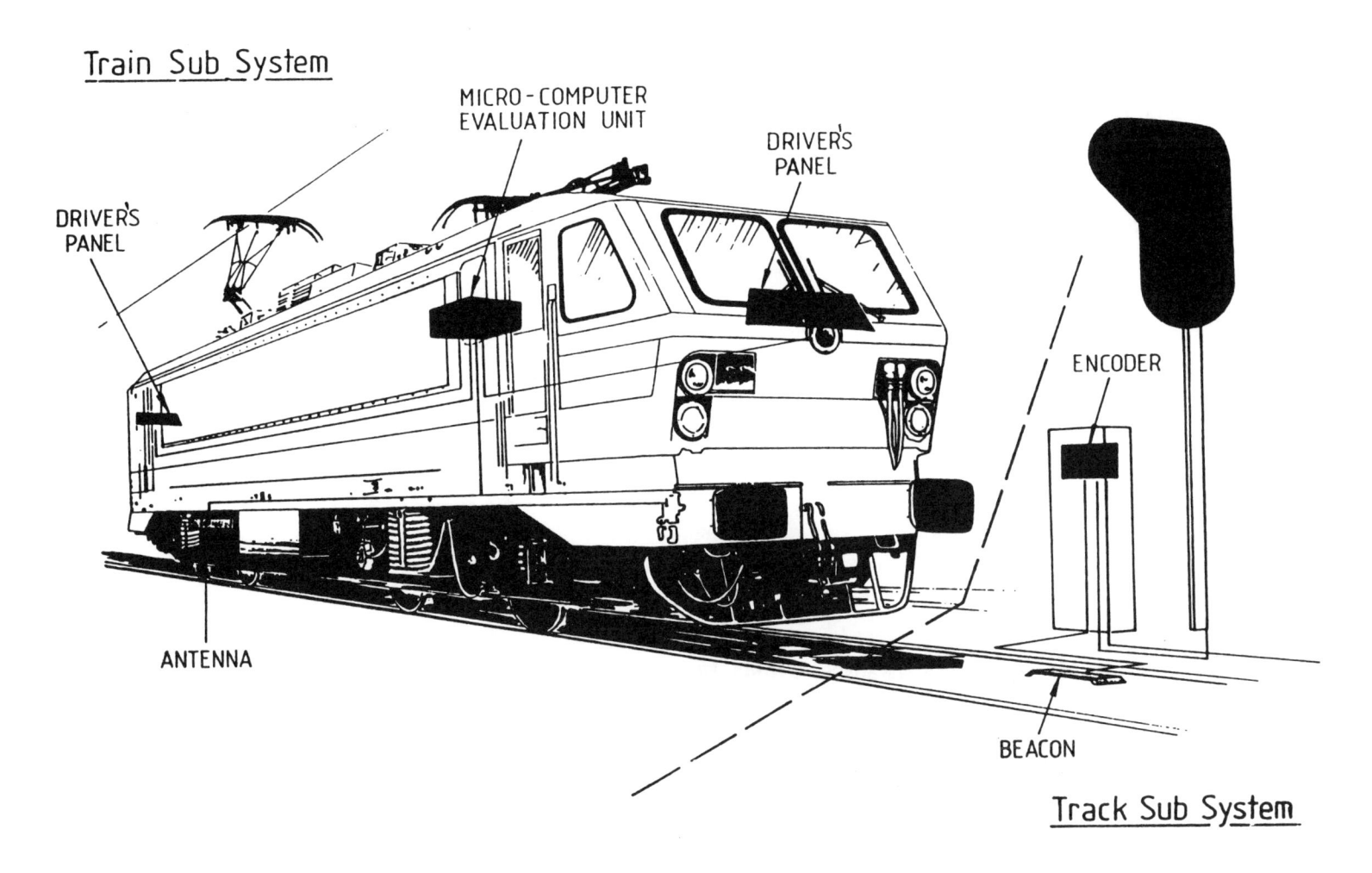

Fig. 10.2 ATP system and architecture

TRAIN SUBSYSTEM

The essential elements of the train subsystem are shown in **Fig. 10.3**. The design of the processor may use techniques similar to those employed by SSI to achieve high levels of safety and availability, as described in Chapter 2. Speed and position are calculated from the output of the tachogenerator, which may be duplicated or triplicated for safety and availability.

Train data relating to braking performance, maximum train speed, etc are entered by the driver at the start of the journey, or permanently programmed in, depending on the type of train.

Track data, relating to signal aspect, speed restrictions, distances, gradients, etc are received from the track subsystem via the antenna.

From the track, signal, and distance data, the processor calculates the permitted speed and braking rate. If the actual values derived from the tachogenerator exceed the permitted values, the driver is first warned via the display system and then a brake application is made.

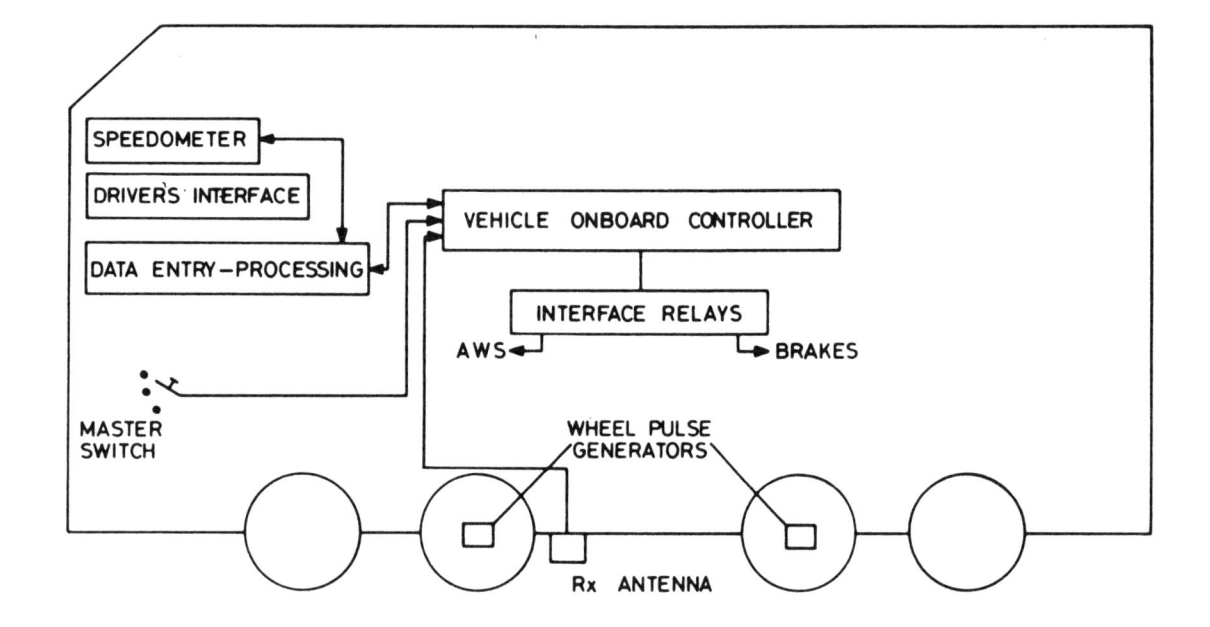

Fig. 10.3 Train subsystem architecture

Fig. 10.4 Track subsystem architecture

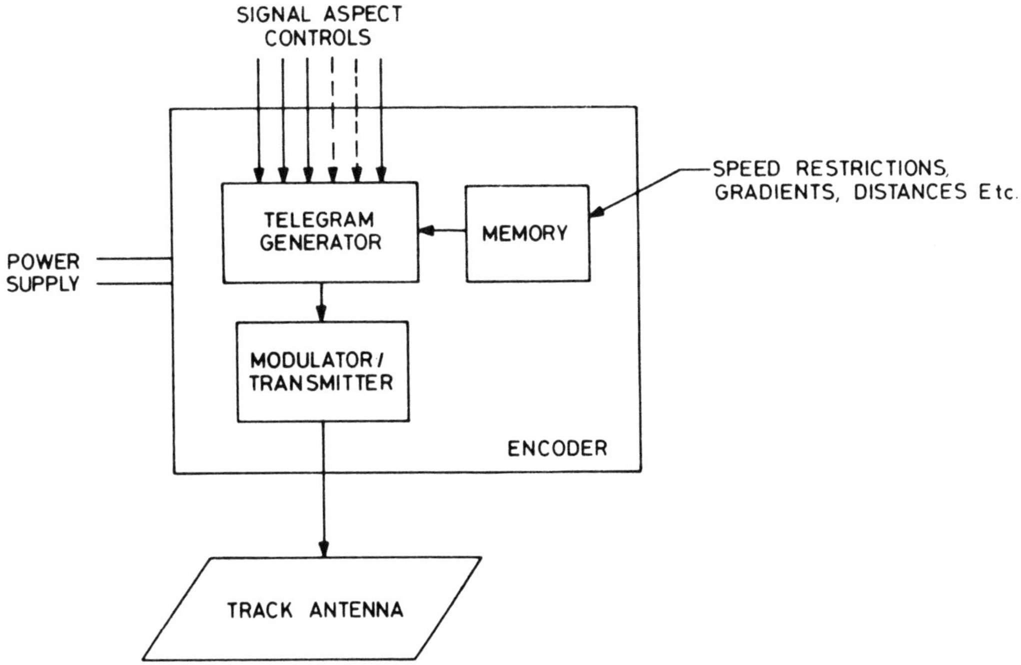

THE TRACK SUBSYSTEM

The track subsystem (see **Fig. 10.4**) comprises lineside encoders, typically one per signal, which generate coded messages for transmission to the train. An interface to the signal control circuitry determines the signal aspect to be transmitted, and selects from the data programmed into the encoder, the track data relating to speed restrictions and gradient appropriate to the route which is signalled. The encoder formats this data into a telegram in a highly secure code and this is modulated on a carrier and fed to the track antenna to be transmitted to the train.

A wide variety of designs of track antenna exist. The rails themselves may be used, or long loops of conductor may be laid throughout the track, either arrangement giving the possibility of continuous transmission to the train.

Alternatively, short loop aerials may be installed to form discrete information points intermittently distributed along the track. These may be active devices, transmitting continuously, or semi-active transponders, replying only when interrogated by the train, and possibly powered by carrier energy transmitted from the train (see **Fig. 10.5**). Continuous and intermittent transmission can be mixed within the same system.

Discrete information points of any form will be called beacons in the remainder of this chapter, although it should be noted that this terminology has yet to be standardised, and in some sources beacon is used to denote an active device as distinct from a transponder.

TRANSPONDER SYSTEM
TRAIN TRANSMITS CONTINUOUSLY
TRANSPONDER TRANSMITS ONLY WHEN INTERROGATED

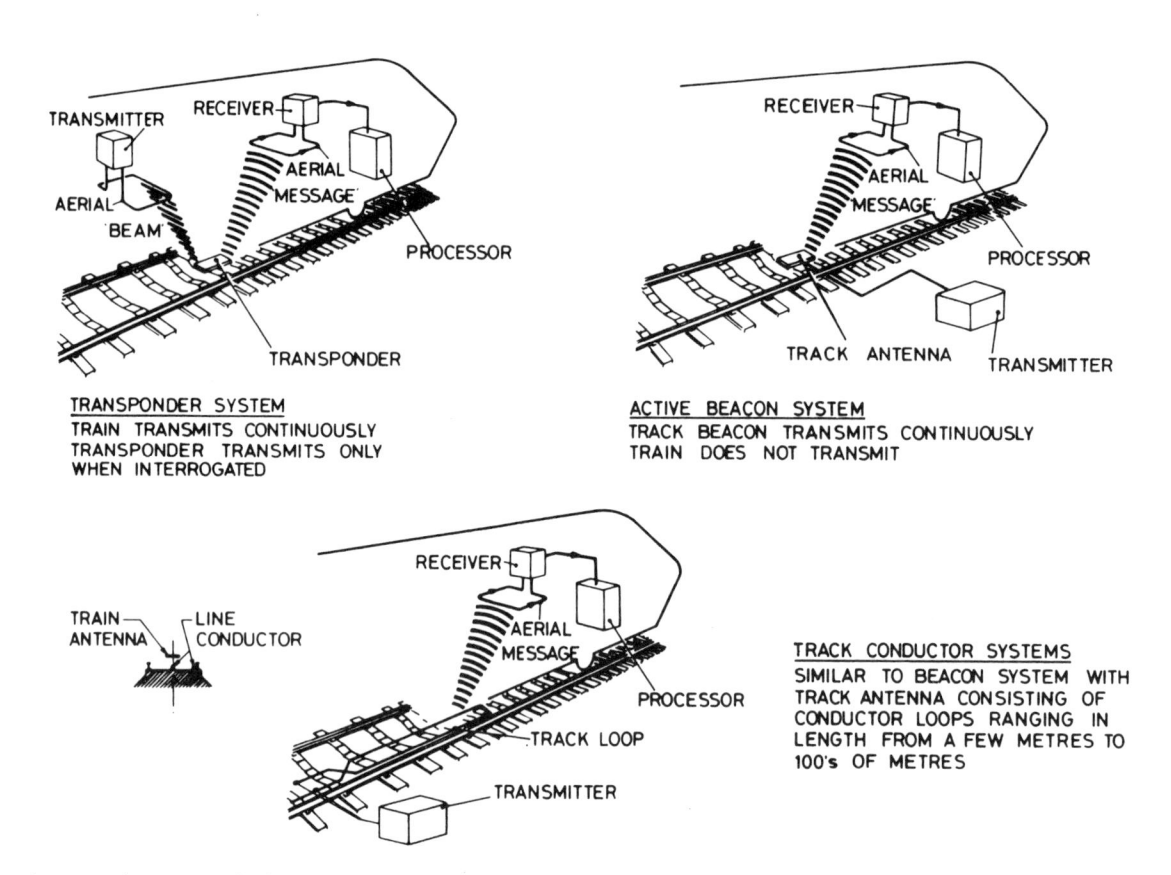

ACTIVE BEACON SYSTEM
TRACK BEACON TRANSMITS CONTINUOUSLY
TRAIN DOES NOT TRANSMIT

TRACK CONDUCTOR SYSTEMS
SIMILAR TO BEACON SYSTEM WITH TRACK ANTENNA CONSISTING OF CONDUCTOR LOOPS RANGING IN LENGTH FROM A FEW METRES TO 100's OF METRES

Fig. 10.5 Track-to-train transmission

Data Transmission and Security

Data transmission between track and train uses a form of short range radio transmission. Although with the short distances involved there should be none of the variable propagation problems sometimes associated with radio systems, there is still the question of protection against interference from other radio users.

The nature of its data transmission puts ATP in the category of low power radio devices according to the classification used by the UK radio licensing authority, the Department of Trade and Industry (DTI). This category includes systems such as radio alarms, metal detectors, anti-theft devices and remote door controls. These devices have been exempted from licensing, but it is still necessary for each design to be type approved by the DTI to ensure that the power levels are not exceeded.

A consequence of working in these frequency bands is the risk of interference from other users of the same frequencies, although the power levels specified by DTI have been chosen to minimise the risk of such interference. The highly secure coding used by ATP systems ensures that even if, under exceptional circumstances, interference did occur there would be no risk to safety. The principles of protection of information by coding are described briefly in relation to SSI in Chapter 2.

Continuous Transmission Systems

Continuous transmission systems have the advantage that information on the train can be immediately updated when any change occurs in the aspect of the signal ahead. This tends to improve line capacity, as a train which has passed a caution signal can respond immediately to the early clearance of the next signal rather than be constrained to follow the braking curve until it reaches the next information point, as in the case of intermittent systems. A continuous system also provides instant response to the emergency replacement of a signal at any time.

With a continuous system, failures of encoders, transmitters, antennae, train receivers, etc can readily be detected by the train processor as a loss of transmission, and the ATP system can default to a safe state accordingly. With intermittent systems, a lack of transmission is a normal state, and more sophisticated arrangements must be made to detect failures.

Continuous systems generally need to be applied comprehensively, and do not lend themselves to selective application to the most critical areas, for example by giving priority to fitting junction areas.

Continuous systems based on track conductors tend to be more expensive on account of the lengths of conductor which have to be installed and maintained. Those using the rails for transmission, based on coded track circuits, tend to have a low information capacity due to the limited bandwidth available. This limited information capacity is enough to provide excellent ATP facilities for railways with a single type of train with fixed characteristics, such as rapid transit systems or the French TGV, but it is insufficient to cater for all conditions encountered in the retrospective application of ATP to a mixed traffic railway.

Intermittent Systems

In comparison with continuous systems, intermittent systems tend to be cheaper and more readily applied selectively. The latter feature allows the most critical locations to be protected at an early stage of any transitional period of introducing ATP. As already noted, the chief weaknesses are in the areas of headway and fault detection. The detection problem is completely soluble, and headway limitations can be eased at the cost of installing extra equipment.

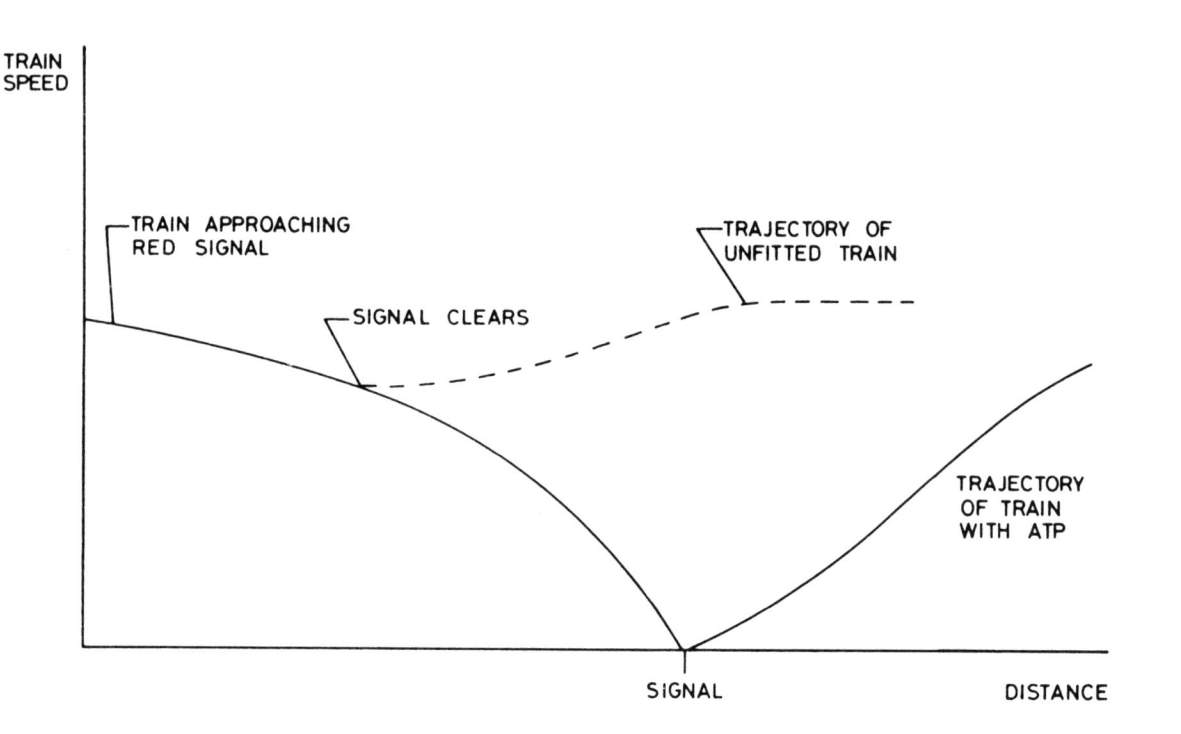

Fig. 10.6 Possible effects on headway

Line Capacity

When a train equipped with an intermittent APT system is approaching a red signal, it must stay within the braking curve leading to a stop at the signal even if the driver sees the signal clear ahead of him. Thus in the situation shown in **Fig. 10.6**, a train not fitted with ATP would have been able to accelerate away as soon as the signal cleared while a train fitted with ATP would have to come almost to a stand at a signal which had cleared, incurring unnecessary delay. Whilst this example clearly shows that intermittent ATP can introduce additional delays, the significance of this should not be exaggerated. Trains running to timetable rarely encounter red signals, and even when disturbances occur, the combinations of train movements where ATP constraints cause additional delays may not be common. A computer simulation of a busy BR junction indicated that simple intermittent ATP would not cause any delay when trains were running to timetable, and would introduce only a few minutes' additional delay in total to a severely disrupted rush-hour period.

Nevertheless, such delays are undesirable and can be reduced by placing additional information points on the approach to the signal so that the train can be released from its braking trajectory if the signal clears. The effect of this is illustrated in **Fig. 10.7**. Several such in-fill beacons can be supplied if required, or an extended loop of track conductor may be used, in which case an approximation to a continuous ATP system is achieved.

With a single in-fill beacon the train will be constrained to trajectory A if the signal clears after the train has passed the beacon, or trajectory B if the signal clears before the train passes the beacon.

With an in-fill loop the train can follow a trajectory such as C as soon as the signal clears.

Release at Red

With an intermittent system, it is generally desirable not to supervise the train speed down to zero on the approach to a red signal. If this were the case, the ATP fitted train in **Fig. 10.6** would never be able to pass the zero point even when the signal cleared. This can be avoided by providing a short section of track conductor loop or other continuous transmission system on the approach to the signal so that the train

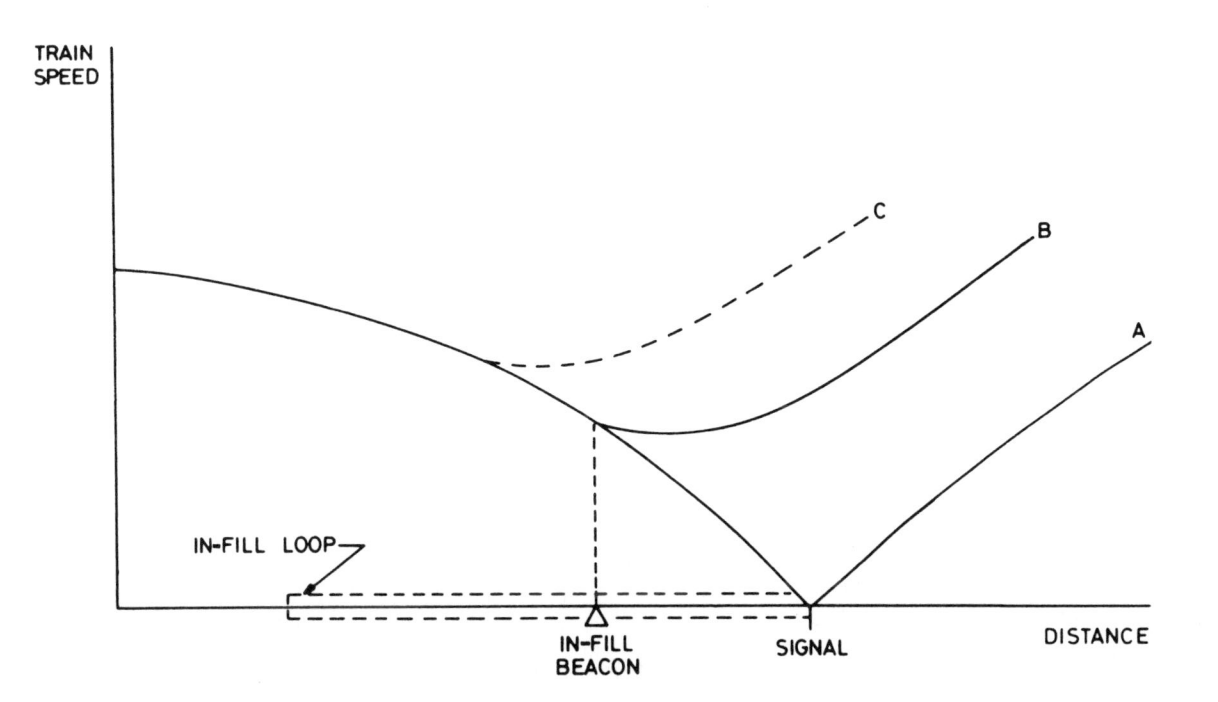

Fig. 10.7 Effects of in-fill

Fig. 10.8 Release speed

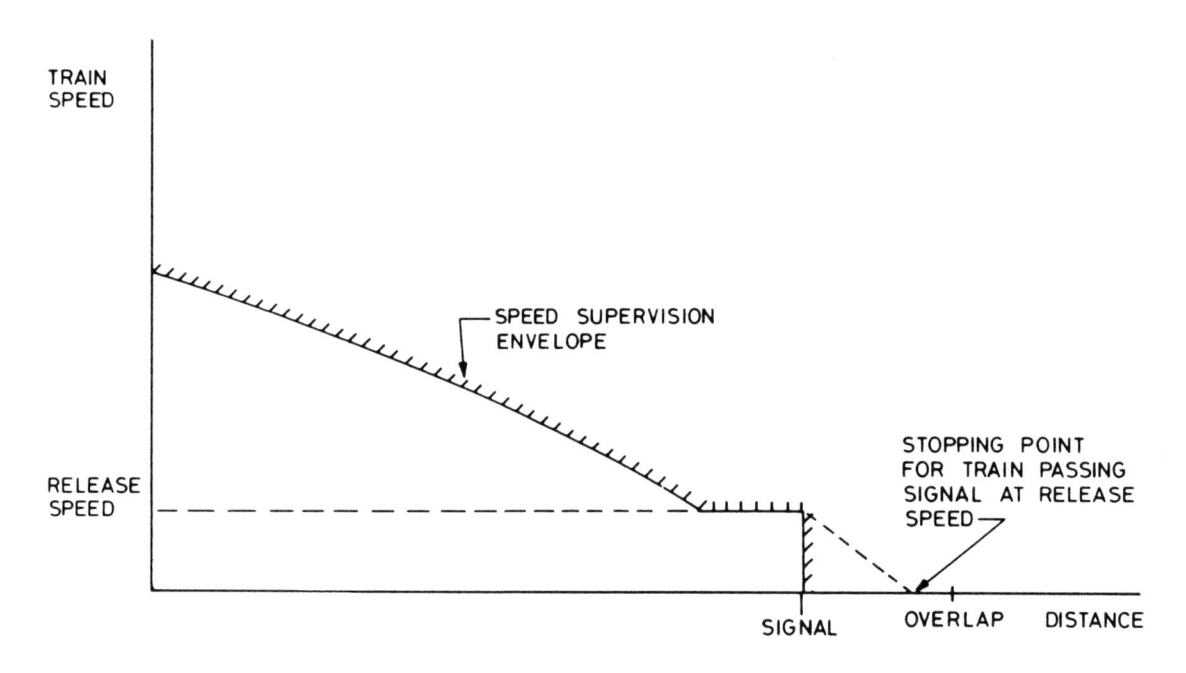

subsystem can always be given a release at or before the zero point when the signal clears. An alternative method needing no additional installations on the track is to supervise the train braking down to a release speed (see **Fig. 10.8**) calculated so that if the train passes the signal at this speed it will stop within the overlap on receiving a trip command from the information point at the signal, which acts as a trainstop.

Beacon Failure

Failure of a beacon or of the receiving equipment on board the train can be detected by including in the message from each beacon, the distance to the next beacon. If the next message is not received when the train has travelled this distance, then the processor deduces that a failure has occurred and gives an appropriate warning to the driver.

The BR ATP System

CONCEPT OF A SYSTEM

The word 'system' is so freely used today that it has become a debased currency. In the present context, references to the BR system of ATP are analogous to references to, for example, the BR system of colour light signalling. The aims, function and architecture of the BR system are described, rather than specific proprietary systems or items of equipment. Indeed, it is intended to implement the BR system as an adaptation of existing proprietary systems, and although a BR-specific system will exist, it will have few BR-specific components.

AIMS

The system is required principally as a safety feature to prevent trains from exceeding speed restrictions or passing signals at danger. It is aimed initially at trains which run up to the current maximum of 200 km/h. However, it will also be required to assist drivers of higher speed trains, when these are introduced, by displaying the maximum permitted speed over the line ahead.

It is to be used on lines with various types of traffic: high speed intercity, high density suburban, low density rural, various types of freight and parcels trains and on lines with mixtures of these. It must be capable both of being added to existing vehicles and signalling equipment with a minimum of disruption and of being applied to all forms of traction and types of train on British Railways.

Initially the system will supplement rather than replace the existing automatic warning system (AWS). However, it is designed so that it can eventually replace AWS.

The system is required to function correctly on lines where trains travel in either direction, without any requirement for the driver to enter direction information.

SYSTEM ARCHITECTURE

BR ATP will be an intermittent system with a beacon at each signal in fitted areas, and with additional in-fill beacons or loops at selected signals to maintain line capacity. This architecture has been decided on the basis of cost studies, computer simulations and reviews of available products.

Each beacon will be allocated a unique identity code which will be included in the data transmitted to the train. Examples of the track data provided by the beacons for some typical ATP sequences are shown in **Fig. 10.9**.

The train subsystem architecture will be essentially as shown in **Fig. 10.3**.

SPEED SUPERVISION PRINCIPLES

Supervision of train speed involves a comparison between the train speed and the most restrictive 'current limit'.

A current limit is a speed limit to which the train speed is currently being supervised. Current limits may be derived from track and train data, or may be pre-programmed into the equipment. In certain special conditions, the train speed is supervised to a release speed which may be regarded as a special form of current limit.

A permanent speed restriction (PSR) or a temporary speed restriction (TSR) current limit may be present in track data in either of two forms:

- Immediate: as a speed value which applies from the positional reference used in the track data.
- Target: as a speed value which is effective as a current limit only when the train has reached or has passed the position associated with the target.

270   AUTOMATIC TRAIN PROTECTION

Fig. 10.9 Typical telegram sequences

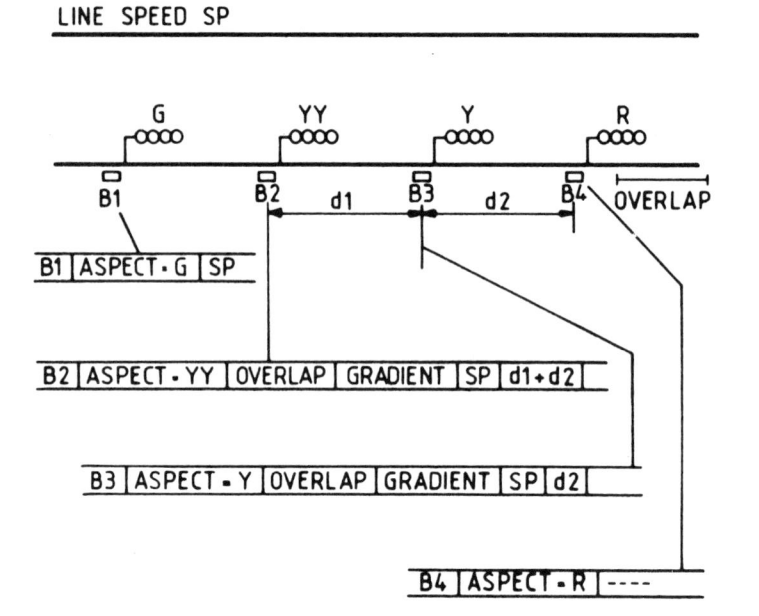

A) STANDARD LINE SPEED – APPROACH TO A RED SIGNAL

B) STANDARD LINE SPEED – APPROACH TO JUNCTION (NO FLASHING YELLOWS)

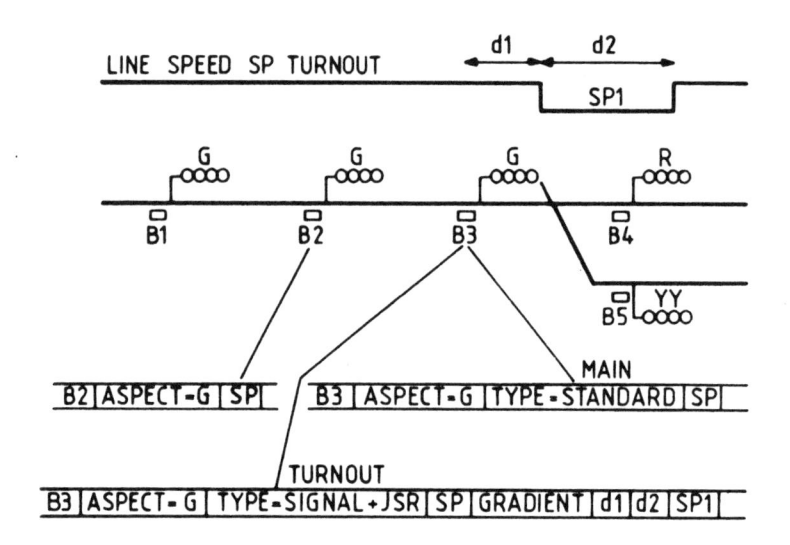

JSR - JUNCTION SPEED RESTRICTIONS

SUPERVISION OF TRAIN BRAKING

A target limit is a speed limit which may currently be exceeded but which must be satisfied when the train reaches some specified point on the track ahead. Target limits are always specified in track data and may be of the following categories: PSR, TSR or red aspect at the next, or next but one signal.

Supervision of train braking involves a comparison between the position of the train speed/distance co-ordinate and the position of various speed/distance curves which represent different braking trajectories to the most restrictive target limit.

For simplicity, it is assumed that the train brakes provide a constant brake force and that the basic braking trajectory is given by a constant deceleration speed/distance curve. The basic curve is given by:

$$Db = (V^2 - Vt^2)/2(B + I) \quad \ldots \quad (1)$$

where Db is the braking distance required to reduce the train speed from the present value, V, to the target speed value, Vt. B is the deceleration of the train on level track attributable to full service braking. I is the deceleration attributable to the inclination of the track and is assumed to be constant throughout the braking. I is positive for rising gradients, and is negative for falling gradients.

Equation (1) is meaningful only for values of V greater than or equal to Vt.

In the case where the train brakes provide a constant brake force but the gradient varies, the train does not follow a constant deceleration curve. Mathematically, the correct approach is to divide the track into sections of constant gradient, and to determine a composite curve from the basic constant deceleration curve of each section. This procedure appears to be unnecessarily complex for a supervisory system. Thus, it is assumed that the ATP system will use only a single basic curve, and that the value of I will be selected to ensure that the safety of the train is not compromised. On some occasions, this will result in the train being supervised to more restrictive conditions than would otherwise be necessary.

Equation (1) is used as the basis for determining when the ATP system must intervene. However, it is necessary to allow for the gradual build up of brake force which occurs when the brakes are first applied. This is taken into account in the 'intervention curve' which is a displaced version of the basic curve. The displacement represents the distance travelled by the train at constant speed during the time that the brake force is building up. The braking distance required for intervention, Div, is defined by:

$$Div = Db + V.Tb \quad \ldots \quad (2)$$

where Db is defined by equation (1), and Tb is the delay in building up the brake force from its present value to that of the full service rate.

The value of Tb varies from the maximum value Tbs, which is the full delay when the brakes are off, to zero when the brakes are fully on. Thus the position of the intervention curve varies, moving closer to the target limit as the retardation due to braking increases. This has the desirable effect of reducing the risk of an intervention when the driver has already taken the necessary action.

A similar curve, the 'warning curve', is defined for Dwa, the distance from the target limit at which a warning is given. It is assumed that the driver is allowed a constant warning time, Twa, in which to make a full service brake application so as to avert intervention by the ATP system. As the brakes will take a time Tb to become fully effective, and intervention is to be avoided, the time displacement between the warning and intervention curves must be Twa + Tb. Thus the warning curve is defined by:

$$Dwa = Db + V.(2.Tb + Twa) \quad \ldots \quad (3)$$

It follows that the position of the warning curve will also vary according to the degree of braking, moving closer to the target limit as the brake force increases. This has the desirable effect of reducing the risk of a warning when the driver has already taken the necessary action.

A further curve, the 'indication curve', is defined for Did, the distance from the target limit at which a target speed value is indicated on the green 'permitted speed' lights on the driver's display described below. It is assumed that the driver is allowed a constant indication time, Tid, in which to prepare for or to anticipate a warning from the ATP system if he has made no brake application. The indication curve is defined by:

$$Did = Db + V.(2.Tbs + Twa + Tid) \quad ... \quad (4)$$

This curve is dependent on Tbs, but not on Tb, so the position of the indication curve does not vary with the brake application. This reduces the risk of the speed indicated on the green permitted speed lights flashing back and forth between a current and a target limit depending on the brake application.

The indication curve has a lower bound determined by the speed value at the target limit, since for train speeds less than Vt, the target limit and Did have no meaning.

The warning and intervention curves have similar lower bounds at the target speed plus, respectively, 5 km/h and 9.5 km/h. These lower bounds ensure consistency between the warning and intervention criteria used for supervision of train braking and those used for the supervision of train speed.

The form of the indication, warning and intervention curves is shown in **Fig. 10.10**.

If Tbs is small compared with Twa, there is no real disadvantage in using Tbs in place of Tb, and the position of the intervention and warning curves are fixed in the worst-case position. If Tbs is large compared with Twa, the use of a fixed value of Tb is likely to give significant ergonomic problems and to enforce an unduly restrictive driving style.

If more than one target limit is present, it is necessary to determine which target limit represents the most restrictive condition. This may be achieved by determining a basic speed/distance curve for each target limit and finding which of these curves is closest to the position of the train or, equivalently, which is closest to the position used as a reference for distance measurement. For this purpose, it is not necessary to take into account Tb, Twa or Tid.

If the basic speed/distance curves are calculated correctly using composite curves to allow for changes in gradient, it is not possible for the curve for one target limit to intersect the curve for another target limit. (However, target limits can share a common curve.) Using the simple constant deceleration curves instead of composite curves, the curve for one target limit can intersect the curve for another target limit if the gradient profile is unfavourable. The implications are that the choice of the most restrictive target limit can depend on the train speed, and that the decision must be continually updated.

If the two or more targets limits are equally restrictive, ie they share a common basic curve, or their basic curves intersect at some particular speed value, then the choice of the most restrictive target limit may not be completely arbitrary.

If the train speed is being supervised to a release speed, then the red signal target limit no longer applies, and the presence of other target limits must be taken into account.

DETERMINATION OF RELEASE SPEED

The calculation of the maximum value of the release speed is based on a simple principle: the speed to which the train is being supervised must be low enough to ensure that, if the train is tripped, the over-run will be contained within the overlap. The usable overlap length must be included in the track data at the previous signal.

The basic equation which relates Do, the minimum usable length of the overlap, to Vmax, the maximum speed at which the train may pass the signal, is:

$$Do = Vmax^2/2(B+I) + Vmax.Tbe \quad ... \quad (5)$$

where Tbe is the time delay associated with an emergency application of the brakes.

AUTOMATIC TRAIN PROTECTION 273

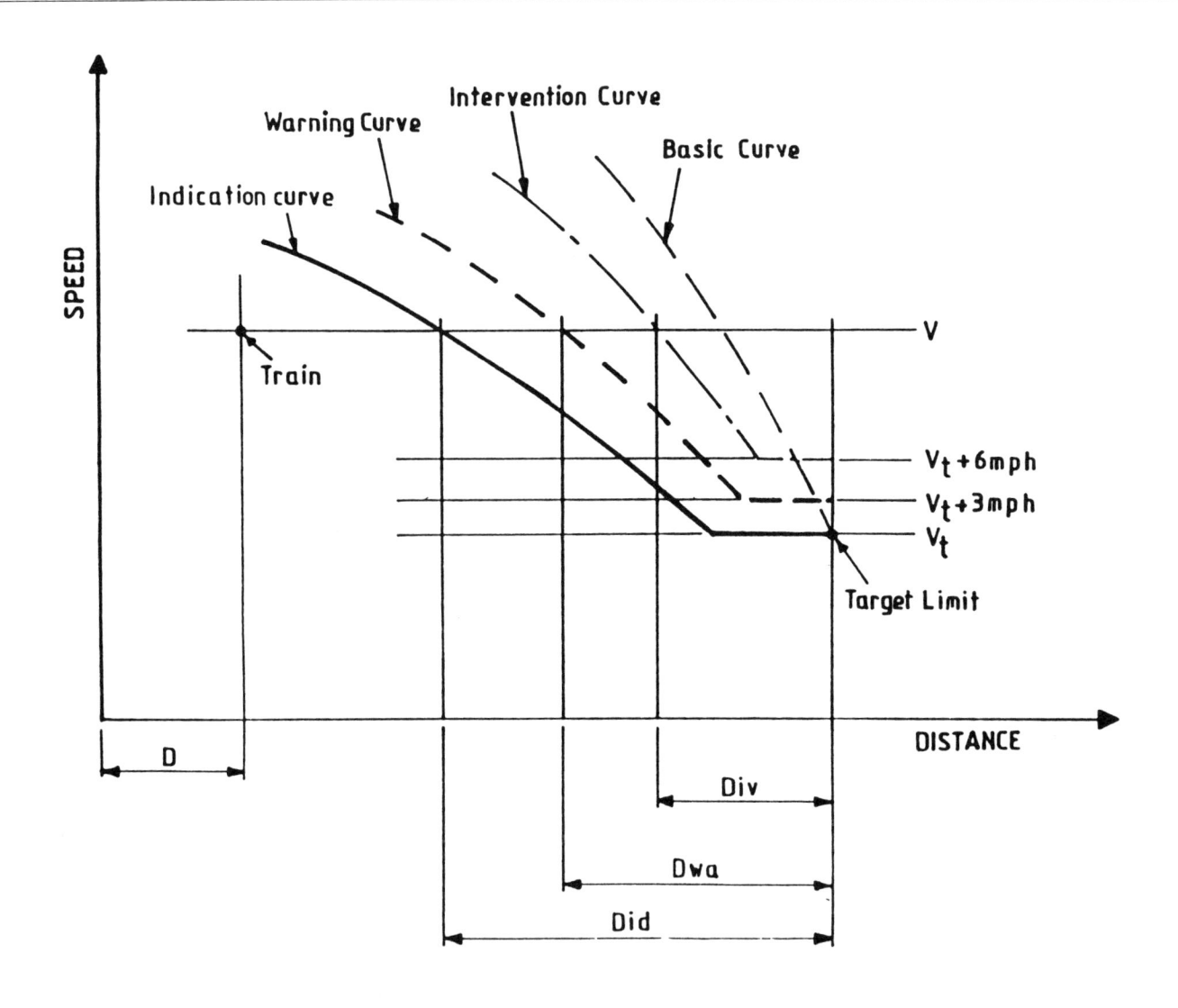

Fig. 10.10 ATP indication, warning and intervention curves

This yields an approximate solution:

$$V_{max} = \sqrt{[2(B+I)D_o]} - 2(B+I)T_{be} \quad \ldots \quad (6)$$

which gives good results if D_o is much greater than $0.5(B+I)T_{be}^2$.

Even if this condition is not satisfied, it can be shown that this equation always gives a safe (low) value of V_{max}.

Very small values of D_o can generate values of V_{max} which are zero or negative. Negative values are treated as zero.

The 'release speed', V_r, is derived from V_{max} taking into account the following factors:

- The train speed can exceed a supervised speed by up to 9.5 km/h before intervention occurs.
- The most restrictive current limit may be lower than V_{max}.
- The release speed can only be displayed in steps of 8 km/h.

A zero value for the release speed is an unusual but legitimate value, generally indicating that the length of any overlap is negligible.

The release point, that is the position at which a release speed may be displayed, is determined by the intersection of the indication curve for the red signal target limit and the release speed.

D_{rp}, the distance between the release point and the signal (which must be the most restrictive target), is given by:

$$D_{rp} = V_{rd}^2/2(B+I) + V_{rd}(T_{bs} + T_{wa} + T_{id}) \quad \ldots \quad (7)$$

It should be noted that supervision of the release speed can only give protection within the overlap that existed at the time the train passed the previous signal. Any signal control which allows the overlap to be cancelled, for example if the train was a long time between signals and hence could be presumed to have stopped, would not be protected by ATP.

TRAIN DATA ENTRY

In order to provide speed supervision, the train equipment must have data describing the following train characteristics:

- Maximum permitted train speed: this is required directly for supervision of the train speed.
- Train category: this is required to enable the train equipment correctly to select from the track data the appropriate PSR or TSR current limit, or the appropriate PSR or TSR target limit where these are differential (ie have speed values which vary depending on the category of the train).
- Train length: this is required so that the train equipment can ensure that a PSR or TSR current limit remains in force while any part of the train is affected by the restriction.
- Braking rate: this is required so that the train can calculate the necessary speed/distance curves for supervision to PSR, TSR and signal targets; the braking rate used is an assumed worst-case deceleration rate achievable with a full service brake application on level track.
- Braking delay: this is required so that the speed/distance curves can be adjusted to allow for the inherent time delays in reaching the full service braking rate assumed above.

In general, it is necessary to enter all the above data at the start of each journey, and to re-enter relevant data after any change which affects the specified characteristics.

The train data required for the supervisory functions is entered via a keypad and a dot matrix or similar display unit. The driver is not required to see the display or use the keypad while the train is moving, but the train data in use can be examined, but not changed, at any time that the train equipment is activated.

Data is entered via the keypad in response to messages on the display prompting for the information as it is required.

The data entry routine is customised to make it appropriate for the particular vehicle to which the train equipment is fitted. Where the data can be largely predetermined for the type of vehicle, such as fixed formation trains, multiple units, then the data entry routine is simplified to minimise the driver's input.

The train equipment is programmed with predetermined limits against which any data which has been entered can be checked for compliance. Any non-compliant data, or data which is inconsistent with other data already entered, is rejected and error messages are displayed indicating the nature of the non-compliance or inconsistency.

When compliant and consistent data has been entered in response to all prompts, the data entry routine displays each item of data in turn, seeking confirmation that it is correct and offering an opportunity to amend it.

OPERATIONAL FEATURES

A good starting point for a consideration of the BR ATP system is the driver's interface, or display shown in **Fig. 10.11**. This has the following features:

- A speedometer in which the train speed is shown by means of a conventional analogue pointer instrument.
- The main display, a three-character alpha-numeric display which is used to advise the driver the extent to which he is being supervised and the reason for any action taken by the system.
- Yellow lights around the periphery of the speedometer dial which are used to indicate the value of the release speed.
- Green lights around the periphery of the speedometer dial which are used to indicate the value of the permitted speed.
- Four push buttons, each capable of being internally illuminated.

Where the system is used at speeds in excess of 200 km/h the three-character numeric indication in the auxiliary display on the left is used to give authority to drive at the higher speed.

The following audible indications are associated with the driver's display:

- Blip: a short, non-alarming unmodulated tone which is given each time there is a change in the status of the display, but where the driver is not required to take any specific action.
- Bleep: a longer, non-alarming unmodulated tone, which is used in conjunction with the auxiliary display only, to draw the driver's attention to a reduction in the authorised speed or to indicate the imminent withdrawal of a speed authority in the cab.
- Warble: a frequency-modulated tone which is sounded continuously to draw attention to a flashing indication on the main display.

The driver's interface supplies rather more information than the minimum required for the basic protection function, which could be provided by the audible warnings and illuminated push buttons alone. The additional information given by the main display and the peripheral lights is intended to ease the driver's task by making him fully aware of the extent and limits of the speed supervision, and to minimise headway constraints by allowing him to take full advantage of in-fill and release speeds.

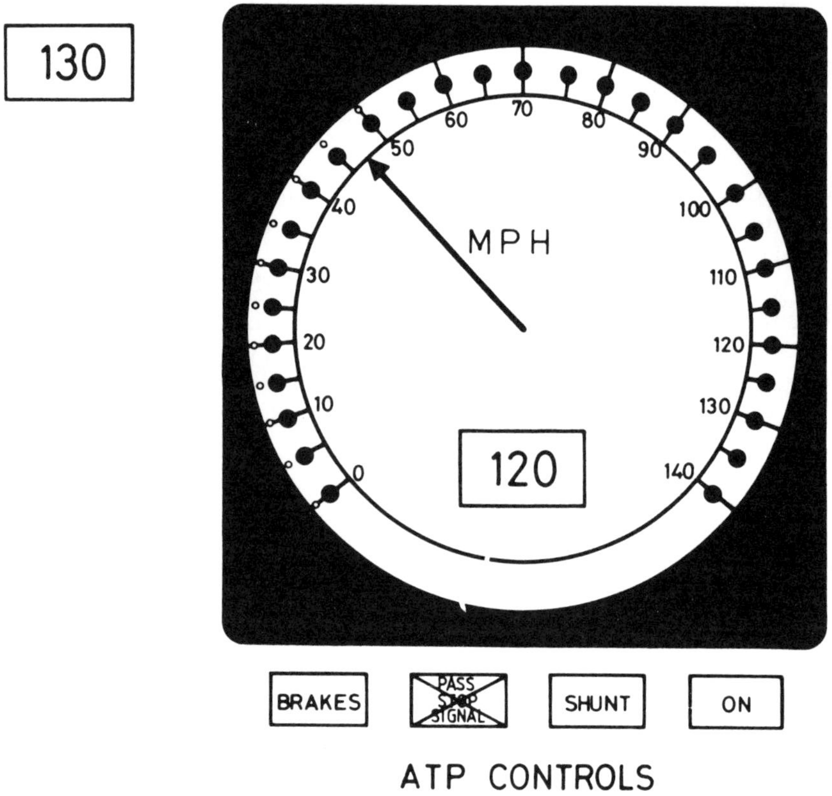

Fig. 10.11 ATP driver's display

The main display indicates the state of the last signal passed, according to the information received from the beacon at the signal:

=== — signifies that the last signal was green
  0 — signifies that the last signal was double yellow
000 — signifies that the last signal was single yellow

DRIVING UNDER RESTRICTIVE SIGNALS
The main display indicates whether the train braking is being supervised to a signal at danger and, if so, whether that signal is the next one ahead. The display shows 0 when the ATP system understands the next signal to be single yellow (where 4-aspect signalling applies), and 000 when it is assumed to be red.

The driver needs to know this information because he cannot assume that the behaviour of the ATP equipment is necessarily consistent with the signal aspects he observes. Inconsistencies may arise when the signal aspect sequence changes: unless continuous track-to-train communication is available, the train equipment cannot be updated at all positions along the track.

In particular, the driver may see signals ahead changing to less restrictive aspects, but the ATP equipment may continue to supervise the train braking to the previous more restrictive aspect sequence. In this situation, the driver has no option — he must continue to drive within the constraints of the previous aspect sequence until he can see from the main display that the train equipment has been updated.

Thus, the main display assists the driver by advising him of any yellow or red target limit to which the train braking is being supervised, even if that target limit is not the most important supervisory constraint. Information about the latter is provided by the green permitted speed lights around the periphery of the speedometer dial.

If there are no target limits, or if none of the target limits present is imminently likely to generate a warning, then the appropriate permitted speed light shows the most restrictive current limit to which the train speed is being supervised.

However, if a warning in respect of a target limit is imminent, then the permitted speed light shows the speed value of the most restrictive target limit to which the train braking is being supervised. If the permitted speed changes to a value which is lower than the actual speed indicated by the speedometer pointer, this effectively gives notice of an impending warning for a new target limit.

The display of a release speed on one of the yellow lights in place of a green permitted speed indicates to the driver that, because of discontinuities in the track-to-train communication link, the ATP equipment is unable to determine whether the next signal ahead is still at danger. To avoid the train being forced to approach this zero speed target limit at a very low speed, even if the signal indication has changed to a proceed aspect, the equipment supervises the train speed to the release speed.

The display of a release speed is not an authority to pass the signal at that speed. If the signal is at danger, the driver must stop the train at or before the signal. If the signal is displaying a proceed aspect, the driver may allow the train to approach and pass the signal, but at a speed no greater than the release speed.

A release speed is displayed only when it is the next signal ahead which is at danger, and only after the train has passed a particular position on the track, the release point, determined by the train equipment. A release speed is not to be displayed while the train equipment is receiving track data from a continuous in-fill device as, in this case, there should be no dubiousness about the aspect conditions ahead.

WARNING AND INTERVENTION

If the train fails to brake adequately after passing a caution signal, so that the train speed exceeds the warning curve in **Fig. 10.10**, the driver will be given an audible warning by the warble and the 0 or 000 in the main display will flash. The fact that the green permitted speed light indicates a value lower than the reading of his speedometer, will be a further warning that he is overspeeding, and will indicate by how much he must reduce his speed.

If the train speed now exceeds the intervention curve a full service brake application is made and the 'brakes' light is lit to provide confirmation that it is the ATP system that has applied the brakes.

When the train speed has been brought down below the intervention curve, the brakes light flashes as a prompt to the driver that he may release the automatic brake application to regain control over the brakes.

The alternative strategy, applied by some railway administrations, of not releasing the automatic brake until the train has been brought to a stand will be reviewed in the light of experience of the pilot ATP scheme.

SPEED RESTRICTIONS

The supervision of braking on the approach to speed restrictions is similar to that for signals. If the warning curve is exceeded, the driver is given an audible warning and the target speed appears as a flashing numerical value in the main display. The intervention and release sequences are the same as for signals.

TRAIN TRIP

On passing a signal at danger, the message received from the beacon at the signal will initiate a train trip sequence. There will be an emergency brake application, the legend SPD will flash in the main display, the warble will sound, and there will be a steady illumination of the brakes light. When the train has been brought to a stand, the brakes light will flash, indicating to the driver that he can regain control of the braking system by pressing the brakes push button. The main display will then show a steady SPD legend, and the train speed will be supervised to 32 km/h until an ATP beacon at a signal with a proceed aspect has been passed.

TRAIN TRIP OVER-RIDE

The driver may over-ride the train trip facility in order to pass a red signal when permitted to do so by the Rule Book. Over-riding is achieved by pressing the 'pass stop signal' push button. The ATP system will only respond if the train is stationary at the time, and will then suppress the train trip for one beacon only. Passing a subsequent beacon or reversing the train will remove the over-ride from the train trip. While the train trip is over-ridden, the maximum speed is supervised to 32 km/h.

ROLL PROTECTION

A train trip sequence will also be initiated if the train is detected to be rolling away in either direction. This is detected by comparing the direction of movement with the position of the driver's master switch.

SHUNTING

The system provides a limited degree of protection for low speed shunting movements without requiring any train data to be entered. The protection provided is speed supervision to 32 km/h, train trip at stop signals, and roll protection.

There are three shunting modes: hauling only, propelling forward, and propelling reverse. The driver may select a shunting mode by pressing the 'shunt' push button when the train is stationary. The mode entered depends on the position of the driver's master switch when the shunt button is pressed.

The reason for having three distinct shunting modes is to have an optional means of selectively disabling the operation of the train trip in either direction of travel. Disabling the train avoids possible spurious operation of the trip when the train is being propelled past an ATP fitted signal. This spurious operation will occur if the leading part of the train causes a signal to be replaced to danger before the vehicle from which the train is being driven has passed the signal.

When the train trip has been selectively disabled in a particular direction of travel, any number of signals at danger may be passed in that direction, even during successive back and forth movements. However, in the opposite direction, the train trip will still remain operative and the train trip over-ride must be used if it is required deliberately to pass a signal at danger.

PARTIAL SUPERVISION

The partial supervision mode of operation provides a limited degree of protection for running movements in non-ATP areas, for example by ensuring that the train maximum speed is not exceeded; and in ATP areas in circumstances where the full supervision mode is not available, for example on opening the cab.

The level of protection depends on the track data available but does not include supervision of train braking in respect of signalling conditions on the line ahead.

Partial supervision is automatically entered when there is some detectable defect in the system, recovery is possible, and isolation of the ATP system is not necessary. Some possible causes of system errors are:

- Poor accuracy in the measurement of distance by the train equipment.
- Electrical interference causing corruption of the track data received by the train equipment.
- Partial or total failure of the ground equipment.
- Insufficient or inconsistent data received by the train equipment.

It may be difficult positively to identify the cause from the symptoms known to the train equipment.

The main display provides a suitable indication. A full service application of the brakes is initiated, or an existing application is maintained. The brake application can be released after the indication has been displayed for 10 s. This delay gives the driver time to adjust to the presence of the system error, and to the possible consequential reduction in the level of protection, without immediately placing the train at risk.

SELF-TEST ROUTINE

When power is applied to the train equipment, the latter goes through a self-test routine to verify its correct functioning and to allow the audible and visual indications to be checked. If the functional test fails to complete satisfactorily, the ATP equipment must be isolated.

SIGNALLING INTERFACE

The lineside encoder needs to be provided with signal aspect and route information for transmission in the data to the train. The encoder is provided with high impedance inputs capable of sensing 110 volt 50 Hz voltages. The encoder may thus be connected in parallel with the feed to a signal lamp from an SSI signal module, as in **Fig. 10.12**. Where relay circuitry is in use, the encoder may also be connected in parallel with the signal aspect feeds, but separate 110 volt feeds may be taken over spare relay contacts as an alternative.

For signals equipped with junction indicators, an additional input to the encoder is required to give a positive indication when the straight ahead route is signalled. This is necessary to prevent the ATP data from defaulting to that for the straight ahead route if the input from the junction indicator becomes open circuited when a diverging route is signalled.

BR APPLICATION STRATEGY

It is BR policy in response to the rising trend in signals passed at danger, reinforced by the recommendations of recent accident enquiry reports, to fit ATP to a large percentage of the network. In order to achieve this in as short a time as possible, the ATP system requirements for BR ATP were drawn up with the capabilities of existing technology in mind.

Fig. 10.12 Signal interface with SSI system

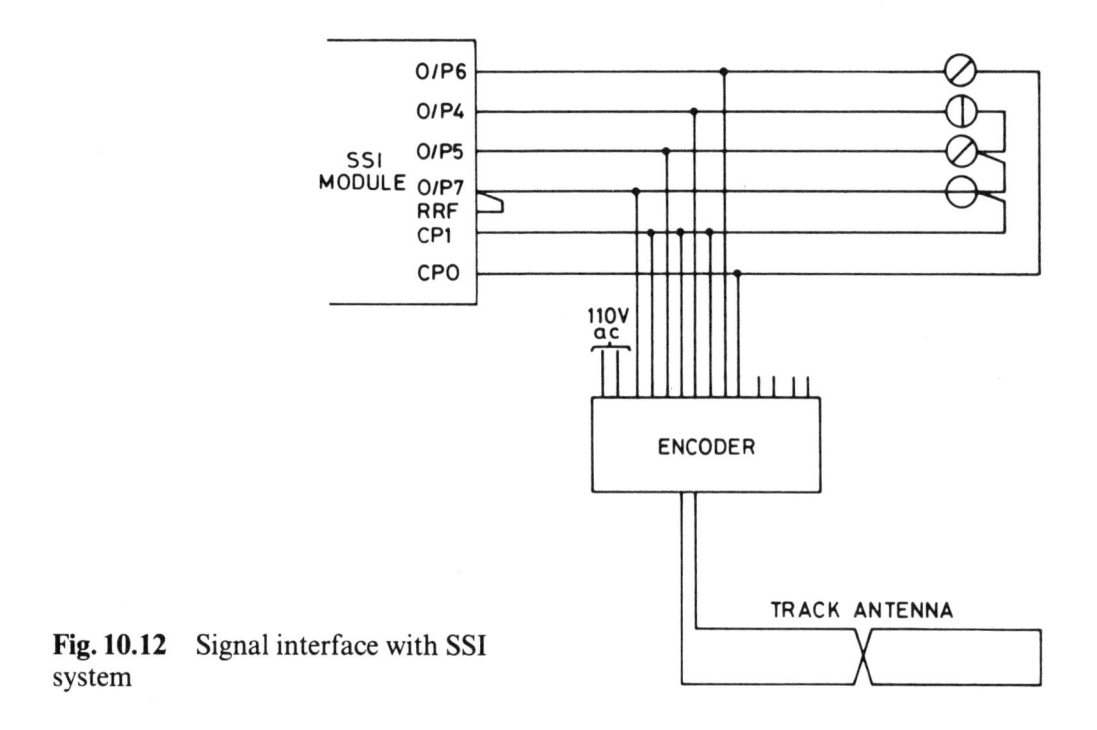

BR must be able to procure ATP equipment providing the system functions described above by competitive tendering from different manufacturers. As a first step, contracts have been let to two different suppliers for pilot ATP schemes on the Chiltern Lines and the main line from Paddington to Bristol.

In each case the manufacturer's current product will be adapted, mainly by software development, to provide the BR system performance as described above.

The pilot schemes will be different in their track-to-train interface. One manufacturer uses track antennae loops of conductor 25, 50 or 300 m in length, depending on circumstances, while the other uses a stainless steel loop antenna about 1 m in length. The latter antenna can be extended by the addition of a track conductor loop where in-fill for improved line capacity is required. The two manufacturers use different carrier frequencies and data formats for track-to-train transmission. One of the results of the pilot schemes will be that a standard track-to-train interface specification can be arrived at to enable competitive procurement of track and train equipment to BR system and performance specifications.

The Future

ATP as described above is expected to solve completely the problem of signals passed at danger and at last bring to an end a period of railway history in which driver error has been a major cause of accidents.

However, the ATP system should not only be looked on as a solution to past problems; it will also provide a secure base for future developments in signalling and train control, and the presence of sufficient capacity for growth is not the least of the features that will be examined in assessing ATP equipment for use on BR.

CHAPTER ELEVEN

The Future

Just as when the first IRSE Textbook was written, the future still holds almost unlimited prospects for enhancing the supervision, control and even the safety of the railways.

The very fact of there being so large a prospect increases the difficulty of forecasting, particularly today when decisions on the choice of technology are more than ever (and quite properly) influenced by financial viability or perceived business needs.

When the previous book was written, British Railways SSI was in its earliest stages and not everyone was satisfied that it could be made adequately safe. Now it is the recognised standard for BR, and is showing its potential by being adapted to the needs of many other administrations. Where will it go from here? Obviously, service experience has shown the need for improvement in some areas and these are being addressed. Work will be carried out to enhance the immunity of the trackside equipment against the growing spectrum of interference. This arises principally from the newer, more powerful, trains with their innovative control equipment, but also from traditional lightning strikes.

Great attention will have to be paid to the overall system engineering aspects in order to produce the most reliable, cost-effective option.

Some of the improvements here will probably come from the growing use of optical fibre communications, already forming part of the long distance network, for the shorter, interlocation channels.

Studies have been undertaken to assess whether a two-out-of-three processor configuration should be applied to the trackside modules to increase their availability. There are many difficulties and it will probably be more cost-effective to provide duplicate modules in key areas where a single failure could cause widespread disruption. The application of BR SSI to lightly loaded routes has so far been through the medium of RETB, as set out in Chapter 3. This technique, in simple or enhanced form, has considerable potential abroad but its expansion in the UK has become limited by the shortage of available, exclusive radio channels and the problems of retaining captive stock. When satellite transmission matures it could be that there will be a resurgence of interest in this area. Ground-based leaky feeder transmission could be used in tunnels.

Work carried out at British Railways Research Division in Derby has shown that it is possible to integrate axle counter heads and their local interfaces directly into a BR SSI system so that the counting logic is carried out within the central interlocking. It is believed that systems of this type can be applied with advantage to lightly loaded lines and the next few years should see major growth in this area. Once confidence has been gained in the reliability of such a system, it could readily be applied on busier lines.

Other special modules are being developed by industry to permit simple serial interface connections between a BR SSI system and another BR SSI system at a different location, or any other safety signalling system, for example, other forms of computer interlocking, transmission-based train control systems, etc. Modules of this type will probably provide assistance in the application of ATP.

One of the strengths of SSI has been the extent to which diagnostic facilities are provided and there should be further enhancement in this area to give even greater assistance to the technician when dealing with complex faults in some of the very large schemes now being installed or contemplated.

In parallel with these detailed developments, work will start on the next generation system. The designers will have many arguments to resolve:

- Should the two-out-of-three and two-out-of-two approach be retained and would it be found necessary to use different hardware and software in each channel?
- Does the increased complexity of such a variant bring about its own problems of implementation?
- Should self-healing or adaptive systems be contemplated?
- Should there be categories of safety lying between the present vital and non-vital: for example, could indications be so arranged that they could be relied on in an emergency?
- Should the system take account of level crossing requirements or should these be met by a separate unit?
- Should the new design have options or alternatives to make it more cost-effective for smaller, and possibly free standing interlockings?
- Will new processors such as VIPER designed and validated with safety applications in mind become viable contenders for major interlockings?
- Should the equipment driving points and signals be mounted in rooms, in lineside cases or directly in the driven equipment?

Whatever arrangement is chosen, one thing is certain: the growing interest in safe software for many industries apart from railways has led to the availability of many more software tools to assist in the lengthy and expensive design and validation processes. It must not be forgotten that however good the tool, it must be properly applied and the greatest asset here is the experience of those in the industry who brought the initial system into existence.

It is reasonable to assume that whatever ongoing system emerges, it will still be data driven. The cost of data preparation is a major factor in any of the current computer-based interlocking systems. More important, the required skills are in short supply and this situation shows little sign of improvement. Knowledge-based or expert systems are being considered for use in this area. These systems are basically computer programs which allow the convenient storage and manipulation of knowledge in such a way that problems sufficiently complex to require a degree of human expertise for their solution can be solved by the program. They will not remove the need for signalling skills, but they will enable existing skills to be better employed. The validation of such systems will need very special attention, but it will be achieved and in a few years they will be in regular use.

Computers, as ever, will receive a lot of attention but what is to happen to track circuits, point machines and lineside signals?

Track circuits bear the full brunt of the growing and interference-laden traction current and, within the UK electrical safety requirements, it has been found extremely difficult to meet the civil engineering wish for jointless track circuits in electrified areas.

Work with metro track circuits using Fourier transform or pseudo-random Gold coding is looking hopeful and may in time be made suitable for the longer main line track circuits. In the much longer term, the use of broad band wave guide systems as a means of train detection and now being tested for metro use, may become viable for main lines. One of the problems in this area is that the increased component counts of sophisticated designs lead to a lowering of the intrinsic reliability. The track circuit is probably the most frequently used piece of trackside equipment and is therefore a very sensitive area in respect of reliability. New designs will become available and will probably make greater use of integrated circuits to maintain availability. It is likely that new designs will feature direct input to SSI or similar systems, rather than making use of an intervening track relay.

In recent years, axle counting systems have been greatly improved to meet the needs of those railway tracks where track circuits could not be employed, due to steel sleepers, flooding, very long sections without intermediate power, etc. They are now, however, beginning to make some inroads into areas where their use is not mandatory. It is not yet clear whether this incursion will increase or decrease the amount of effort directed towards track circuit development.

Clamp locks and point machines will probably undergo detailed changes to increase their reliability further, but there is unlikely to be any fundamental change in their design.

Trailable points, favoured in some countries of mainland Europe, are unlikely to come into use on BR main lines but may find growing favour in yard areas.

Running signals have been dispensed with on a number of metros and on some of the European dedicated high speed lines where reliance is placed on cab signals. The extent of mixed traffic working on most BR lines make it unlikely that this practice will be adopted, but the London, Tilbury and Southend Line could be a possible candidate.

There has, however, been a firm decision to proceed with the provision of ATP and it is expected that within five years, approximately 30% of BR will have been equipped covering the order of 80% of the traffic.

Two pilot schemes currently authorised will have been evaluated by the time this book is published. Any comment at this stage on the likely outcome would be invidious, but the decision will be one of the most important made by BR in recent times.

So far this chapter has concentrated on the safety aspects of signalling where there is undoubtedly an ongoing climate of change, but the nature of that change, after the dramatic shift to SSI, is now likely to be a steady evolution.

In the control and supervision area, change has been both rapid and widespread and it shows signs of continuing in that way. The introduction of IECC has enabled VDUs to be used for control on a scale not previously thought possible. There is, however, a school of thought that for major, complex areas, the retention of a vestigial overview panel conveys some advantage when strategic operational decisions have to be made during interruptions and emergency working conditions.

Automatic route setting has now grown in stature and will continue to grow. Forecasting the effect of perturbations over an extended period is an area receiving considerable attention both within BR and in overseas applications of alternative equipment by the contractors; from this, there should be useful feedback.

The use of the management features of IECC to produce traffic trend analysis and fleet maintenance analysis will be extended. It should also be possible to provide a degree of energy management by applying coasting control to selected vehicles via track circuits, radio or beacons.

'Signalling the passenger', a phrase coined some years ago, will become ever more meaningful. The presence of much more real-time information about the traffic and its performance, will enable much more passenger information to be given automatically with details of delays — which are never likely to disappear completely — and possibly preferred alternatives. Expert systems are likely to find application in this work on the basis that they will learn (and remember) which of their forecasts was correct and make use of these again in future.

Synthesised public address announcements, automatically generated from the passenger information system and multilingual where appropriate, will almost certainly appear before long.

Some of the foregoing predictions have quite an air of glamour about them but there are many things very necessary in the next few years which, whilst largely unsung, remain crucial to the future.

There must be an improvement in the systems training of signal engineers so that they can integrate the many different elements which will be available to them with confidence and accuracy, and properly exercise engineering judgement. New systems will require new methods for assuring safety and these must be developed in parallel and with meticulous attention to detail. It is insufficient in today's climate, and more so in tomorrow's, just to get it right. It has to be seen to be right and much work is needed to design adequate, but user friendly procedures for these activities. If the procedures are not user friendly then the effort required to complete them will detract from that available for problem solving. Scarce resources must be applied where they are most appropriate.

With the advent of the common market in 1992, it will not only be the commercial barriers which are removed. There will be an enormous amount of work both prior to and after that date to harmonise standards, some of which are in very difficult areas such as electromagnetic compatability (EMC) and possibly the track-to-train interface for ATP and similar systems.

Cross-fertilisation, which was previously interesting and useful but often sprang from casual encounters, may in some cases become mandatory within the EC, especially where through running is concerned and this could provide one of the greatest challenges in the future of signalling.

In summary, what then does the future hold? Certainly a period of consolidation in the SSI field with detailed improvements, a widening of application and close integration with the chosen ATP scheme. There will be an increase in the use of IECC or similar systems with enhanced automatic route setting facilities and in particular, the provision of real-time management and passenger information to a much greater level than now exists.

In parallel a great deal of work will be carried out in areas such as expert systems, safety validation technology, interference assessment, system engineering technology and reliability enhancement. The importance of reliability cannot be overstated and it will be a keynote in all future work.

By the end of the 1990s it is likely that the next generation SSI will be in place on BR. Hopefully the European influence on all administrations will be such that there will be co-operation in its development and a larger market to provide a better cost base.

Prediction is notoriously difficult as everyone concerned with forward planning will agree. One thing is certain and that is that if only half of what is mentioned in this chapter comes to pass, there will be an interesting, exciting and very demanding future for those engineers already in, or choosing to join, this very special profession.

Chapter-by-chapter Index

1 Recent Changes in Signalling Philosophy
Junction signalling   1
Combining of berth and overlap track circuits   3
Release of approach locking   3
Emergency replacement of automatic signals   4
Time of operation locking   4
Subsidiary and shunt signals   4
Point to point locking   4
Warning class routes (delayed yellow)   5
Simplified bidirectional signalling   5
Staff warning systems: TOWS and ILWS   7
Control tables   9

2 Solid State Interlocking
Introduction   18
Safety and availability techniques
  Safety by redundancy   18
  Availability   22
  Safe data transmission   25
System description
  Overview   26
  The control centre   27
  Interlocking software   31
  Trackside data links   35
  Trackside functional modules   39
Scheme design and data preparation
  Design considerations   46
  Allocation of functional modules   48
  Design workstation   48
  Data preparation   50
SSI data examples   50

3 Single Line Signalling
Introduction   59
General philosophy   59
Requirements of the Department of Transport for the operation of single lines   59
  Train staff and ticket   60
  Divisible train staff   60
  One train working   60
  Electric token   60
  Direction lever and track circuit   61
  Other methods   61
Electric key token
  General description   64
  Explanation of phase   66
  Operation of system   67
Reed communication between instruments using public telecommunications lines
  General description   72
  Transmission equipment   72
  Transmission medium   72
  Mode of operation   75
  Security   75
Simplified infrastructure
  Level crossings   77
  Running connections   77
  Siding connections   77
  Fixed signalling   77
  Other features   80
Radio electronic token block working
  Introduction   80
  Detailed description   83
  Traction unit   83
  Definition of limits   83
  Types of token   83
  General   84

No-signalman remote key token working
  General description 84
  Mode of operation using FDM 86
  Mode of operation using physical line circuits 91
  Note on line circuits 91
British Railways tokenless block system
  General description 92
  Mode of operation 94
British Railways tokenless block with reed transmission
  General description 97
  Transmission system 97
  Mode of operation 97

4 Immunisation and Earthing of Signalling Systems
Introduction 99
Categorisation of interference to signalling equipment 99
Mechanism of interference from external electrical systems 100
Principles of immunisation 101
Electrical interference from electric traction systems
  DC electric traction 102
  AC electric traction 102
  Dual traction systems 103
  Semiconductor traction control systems 103
Operation of signalling equipment in electric traction areas – track circuits
  DC areas 104
  AC areas 105
  Dual areas 105
  Traction return circuits 105
Operation of signalling equipment in electric traction areas – tail cables 106
Operation of signalling equipment in electric traction areas – lineside circuits 108

Electrical interference from other electrical equipment
  Rail connected equipment 110
  Train-borne systems 110
  Electric potentials in the ground 110
  Power transmission lines 110
  Radio interference 111
  Power supply transients 111
Earthing of signalling equipment 111
Protection of signalling equipment from earth faults 111
Lightning protection 113
  Effects on signalling equipment 114
  Preventative methods 115

5 Train Detection
Introduction 117
Type TI 21 track circuits
  Introduction 117
  Principles of operation 117
  Transmitter 117
  Receiver 119
  Typical installations 120
  Information transmission via the track circuit 120
Axle counters
  Introduction 122
  System overview 123
  Trackside detection equipment 123
  Relay room equipment 125
  Multi-section and dead-end track 127
  Switch and crossing layouts 127
  Relative merits of axle counters and track circuits 128
  Use of axle counters in the UK 128
Transponders
  Introduction 128
    (*continued*)

Transponders (*continued*)
  Principles of operation 130
  Practical application for train detection 131
  Use of transponders for train detection in the UK 132

6 Level crossings
Introduction 134
Automatic half barriers
  General description 139
  Crossing initiation 140
  Description of operation 140
  Crossing with signal staging 149
Automatic open crossings locally monitored
  General description 152
  Description of operation 153
Automatic half barriers locally monitored
  General description 158
  Description of operation 161
Manned barrier crossings
  General description 166
  Types of crossing 167
  Modes of operation 167
  Protecting signals 168
  Operation of CCTV 168
  Description of operation 170
  Operating of crossing with local control unit 181

7 Equipment
BR930 series relays
  General construction 183
  AC immune relay 184
  Biased AC immune relay 184
  Slow pick-up neutral relay 185
  Slow release neutral relay 185
  Magnetically latched relay 185
  Polarised magnetic stick relay 185
  DC track relay 187
  Lamp proving relay 187
  Biased contactor 187
  Twin relays 187
  SSI interface relay 187
Main signals 188
Subsidiary signals 188
Limit of shunt signals 188
Repeater signals 188
Position light signals 190
Fibre optics 190
  Theatre-type route indicator 192
  Stencil-type indicator 192
  Repeater signal 193
  Future developments 193
Clamp locks 195
Train-operated points system 195
Level crossing barriers
  Operational requirements 197
  Construction 198
  Power unit 199

8 Operator Interface
Integrated electronic control centre
  Hardware 201
  Signalman's display system 205
  Automatic route setting 213
  Timetable processor 220
  Gateway system 222
  IECC system monitor 222
  Program, data and data preparation 223
  Testing 224
  The future 225

Panel processors   226
　Input file   228
　Logic module file   228
　Function file   228
　Output file   229
Centralised traffic control
　Introduction   229
　Origin   229
　Time coding   230
　Office control machine   230
　Push button operation   230
　Computer control   231
　Automatic routing   232
　Train graph   232
　Meaningful warning messages   232
　Operator prompts   232
　Operator displays   232
　The future   234

9　Signalling the Passenger
Introduction   235
Display techniques – mechanical
　Flap indicators   236
　Electromechanical dot matrix   237
Display techniques – electronic
　CCTV   238
　Light emitting diode   241
　Liquid crystal display   242
　Electrochemical display   242
Audio techniques
　Local public address   242
　Recorded announcements   244
　Long line public address   245

Control techniques
　Control and automation   247
　Software   247
　Area control (long line systems)   249
Applications   250
The future   254
Conclusion   257

10　Automatic Train Protection
The AWS system   258
Train stops   248
Speed supervision   259
Data requirements for ATP   260
ATP system architecture   260
　Train subsystem   262
　Track subsystem   263
Data transmission and security   265
Continuous transmission systems   265
Intermittent systems   265
Line capacity   266
Release at red   267
Beacon failure   269
The BR ATP system
　Concept of a system   269
　Aims   269
　Systems architecture   269
　Speed supervision principles   269
　Supervision of train braking   271
　Determination of release speed   272
　Train data entry   274
　Operational features   275
　Driving under restrictive signals   276
　Warning and intervention   277
　　(*continued*)

The BR ATP system (*continued*)
  Speed restrictions 278
  Train trip 278
  Train trip over-ride 278
  Roll protection 278
  Shunting 278
  Partial supervision 279
  Self-test routine 279
  Signalling interface 279
  BR application strategy 279
The future 281

11 The Future 282